Aidan Dodson is a Senior Research Fellow at the University of Bristol. Prior to returning to full-time academia he was also a Civil Servant, working on a number of naval programmes, including as project leader for the current Falkland Islands Patrol Ship, HMS *Clyde*. The author of nineteen books and some three hundred articles, his latest book, *The Kaiser's Battlefleet*, was published by Seaforth in the spring of 2016.

Hans Lengerer is an acknowledged authority on the Imperial Japanese Navy, and has written a number of books on the subject, including the *Technical and Operational History* and *… at War* series. His latest book with Lars Ahlberg, *The Yamato Class and Subsequent Planning*, was published by Nimble Books in December 1914.

Peter Marland is a former RN Weapon Engineer Officer with sea jobs in HM ships *Blake, Bristol* and *Euryalus*, and postings ashore in research and in procurement. He now works as an Operational Analyst, and has contributed a number of articles to *Warship* on post-war Royal Navy weapons and electronics. He is a Chartered Engineer, with post-graduate qualifications in Project Management and in teaching.

Stephen McLaughlin is a librarian at the San Francisco Public Library. Besides contributing regularly to *Warship*, he is the author of *Russian and Soviet Battleships* (Naval Institute Press, 2003). His most recent publication is 'Battlelines and Fast Wings', in the *Journal of Strategic Studies*, Vol 38 No 7. He is co-editor of an annotated version of the controversial *Naval Staff Appreciation of Jutland*, published by Seaforth in the spring of 2016.

Vincent P O'Hara is the author, co-author or editor of nine books including the recently-published *In Passage Perilous: Malta and the Convoy Battles of June 1942* (Indiana University Press 2012) and *Torch* (USNIP 2015), a detailed study of the Allied landings in North Africa in November 1942. He is also a contributor to journals such as *Naval War College Review, MHQ,* and *Naval History*.

Conrad Waters is a barrister by training but a banker by profession. He is the author of a number of articles on modern naval history and current affairs, and editor of the annual *World Naval Review* (Seaforth). He is currently preparing a study of naval developments since the end of the Cold War entitled *Navies in the 21st Century* for Seaforth Publishing.

Richard N J Wright is a former Lieutenant-Commander RN and a retired Queen's Messenger. He has a wide range of naval interests and is the author of the seminal study *The Chinese Steam Navy* (Chatham Publishing 2003).

Richard Worth and **Vladimir Yakubov** are the authors of *Raising the Red Banner* (Spellmount 2008) and are currently compiling an encyclopaedic volume on Soviet warships in the Second World War. In addition to his writings, Vladimir Yakubov is also an award-winning ship modeller. Richard Worth's credits include *Thunder in Its Courses* (Nimble 2011) and *Fleets of World War II* (Nimble 2015).

WARSHIP 2016

WARSHIP 2016

Editor: **John Jordan**

Assistant Editor: **Stephen Dent**

CONWAY
BLOOMSBURY
LONDON · OXFORD · NEW YORK · NEW DELHI · SYDNEY

Title pages: The French colonial sloop *Amiral Charner* arriving in Melbourne, Australia, on 4 November 1934 for the Centenary Celebrations. *Amiral Charner* and her sisters are the subject of an article by the Editor published on p.8 of this year's annual.
(Allan C Green collection, State Library of Victoria)

Conway
An imprint of Bloomsbury Publishing Plc

50 Bedford Square	1385 Broadway
London	New York
WC1B 3DP	NY 10018
UK	USA

www.bloomsbury.com

CONWAY™ is a trademark and imprint of Bloomsbury Publishing Plc

First published 2016

© Bloomsbury 2016

All rights reserved. No part of this publication may be reproduced or transmitted in any form or by any means, electronic or mechanical, including photocopying, recording, or any information storage or retrieval system, without prior permission in writing from the publishers.

No responsibility for loss caused to any individual or organization acting on or refraining from action as a result of the material in this publication can be accepted by Bloomsbury or the author.

British Library Cataloguing-in-Publication Data
A catalogue record for this book is available from the British Library.
Library of Congress Cataloguing-in-Publication data has been applied for.

ISBN: HB: 978-1-8448-6326-6
ePDF: 978-1-8448-6438-6
ePub: 978-1-8448-6437-9

2 4 6 8 10 9 7 5 3 1

Typeset by Stephen Dent
Printed and bound in Great Britain by CPI Group (UK) Ltd, Croydon CR0 4YY

To find out more about our authors and books visit www.bloomsbury.com. Here you will find extracts, author interviews, details of forthcoming events and the option to sign up for our newsletters.

CONTENTS

Editorial	6

Feature Articles

The Colonial Sloops of the *Bougainville* Class — 8
John Jordan looks at the French *Bougainville* class, built during the interwar period specifically for the policing of the country's overseas possessions.

The Colonial Sloop *Eritrea* — 30
Michele Cosentino writes about *Eritrea*, the prototype of a force of purpose-designed sloops to be employed in the 'presence' and showing-the-flag roles.

The Japanese Destroyers of the *Asashio* Class — 42
Hans Lengerer looks at the design characteristics of these large destroyers and, in particular, the problems they experienced with their turbines.

The Naval War in the Adriatic Part 2: 1917–18 — 62
Enrico Cernuschi and **Vincent P O'Hara** cover the years 1917–18, which saw the triumph of the Entente forces and the defeat of the Imperial Austro-Hungarian Navy.

Post-War AIO and Command Systems in the Royal Navy — 76
Peter Marland describes the sequence of AIO and Command Systems during the post-war era.

The Soviet *Fugas* class Minesweepers — 99
Vladimir Yakubov and **Richard Worth** provide an account of the conception and construction of these interwar minesweepers, together with a service history of the ships completed.

Divide and Conquer? Divisional Tactics and the Battle of Jutland — 111
Stephen McLaughlin examines how the tactic of dividing a fleet into small, independently manoeuvring formations might have worked at the Battle of Jutland.

Modern Littoral Surface Combatants — 126
Conrad Waters looks at the new breed of surface ships designed specifically for littoral warfare.

The Chinese Flagship *Hai Chi* and the Revolution of 1911 — 143
Richard Wright attempts to resolve a mystery concerning the activities of the Chinese flagship *Hai Chi* during 1911–12.

The Battleship *Courbet* and Operation 'Substance' — 152
Stephen Dent investigates an almost forgotten episode in the long career of one of France's first dreadnoughts.

The 'Flat-Iron': The Coast Defence Battleship *Tempête* — 161
Philippe Caresse tells the story of this unusual vessel, one of a number of coast defence monitors built during the period which followed the American Civil War.

Warship Notes — 175

Reviews — 187

Warship Gallery — 205
A D Baker III provides a detailed drawing and brief technical description of the dreadnought HMS *Colossus*.

EDITORIAL

Readers will notice a number of changes in the look of the annual this year. For some years our Assistant Editor, Stephen Dent, has been looking to improve the appearance of the internal pages by adopting more modern typefaces and more attractive, flexible layouts. Conway's new parent company, Bloomsbury Publishing, has given us every encouragement and support, and has provided a new cover to complement the internal redesign. The changes have already attracted favourable comment from our contributors, who particularly liked the new table format, and we hope that they will be equally well received by the readership. Throughout the redesign process our primary concern has been to ensure that the high quality of writing and presentation which readers have come to associate with the annual is not adversely affected; we think it is actually enhanced.

The evolution of *Warship* has been underpinned by continuity in the editorial team, who have been together now for twelve years. Some of the established contributors remain, while many new contributors have been introduced during that period. Stephen Dent, who has often contributed Warship Notes and edited the Warship Gallery in the past, has this year contributed his first feature article, on the little-known employment of the old French battleship *Courbet* for trials of a British secret weapon during the Second World War. The article sheds light on the often fraught wartime relationship between the Royal Navy and the Free French naval forces (FNFL), which came to a head in this particular episode over which flag (or flags!) she should fly during her unorthodox mission. We also welcome on board A D Baker III, former editor of the authoritative *Combat Fleets*, with whom many readers will also be familiar due to his work as an illustrator for the books of Norman Friedman. This year sees the publication of the first of what we hope will be a series of contributions centred around the artist's superb line drawings, based on the official plans of the ships. This year's contribution, on HMS *Colossus*, is the subject of our Warship Gallery.

Readers cannot have failed to notice the anticipation already building for the centenary of the Battle of Jutland, which took place on 31 May 1916. Besides a major exhibition featuring the largest collection of artefacts, photographs and personal testimonies ever assembled which will open in the Spring at the National Museum of the Royal Navy in Portsmouth, a number of new studies are scheduled for publication during the early part of next year, the most significant of which will be reviewed in *Warship 2017*. In the past, much of the criticism directed at the C-in-C Grand Fleet, Admiral John Jellicoe, for his failure to inflict a crushing defeat on the German High Seas Fleet has focused on his predilection for an over-rigid centralisation of command. As a follow-up to his ground-breaking article on Jellicoe's deployment at Jutland published in *Warship 2010*, it was suggested to Stephen McLaughlin that he might like to consider an article on the alternative 'divisional' tactics considered and developed by the Royal Navy before the Great War and subsequently championed by some of Jellicoe's critics as an alternative path to decisive victory in a battle involving large and inherently unwieldy battle-fleets. Happily Stephen agreed, and the results are published in this year's annual.

Our coverage of the Great War is completed by the concluding part of 'War in the Adriatic' by Enrico Cernuschi and Vincent O'Hara, which focuses on the period 1916–18. Besides filling a pronounced gap in the history of naval operations during the period in question, the article raises some fundamental questions about the use of seapower in confined waters. Neither the Austro-Hungarian Navy's overall strategy nor the tactics it employed using the resources at its disposal emerge with much credit from this analysis.

With the scale of the battlefleets much reduced following the Washington Naval Arms Limitation Treaty of 6 February 1922 and the institution of a ten-year 'battleship holiday', the attention of those major powers with world-wide interests to defend turned to 'trade protection' cruisers, long-range submarines capable of operating from overseas bases, and small (but capable) surface units for overseas deployment which could remain on station either permanently or for extended periods of time, and which could undertake local sea control and policing duties. The 'trade' cruisers have received extensive coverage in English-language sources; in contrast, the 'colonial sloops' built by Britain, France and Italy from the late 1920s have been almost entirely neglected because of their marginal contribution to 'fleet-on-fleet' action and to oceanic antisubmarine warfare during the Second World War. Yet during the interwar period considerable resources were invested in these ships, particularly by France which, despite the need to police a world-wide empire second only to that of Britain, had only a small force of cruisers compared to the Royal Navy. The French colonial sloops of the *Bougainville* class, laid down from 1929, were powerful, well-armed ships which incorporated features specifically intended for independent operation in tropical and equatorial climates. The design not only influenced the characteristics of the 'treaty-exempt' category defined by the London Treaty of 1930, but inspired similar ships built or designed by the Italian Regia Marina to police its newly-acquired empire in East Africa and the Red Sea. This year's annual begins with a detailed study by the Editor of the French *Bougainville* class, which is followed by an article by Michele Cosentino on the Italian *Eritrea*; both of these articles include much previously unpublished material.

EDITORIAL

The former German light cruiser *Stralsund* at Cherbourg in 1920; she subsequently became the French *Mulhouse*. *Stralsund* and the other German light cruisers which survived the First World War will be the subject of a major feature article in *Warship 2017*. (Jean Moulin collection)

Our third feature article on warships of the interwar period is Hans Lengerer's study of the IJN destroyers of the *Asashio* class. The *Asashio* represented a return to the 'super-destroyers' of the 'Special Type' following the constraints on destroyer displacement imposed by the London Treaty of 1930. However, the introduction of a new intermediate-pressure turbine proved problematic, with a number of instances of blade failure due to vibration and other stresses. Building on the research of Japanese author Hori Motoyoshi, Hans Lengerer gives a detailed account of the lengthy and involved process of investigating and remedying these defects. The *Asashio* was also the first IJN destroyer to feature the fully-developed torpedo fire control and reloading apparatus and procedures which set the standard for the later *Kagero* and *Yugumo* classes, and which made a considerable impression on the US Naval Technical Mission to Japan in 1945–46; these features are fully documented in the article.

Coverage of the interwar period is concluded by an article on the Soviet minesweepers of the *Fugas* class by regular contributors Richard Worth and Vladimir Yakubov. These were the first minesweepers designed and built for the Soviet Navy; no fewer than 42 ships were completed between November 1936 and April 1941, and the original design was subject to several upgrades in configuration and capability. Although the primary role of the *Fugas* class was minesweeping, once war broke out these ships were employed as maids of all work, taking part in escort and fire support missions, landings and evacuations; many suffered serious damage or loss. The article provides a full listing of their wartime activities.

The period from 1870 to the Great War is represented by two feature articles: Philippe Caresse's study of the French coast defence battleship *Tempête* exposes the technical and tactical limitations of this type of ship, while Richard Wright, author of the seminal book *The Chinese Steel Navy*, returns after a lengthy absence to probe the mystery surrounding the movements of the protected cruiser *Hai Chi* during the period 1911–12, when a pre-planned world cruise effectively removed her from Chinese waters in the build-up to the Revolution of October 1911.

Conrad Waters' regular feature on modern warship developments focuses this year on the craft specially developed by a number of Western navies for littoral warfare. These vary from the frigate-sized Littoral Combat Ships (LCS) currently being built in two parallel variants for the US Navy to the Swedish *Visby* and the Norwegian *Skjold* classes of fast patrol boat. All these types feature low-visibility features such as camouflage and hull-forms and superstructures designed to minimise radar signature; composite materials, modular payloads and waterjet propulsion are also key features of many of the designs. The article demonstrates how concepts of naval warfare are evolving, with shallow-water operations becoming as important as ocean warfare.

In this year's Warship Notes Aidan Dobson provides an account of the twilight career of the German battlecruiser *Derfflinger*, Ian Johnston contributes an eye-witness account of the sinking of the submarine *K13* while conducting diving trials on the Clyde in January 1917, and Henk Visser looks at the design of the first class of destroyer built post-war for the Royal Netherlands Navy, the Type 47A.

Next year's annual is already in preparation and will have the customary range of subjects, navies and historical periods. Major feature articles will include a study of the first Japanese all-big-gun battleships, *Kawachi* and *Settsu*, by Kathrin Milanovich, a detailed account of the service careers and fate of the ex-German cruisers which served with the *Reichsmarine* or with foreign navies after 1918 by Aidan Dodson, an article by David Murfin on the treaty cruiser designs drawn up by the Royal Navy during the late 1920s, and an account of the Australian Navy's ill-fated attempt to design and build a modern 'light destroyer' (DDL) during the early 1970s by Mark Briggs.

John Jordan
April 2016

THE COLONIAL SLOOPS OF THE *BOUGAINVILLE* CLASS

The interwar period saw the construction by the major imperial powers of vessels specifically intended for the policing of their overseas possessions. **John Jordan** looks at the French *Bougainville* class, the characteristics of which set the standard for the 'treaty exempt' category at the London Conference of 1930.

Of the five major powers which attended the Washington Naval Conference of 1921–22 only two, Britain and France, had major empires which extended around the globe. The United States had inherited a number of former Spanish colonies in the Caribbean and the Western Pacific, and Italy and Japan had ambitions to create new empires of their own for reasons of national prestige; however, none of these countries could be said to carry the weight of 'imperial burden' to the same extent as Britain and France, and none maintained substantial naval squadrons overseas. These differences in perceived national security requirements were to result in serious problems for the naval arms limitation process both during the Washington conference negotiations and for the fifteen years (1922–36) during which the treaty system was in force.

When France was presented with a *fait accompli* at Washington with regard to her allocated capital ship tonnage, the French delegation made it clear that this derisory figure could be accepted only on condition that no attempt was made to extend the 5:3:1.75 ratio to 'auxiliary' vessels or to abolish the submarine. France needed cruisers and submarines not only for its battle fleet but to protect its widely-separated possessions overseas. Similar considerations were at the root of the Anglo-US dispute over cruiser numbers which threatened to wreck the treaty system during the late 1920s, and which was resolved only when the British Prime Minister Ramsay MacDonald was prepared to make substantial concessions in the lead-up to the 1930 London

An aerial view of the nameship of the class, *Bougainville* (A1), at Majunga on the northwest coast of Madagascar, during her first deployment to the Indian Ocean. She has yet to be embark a reconnaissance aircraft or the planned minesweeping paravanes amidships, so the after part of the shelter deck is unusually uncluttered. Note the wooden planking on the forecastle and shelter decks, and the canvas awnings over the quarterdeck and the bridge decks. (Forces Aériennes Françaises de Madagascar, courtesy of Jean Moulin)

Bougainville: Profile & Plan

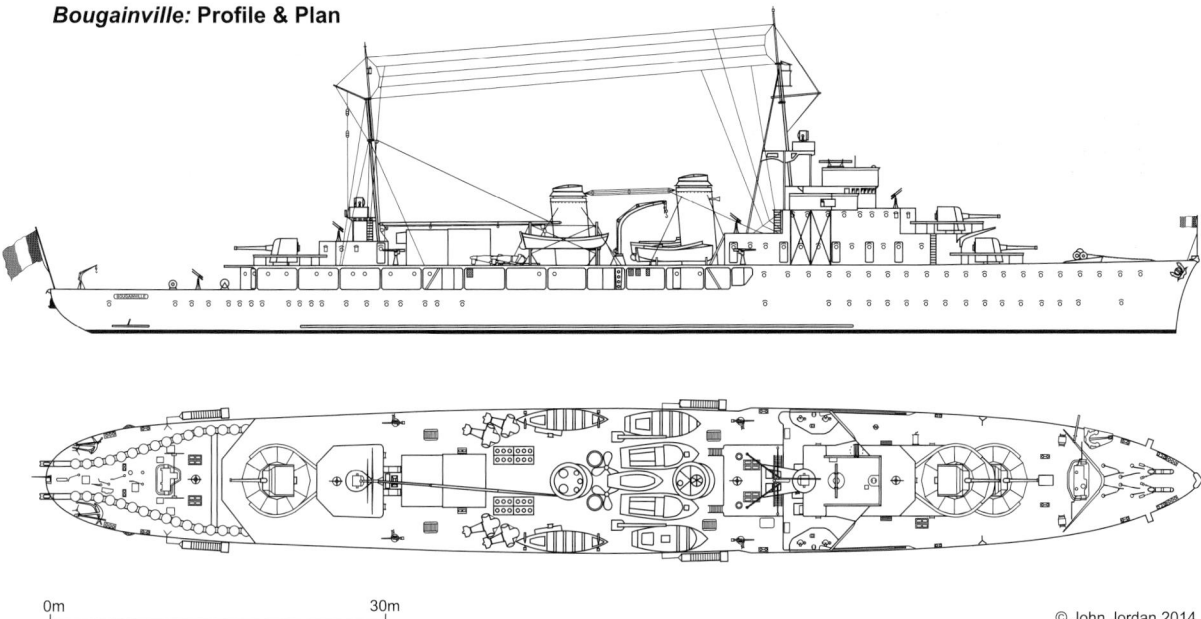

© John Jordan 2014

Conference in the interests of world peace. The British, like the French, ideally would have liked to establish a theoretical division between cruisers for the fleet and cruisers for trade protection, with only the former being counted under the quantitative limits proposed. However, the Americans rejected this proposal out of hand, arguing that cruisers designed for trade protection could still be used to supplement 'the fleet' in the event of conflict between nations and possibly tip the balance.

One of the ways out of this potential *impasse* was to develop a class of warship specifically for policing duties overseas which, while having substantial defensive capabilities, would be poorly suited to operation with the fleet. Clearly there would need to be limits on size and firepower (ie gun calibre <u>and</u> number); the hull of a small cruiser or a large destroyer would suffice. More importantly, offensive qualities such as high speed and the ability to launch torpedoes were out of the question, with endurance (for lengthy transits in the Indian Ocean and the Pacific) and habitability (for the hot, humid climates of the tropics) being more important considerations. A key attraction for all the major powers was that there was no compelling reason why such a vessel should be counted within quantitative treaty limits, as it was purely defensive in nature and posed no aggressive threat to any other power.

Traditionally the French Navy had used older sloops and cruisers for overseas policing duties. However, in the wake of the Washington Treaty, the French 1922 Programme envisaged ten ships specifically designed to serve on foreign stations and to be able to operate in extreme climates. The initial classification of these ships was *aviso pour campagnes lointaines* (sloop for overseas deployments); later the term *aviso colonial* (colonial sloop) was frequently used. The massive investment in terms of funding and resources necessary to rebuild the fleet following the Great War resulted in a five-year delay in the ordering of these ships, which allowed a lot of thought to go into the design.

In the lead-up to the Geneva Conference of 1927, which was to focus on resolving the Anglo-US cruiser issue, Britain formally proposed a new type of 'treaty exempt' vessel for the overseas policing role. The qualitative limits of these ships were to be as follows:

– maximum displacement: 2000 tons (2032mt) standard
– maximum gun calibre: 5-inch (127mm)
– maximum speed: 18 knots
– no torpedo tubes

The proposals were favourably received, and during the 'Coolidge negotiations' between the British, American and Japanese delegations – the French and the Italians declined to attend the conference – an increase in gun calibre to 6.1-inch (155mm) and a maximum number of four guns above 4-inch (102mm) was agreed. However, the failure to resolve the cruiser issue meant that the proposed 'treaty-exempt' category was effectively put on the back-burner until the London Conference of 1930, by which time the British had embarked on a replacement programme for the 'Flower' class sloops, which were often used as station ships overseas, and the French had embarked on the sloops of the *Bougainville* class.

The British sloops of the *Bridgwater* class – they would be followed by the similar *Hastings* and *Shoreham* classes – were relatively small ships of 1045 tons, armed with two 4-inch guns and with a maximum speed of 16.5 knots. The French *Bougainvilles*, on the other hand, were more substantial ships which corresponded closely to the maximum characteristics of the 1927 'treaty-exempt' proposals; they had a standard displacement of 2000 tons, a designed maximum speed of 16 knots,

Bougainville class Colonial Sloops

Estimates	Number	Name	Shipyard	Laid down	Launched	In service
1927	A1	*Bougainville*	FC Gironde, Bordeaux	25 Nov 1929	21 Apr 1931	15 Feb 1933
	A2	*Dumont d'Urville* [S]	AC Maritime du Sud-Ouest	19 Nov 1929	21 Mar 1931	4 June 1932
1929	A3	*Savorgnan de Brazza* [S]	AC Maritime du Sud-Ouest	6 Dec 1929	18 Jun 1931	21 Feb 1933
	A4	*D'Entrecasteaux*	AC de Provence, Port-de-Bouc	29 Jan 1930	21 Jun 1931[1]	6 May 1933
1930	A5	*Rigault de Genouilly*	FC Gironde, Bordeaux	7 Jul 1931	18 Sep 1932	14 Mar 1934
	A6	*Amiral Charner*	AC Maritime du Sud-Ouest	27 May 1931	1 Oct 1932	20 Apr 1934
1931	A7	*D'Iberville* [S]	AC de Provence, Port-de-Bouc	13 June 1932	23 Sep 1934	22 Sep 1935
1937	A8	*Ville d'Ys* (*La Grandière*)	AC de Provence, Port-de-Bouc	23 Feb 1938	22 Jun 1939	20 Jun 1940
	A9	*Beautemps-Beaupré*[2]	FC Gironde, Bordeaux	3 May 1938	20 Jun 1939	–
1938bis	A10	*La Pérouse*[2]	FC Gironde, Bordeaux	[suspended Apr 1940, cancelled Jun 1940]		

Notes:
1. Aborted launch 07.06.31
2. Redesigned with 2 x twin 100mm, enhanced AA, single funnel and modern-style pole mast

Characteristics		Weights	
Displacement:	1969 tons standard; 2126 tonnes normal; 2600 tonnes full load	Hull:	856t
		Protection:	21t
Dimensions:	length pp 98.0m; length oa 103.7m; beam 12.7m; draught 4.15m max.	Propulsion:	460t
		Guns:	174t
		Aviation:	10t
Propulsion:	two Sulzer [S] or Burmeister & Wain diesels on two shafts; 3200bhp = 15.5 knots (designed)	Weight proportional to displacement:	321t
		Weight proportional to 2/3 power:	316t
Fuel:	diesel 220 tonnes; oil fuel 60 tonnes	Special installations (W/T, boats, nav):	42t
Endurance:	9000nm at 14 knots	Reserve:	6.4t
Armament:	three 138.6/40 Mle 1927 (3 x I)	**Washington displacement:**	**2206.4t**
	four 37/50 Mle 1925 (4 x I)	**Normal load**	
	six 13.2/60 Mle 1929 MG (3 x II)	Heavy oil	30t
	50 mines	Diesel	110t
	GL.832 recce floatplane	RFW	10t
Complement:	14 officers + 121 men (peacetime)	**Normal displacement:**	**2356.4t**
		Deep load	
		Heavy oil	30t
		Diesel	110t
		RFW	10t
		Exercise rounds	10.8t
		Various	22t
		Deep load displacement:	**2539.3t**

and an armament of three 138.6mm (5.4-inch) guns.

The characteristics of the new French ships served to concentrate minds at the London Conference, and qualitative limits for the 'treaty-exempt' category were finally agreed as follows:

– maximum displacement: 2000 tons (2032mt)
– maximum gun calibre: 6.1-inch (155mm)
– maximum no. of guns above 3-inch (75mm): four
– maximum speed: 20 knots (increase proposed by the Japanese)
– no torpedo tubes

Alternatively, each of the contracting powers could build ships below 600 tons in which neither armament nor speed was limited; this led to a series of small torpedo-boat designs in France, Italy and Japan, and the *Kingfisher* class coastal sloop in Britain.

The *Bougainville* Design

When the characteristics of the new French sloops were first discussed, the principal staff requirements were established as:

– 25 days sailing at 10 knots;
– an endurance of 10,000nm;
– the ship to be capable of being maintained and repaired at (limited) overseas facilities;
– the hull to be accommodated in existing overseas docks (both French and foreign);
– a high level of comfort for the crew;
– the ability to accommodate new weapons;
– the ability to embark an admiral and his staff for a lengthy transit.

Subsequent discussions centred on:

THE COLONIAL SLOOPS OF THE *BOUGAINVILLE* CLASS

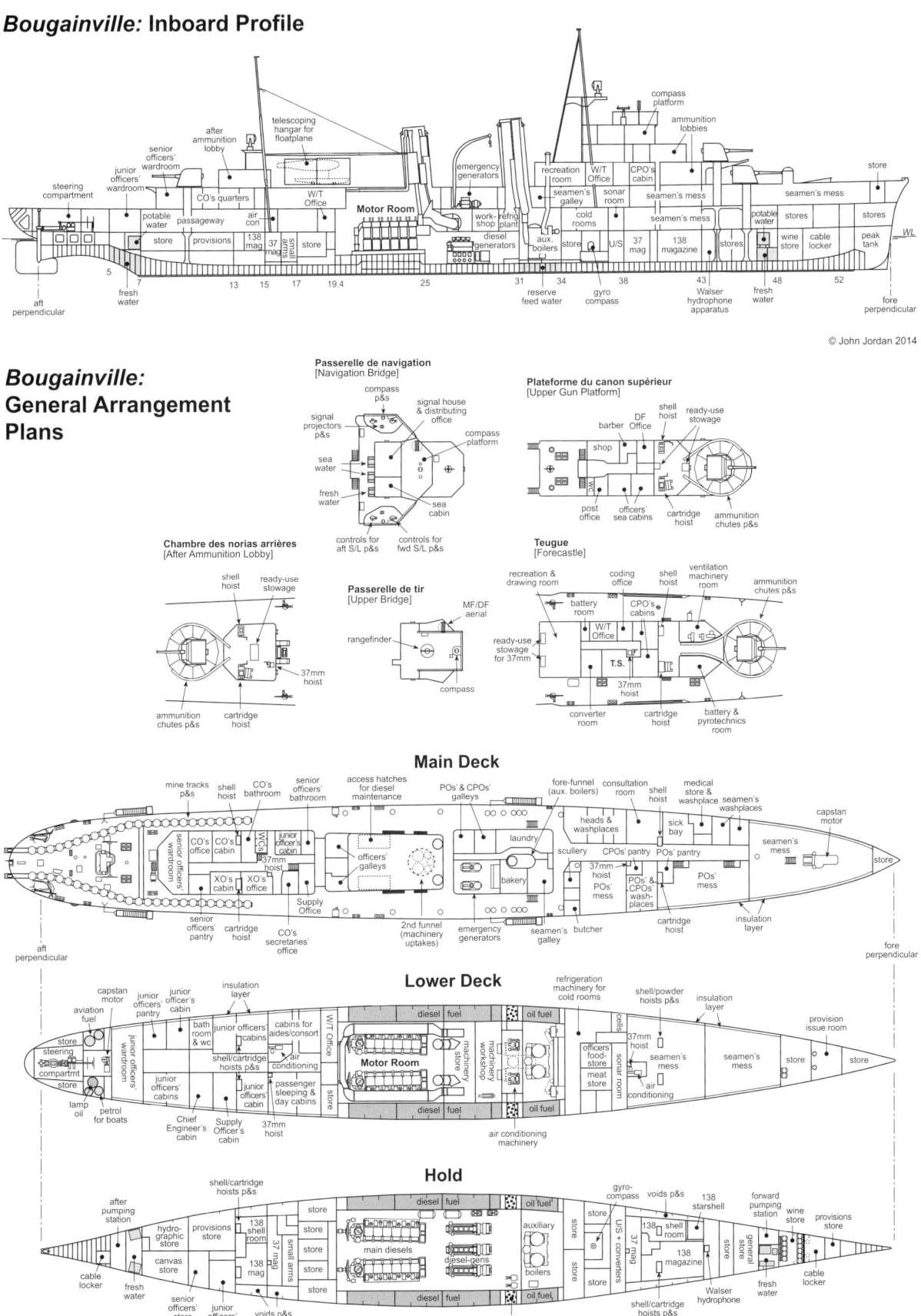

Bougainville: Sections

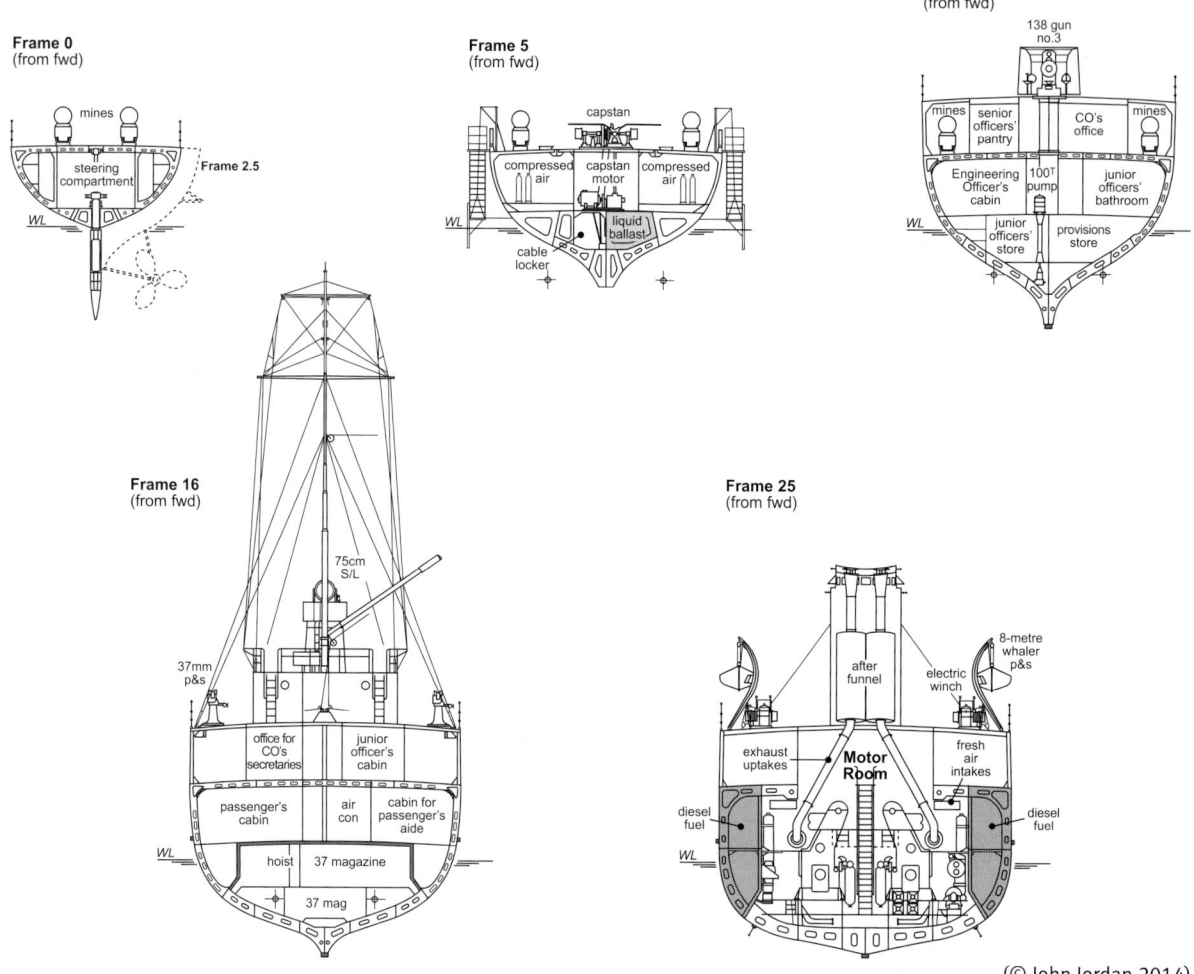

(© John Jordan 2014)

- the type of machinery (diesels were considered for the first time);
- whether aircraft would be embarked (a catapult was not considered feasible for a ship of this size because of the topweight/stability implications);
- the main armament.

It was envisaged that the ships would generally be on station, interrupted by brief port visits to 'show the flag' and periods of maintenance. High speed was considered unimportant for the representative mission, but endurance was crucial. Diesel propulsion was therefore particularly attractive; diesel engines had good availability and a rapid start-up (20–40 minutes) compared to steam machinery, made fewer demands on hull volume, and had greater fuel economy. However, French industry could only build diesels under licence, and these generally had an unfavourable weight/horsepower ratio (60kg/ch vs 34kg/ch for the latest German cruisers).

Given the vast ocean areas which the sloops might have to patrol, it was decided that a small reconnaissance aircraft would be useful. By the time the first units of the class were laid down, a lightweight aircraft was under development for the light cruisers of the *Duguay-Trouin* class, which would likewise spend much of their service lives on foreign stations. A lightweight telescopic hangar would also be a feature of the initial design, to protect the aircraft from the elements.

Although the sloops were slightly shorter than the new destroyers of the *Bourrasque* and *L'Adroit* classes, they had significantly greater beam, thereby enabling them to accommodate the heavier 138.6/40 gun of the latest *contre-torpilleurs*. Three main guns were to be fitted, two forward and one aft, and the light anti-aircraft armament (see below) was powerful for its time.

The architecture of the ships was inspired by the training cruiser *Jeanne d'Arc* of the 1926 programme, which because of her annual round-the-world cruises was designed to operate in the tropics. The design featured relatively spacious mess decks and officer accommodation, with air conditioning throughout and insulation around the accommodation spaces. There was a raised forecastle for good sea-keeping, and the shelter deck amidships was extended to the sides to form a covered way over the upper deck to port and to starboard; these were called *passavents*, the same term used for the prom-

THE COLONIAL SLOOPS OF THE *BOUGAINVILLE* CLASS

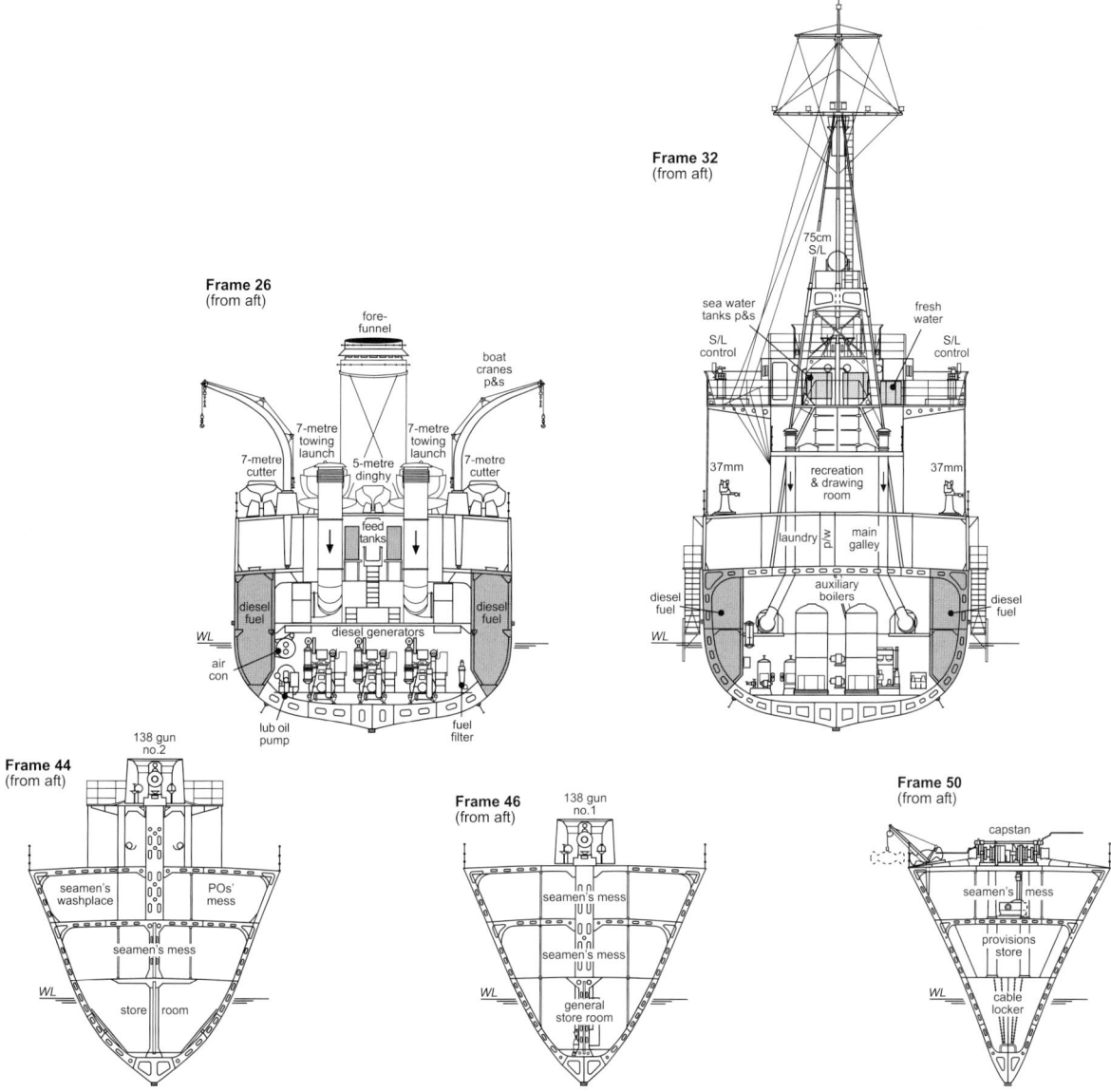

enade decks on ocean liners. A complete set of canvas awnings was also provided; these were normally stowed in a store on the starboard side of the hold aft.

The shelter deck rested on three broad centre-line deck-houses, and there were thirteen pillars per side. The three deckhouses accommodated:

- the ship's 'services' (emergency electro-generators, main galley, laundry);
- the machinery access trunking and officers' galleys (after end);
- the senior officers' accommodation and wardroom, with offices at the forward end.

Orders and Construction

The French naval dockyards of the period were fully occupied with other work, so it was decided that orders for the new sloops would be shared between three private shipyards: Forges et Chantiers de la Gironde (Bordeaux), which was to build A1, A5 and A9; Ateliers et Chantiers du Sud-Ouest & Bacalan Réunis (also Bordeaux), with which the orders for A2, A3 and A6 were placed; and Ateliers et Chantiers de Provence (Port de Bouc – west of Marseille), which built A4, A7 and A8.

The first two ships (A1/2) were authorised under the 1927 Estimates, a further four under the 1929 (A3/4) and 1930 (A5/6) Estimates, and a seventh (A7) under the 1931 Estimates. The French naval budget then came under serious pressure with the construction of the two battleships of the *Dunkerque* class and the six 7600-ton cruisers of the *La Galissonnière* class, and it would be 1937 before the next pair of sloops (A8/9) was authorised. A9 would be sabotaged at Bordeaux when 70% complete in June 1940 – she had yet to be fitted with her machinery, and A10, authorised on 12 April 1939 under the 1938 *bis* supplementary estimates and allocated to A C Gironde, would never be laid down. These last two

13

ships were to have been built to a revised design as survey ships, with a modified armament of two twin 100mm guns. On 1 April 1940 *Ville d'Ys* (A8) was renamed *La Grandière*.

Unlike their smaller British counterparts, which were built to provide numbers at the lowest possible cost, the French colonial sloops were relatively sophisticated ships. The contract price for the 1931 ship was 36.5m FF, a figure not far short of the contemporary *contre-torpilleurs* of the *Aigle* class, which cost c.46m FF.

Hull and Fittings

The hulls of French warships of the period were generally constructed of 50kg mild steel. The *avisos* of the *Bougainville* class were unusual in having the lower part of the hull and keel of 50kg steel but the upper plating of high-tensile 60kg steel (similar to the 'D' steel used in British ships of the period). Bullet-proof special steel of chrome-cobalt-molybdenum (designated *qualité masque* or 'gunshield quality') was used not only for the gunshields, but also for the bridge block, the walls of the three main deckhouses and the lower part of the after funnel, which housed the uptakes for the main diesels. The decks were also of 60kg steel, with a thickness of 6mm over the bridge and the ammunition lobbies and 5mm elsewhere. These figures were comparable to those of French destroyers, but the use of HT and special steels for the upperworks gave greater protection against machine gun fire, as their overseas policing role meant that they could be taken under fire from land.

Although the naval dockyards were beginning to experiment with electric welding, for the *avisos* traditional riveted construction was retained throughout. There were 'double walls' outboard of the mess-decks and officer cabins on the lower and main decks with inert rockwool insulation between.

Subdivision was on a par with contemporary destroyers. Frame spacing was 1.8 metres, and the hull was subdivided into eleven watertight compartments designated A–K by bulkheads which extended from the keel to the main deck. The greater beam of the *avisos* made it possible to work in fuel tanks outboard of the machinery spaces amidships (see General Arrangement plans).

Due to the need to operate in tropical climates, it was decided to fit wooden planking of 40–50mm Siam teak to the upper deck and upper bridge. (This was a feature normally confined to cruisers and capital ships.) There was a special dark grey non-slip paint on the ammunition lobbies and the bridge, while the internal decks were lined with the standard red-brown linoleum secured with brass battens.

A fine study of *Rigault de Genouilly* (A5) entering Melbourne on 16 January 1937. Ships deployed to the French Pacific Naval Station, which embraced New Caledonia and French Polynesia, regularly visited the east coast of Australia during the 1930s. By this time a GL.832 floatplane, visible abaft the second funnel, was regularly embarked. (Allan C Green collection, State Library of Victoria)

Machinery

The two main propulsion diesels were located at the after end of a large motor room amidships, and their exhausts were led up into the second funnel. They were of two different types:

- Burmeister & Wain Type 655 MTF 90 4-stroke, 6-cylinder under licence to Schneider and built by Penhoët; nominal horsepower was 2191CV for 1600/1800bhp; the weight/power ratio was 80kg/CV.
- Sulzer SRK4 2-stroke, 6-cylinder built under licence by A C Loire (St Denis) and F C Méditerranée (Le Havre); nominal horsepower was 1660/2100CV for 1250/1600bhp; the weight/power ratio was c.90kg/CV.

Six of the nine ships were fitted with the Burmeister & Wain 4-stroke diesel, while A2, A3 and A7 had the Sulzer 2-stroke model.

On two diesels the ships had a designed speed of 15.5 knots (16.4 knots max.); on a single engine they were capable of 9.7 knots (10.9 knots max.). It was estimated that maximum speed would decline by 0.75 knots after three months in the tropics.

On trials in July 1932, at a displacement of 2193 tonnes, *Bougainville* attained 17.5 knots on two shafts and 12.5 knots on a single shaft. Comparable figures were achieved in September: 17.3 knots on two shafts and 14.7 knots on a single shaft, with shaft revolutions of 170–200rpm; some vibration was experienced between 120rpm and 140rpm. The time to get underway was 20 minutes when warm, 40 minutes if the engines had been stopped for more than 24 hours. The diesels could be maintained by the crew on station without returning to France.

The two main diesels drove two shafts with 2.8m-diameter three-bladed propellers (3.15m-diameter three-bladed propellers in the Sulzer ships). The original propeller guards were found to be inadequate and had to be reinforced. There was a single balanced rudder with a surface area of 13.33m^2; it could be turned through a maximum angle of 32° (20–25° normal); and was powered by two Sautter-Harlé servo-motors. The ships proved very stable, and had a low angle of heel when turning.

The three main generators were housed in an extension of the motor room forward and also exhausted through the second funnel. The B&W ships had three MAN 6 GVU 33 Ricardo 4-stroke diesel-generators from Schneider rated at 85kW (120V, 700A); the three Sulzer ships had Sulzer 5 RKH 25 2-stroke diesel-generators rated at 120kW (118V, 1017A). The principal disadvantage of this arrangement was that it resulted in a large machinery space 21m by 9m which was vulnerable to flooding.

To provide ship's services when underway and alongside there were two Riley vertical boilers rated at 10kg/cm^2 built by A C Loire. These were housed in a separate compartment forward of the motor room and exhausted through the forward funnel.

The two Bettus-Loire 4-cylinder diesel-powered electro-generators for emergency use were located in the forward deckhouse on the main deck (see GA plans), their exhausts being led up into the after part of the fore-funnel. They were supplied by A C Loire and Schneider and fuelled by lamp oil. Each could deliver 22kW (46kW max.) at 115/118V (188/225A).

The two air conditioning units were housed in a separate compartment abaft the auxiliary boilers. There were refrigeration/cooling units for the magazines (two: fore and aft), the diesels (two) and the cold rooms (two), plus eleven ventilators.

In contrast, emergency pumping capabilities were less complete than in fleet French flotilla craft of the period. There were just five pumps with a capacity of 100 tonnes/hour (the destroyers of the *Bourrasque* class had nine, plus two smaller): two for the magazine compartments (fore and aft), two for the motor room, and one for the boiler room. It was considered that these ships would be more likely to face attack from the surface or from the air, but *Rigault de Genouilly* would be torpedoed by the British submarine *Pandora* off Oran on 4 July 1940, and sank with the loss of twelve men.

Liquids & Coal

Outboard of the main machinery spaces there were side tanks capable of holding 260 tonnes of diesel (*gas-oil*) for the engines and the main generators. Forward of these, separated by void compartments, were bunkers with 5 tonnes of coal for the galleys and tanks holding 60 tonnes of furnace fuel oil (*mazout*) for the auxiliary boilers. Range was estimated on trials as 12,700nm maximum, 9700nm at 10 knots on a single shaft.

All combustible fuels were stowed in cylindrical tanks on the accommodation deck aft. The lamp oil for the Bettus-Loire emergency generators was carried in a single 900kg cylindrical tank to starboard, and in the same compartment were two 450kg cylindrical tanks which held the petrol for the boats. The aviation fuel for the single reconnaissance aircraft was stowed in two cylindrical tanks, the larger of which had a capacity of 1000kg, in a separate compartment to port. Both of these compartments were lined with insulating material and equipped with a CO^2 fire containment system.

Lubrication oil was carried in tanks in a compartment abaft the machinery spaces; normally 7650kg of MT1 oil (10,000kg at deep load) and 1660kg of MT2 were carried. Other liquids included 20 tonnes of reserve feed water (RFW) for the auxiliary boilers, 30 tonnes of fresh water for the crew, and 10 tonnes of drinking water.

Armament

The main gun selected for the *avisos coloniaux* was the 138.6mm Mle 1927 fitted in the latest *contre-torpilleurs* of the *Aigle* class. A 40-calibre weapon, it was the first French destroyer gun with a German-style sliding breech, which was much faster in operation than the Welin inter-

138.6mm/40 Mle 1927

Profile

Rear

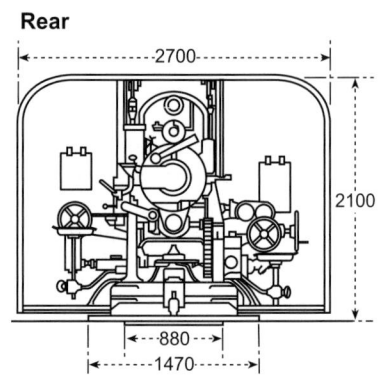

Plan View

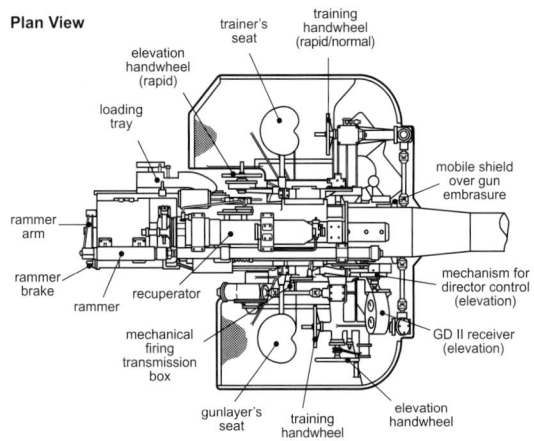

(© John Jordan 2013)

138.6mm/40 Mle 1927

Gun Data

Construction	monobloc autofretted barrel
Weight of gun	4.8 tonnes
Breech mechanism	horizontal sliding block
Ammunition type	separate
Projectiles	OPFA Mle 1924 (39.9kg)
	OEA Mle 1928 (40.2kg)
Propellant	9kg BM7 in cartridge Mle 1910
Muzzle velocity	700m/s
Range at 28°	16,600m

Gun Mounting

Designation	Mle 1927
Protection	3mm shield
Weight of mounting	13.0 tonnes
Elevation	–10° / +28°
Loading angle	any angle
Firing cycle	8–12rpm

Notes:

Mle	*Modèle*	Model
OPFA	*Obus de Perforation en Fonte Aciérée*	Semi-Armour Piercing (SAP)
OEA	*Obus Explosif en Acier*	High Explosive (HE)

rupted screw breech of earlier guns. The theoretical firing cycle was up to twelve rounds per minute – twice that of its predecessor, the Mle 1923 – although 4–5 rounds per minute was more commonly achieved in practice.

A further modification compared to the earlier Mle 1923 was the lowering of the trunnions from 1.34m to 1.25m, making the gun easier to load at low angles of elevation. In order to achieve this it was necessary to sacrifice range: the maximum elevation of the Mle 1927 was only 28 degrees (*vice* 35 degrees), and maximum range, using the same SAP ammunition as the Mle 1923, was only 16,600m (*vice* 18,200m). However, early trials with ships equipped with the basic 3-metre coincidence rangefinder had shown that effective fire control was only possible to 12–13,000 metres. The relatively short barrel of these guns also meant that dispersion was excessive, particularly at longer ranges: 400m at 12,000m and 200m at 7000m. In terms of response times, it took some eight minutes to get the mounting into action in daytime, and ten minutes at night.

The three 138.6mm gun mountings on the *Bougainville* class were fitted with a modified shield which had a squarer, more angular profile than that of the *contre-torpilleurs*, and provided greater protection for the gun crew; each mounting weighed around 13 tonnes complete.

As in the latest *contre-torpilleurs* there were four single 37mm Mle 1925 anti-aircraft guns, which were disposed at the four corners of the shelter deck. The 37mm/50 Mle 1925 was a reliable, if unspectacular gun, which could elevate to 80° and deliver 30 rounds per minute – a creditable enough figure against the slow biplanes of the 1920s, but one which would prove totally inadequate against the high-performance monoplanes of the late 1930s.

For self-defence against strafing aircraft there were to be eight 8mm Hotchkiss MG Mle 1914 in twin mountings Mle 1926. The machine guns were normally stowed below-decks, and when required were affixed to pedestal mountings. The official plans show all four pedestals on the centreline: one atop the forward ammunition lobby, one abaft the bridge, one atop the after ammunition

THE COLONIAL SLOOPS OF THE *BOUGAINVILLE* CLASS

lobby, and a fourth on the quarterdeck. The pedestal abaft the bridge seems never to have been fitted, and some ships later had paired pedestals atop the forward ammunition lobby.

Fire control

For control of the main guns there was a single OPL 3-metre stereo rangefinder Mle 1932 atop the bridge, and a transmitting station equipped with a computer *type aviso*; tangent angle and deflection were calculated mechanically, while a graphical plot supplied range and bearing.

For night firing there were two 75cm BBT Mle 1933 searchlight projectors, mounted fore and aft as in the latest French destroyers. The Mle 1933 had a nominal power of 190,000W and an effective range out to 4500m. As in the *contre-torpilleurs*, the searchlights were remotely controlled from consoles in the bridge wings: there were two consoles per side, the forward consoles (port & starboard) to control the forward searchlight and the remaining pair the after searchlight.

Two smaller OPL 0.8-metre or 1-metre rangefinders were provided for HA fire; the official plans show them mounted on projecting platforms to port and starboard at the after end of the upper bridge. These were not fitted in the earliest ships to be completed.

Ammunition

There were paired projectile and cartridge hoists for the 138.6mm guns fore and aft; these emerged behind no.2 and no.3 gun mountings, with intermediate positions behind no.1 gun. Like the *contre-torpilleurs* of the period the *Bougainvilles* were designed with 'partial' ammunition chutes, fed from the ammunition lobbies, which emerged from the lobbies on either side of main guns. They were generally mounted only when action was in prospect and rarely feature in early photographs of the ships; each of the chutes was in four sections which could be stowed below when not in use. Of the later ships, only *La Grandière* had a completely circular chute (no.2 gun).

The ammunition for the 37mm guns, in six-round magazines, was handled by separate hoists fore and aft which emerged at forecastle deck level close to the mountings.

The magazines were grouped fore and aft. The forward group comprised the shell room and magazine for the forward 138.6mm guns and the magazine for the forward 37mm AA guns; the after group the shell room and magazine for the single after gun, the magazine for the after 37mm guns, and the black powder and small arms magazines. Starshell (for the forward guns only) was in a separate magazine forward (see GA plans). There was sufficient capacity in the main magazines for 709 combat rounds (c.235 rounds per gun) and 76 exercise rounds. Ammunition provision for the 37mm AA guns was c.350 rounds per gun, of which approximately 500 were piercing and a similar number tracer.

Underwater weapons

The ability to lay mines to protect ports and harbours was an important requirement, and was a factor in the decision to locate no.3 gun at the higher level, which effectively cleared the quarterdeck. There were rails for 25 mines per side, as in the latest *contre-torpilleurs* of the

Standard Ammunition for the 138.6mm Mle 1927

138.6mm OPFA Mle 1924
Shell weight: 39.91kg
Burster: 2.30kg Mélinite

138.6mm OEA Mle 1928
Shell weight: 40.20kg
Burstier: 3.45kg Mélinite

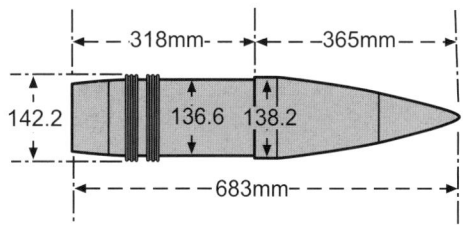

Note: The nose of the OPFA Mle 1924 was subsequently modified to incorporate a dye bag and fuze (*dispositif K*). The modified shells were redesignated OPfK Mle 1924, and were slightly heavier (40.60kg). As these shells were primarily intended for formation firing by divisions of *contre-torpilleurs* it is unlikely they were embarked in the *avisos coloniaux*, which retained the OPf Mle 1924 in its original form.

Cartridge for 138.6mm Mle 1910, 1923, 1924 & 1927
Weight empty: 12.95kg
Weight incl. BM7 charge: 22.00kg

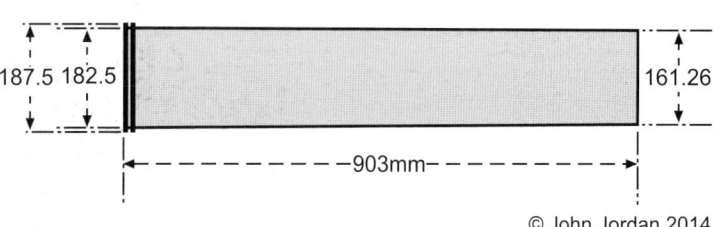

© John Jordan 2014

A later view of the *Rigault de Genouilly*, again with a GL.832 reconnaissance aircraft embarked. The aircraft handling derrick is of the modified braced type and the mainmast has received reinforcing stays; Brest-type life rafts have also been fitted. It is possible that this photo was taken in North Africa shortly before her loss. Note the two twin 8mm Hotchkiss MG mountings forward of the bridge. (Courtesy of Jean Moulin)

Vauquelin class. It was envisaged that the Breguet B4 mine (535kg, 60kg charge, 300m cable) would be embarked, but the larger Harlé H4 (1100kg, 200kg charge, 200m cable) could also be accommodated in smaller numbers. The ships were rarely used for minelaying until 1942.

Another important requirement was the ability to sweep enemy mines. Four Mle 1935 paravanes were carried on the shelter deck amidships (see plan drawing) and handled by the boat crane. They were towed at 20–25m on either beam at around 10 knots.

There were no stern chutes for depth charges; instead the minelaying rails could be used to launch the 200kg Guiraud depth charge. The latter were moved to the stern on trolleys; eight could be embarked per side. Depth charges were not generally embarked until the outbreak of war in September 1939.

As regards underwater detection, in the original design there was provision for the Walser underwater passive detection apparatus and for an ultrasonic active 'pinger', but there is no evidence that either of these was fitted. There was, however, a CET Mle 1931 ultrasonic depth sounder, and a Warluzel echo sounder, which was lowered over the side of ship, was located in the after bridge wings to port. The CET Mle 1931 could register depths up to 2000m in fresh water, and would have been particularly useful in uncharted waters.

Aviation

The floatplane initially embarked was the Gourdou-Leseurre GL.832 two-seat float monoplane, a smaller version of the GL.810/811/812 series which equipped the 10,000-ton cruisers. It entered service in 1934, and was first embarked in a ship of this class in 1935. Like most floatplanes of the period, it could take off and land only in the most favourable sea conditions, but it could be launched from a sheltered anchorage in the tropical climes where these ships saw service. Endurance was approximately 550km at cruising speed. During the late 1930s some ships operated the Potez 452, a high-wing seaplane with a similar 13-metre wingspan and a slightly greater cruising radius.

Originally the aircraft was to have been stowed broken down in a telescoping hangar located on the shelter deck behind the second funnel, as in the contemporary cruiser *Emile Bertin* (see Aviation drawing). However, it was

***Bougainville:* Aviation Arrangements as designed**

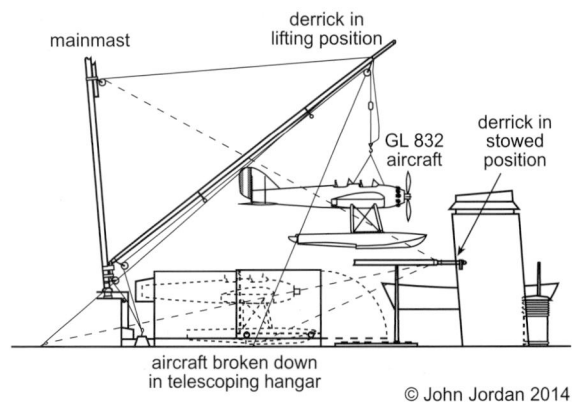

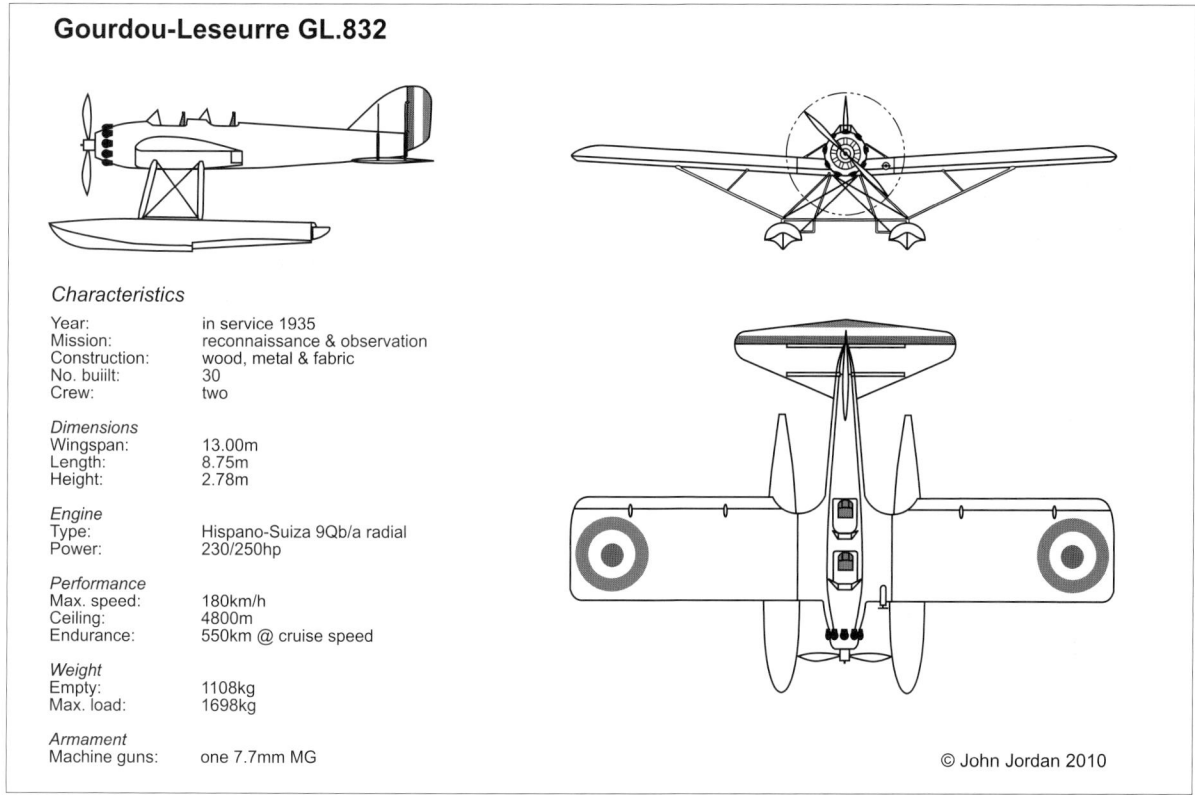

found more practical to stow the aircraft on deck with wings deployed – the aircraft was angled at 45 degrees so that the wing-tips did not overlap the sides of the ship – and to use a specially-designed canvas 'tent' to protect it from the elements.

The aircraft was lowered onto the water and lifted on board by a tubular derrick hinged to the mainmast. This proved too lightly-designed to support the weight of the aircraft; the tubular derrick was replaced by a braced model, and stays were added to the mainmast, which eventually became a light tripod.

When the aircraft was embarked, it was accompanied by a maintenance group of seven men commanded by a lieutenant or a sub-lieutenant.

Boats

Deployment to overseas stations would necessarily entail the use of open anchorages in undeveloped French colonial ports. The design of the *avisos pour campagnes lointaines* therefore featured a comparatively large complement of boats, all of which were carried on the shelter deck forward.

On crutches, and handled by twin free-standing boat cranes, there were:

- two 7-metre motor boats (abeam first funnel)
- one 7-metre motor launch (outboard, to std)
- one 7-metre pulling cutter (outboard, to port)
- one 5-metre dinghy with 3-metre punt inside (centreline)

Outboard of the second funnel, two 8-metre whalers were suspended from davits.

The motor boats were for use by an embarked admiral, the ship's CO and senior officers, the other boats were for ferrying members of the crew ashore (*service de rade*) or, in the case of the whalers on davits, for rescue at sea (*sauvetage*).

The distinctive free-standing 'goose-neck' boat cranes were 6 metres high with a 2-metre reach over the side; each could lift 3.55 tonnes. Two hinged 7-metre boat booms were fitted on the forecastle abeam the bridge for handling the boats when alongside.

Accommodation & Stores

In peacetime the ships of the *Bougainville* class were designed for 35–90 days endurance with a crew of 130. Independent operations on moderately-equipped overseas stations required a greater quantity and variety of stores to be stowed on board, and full advantage was taken of the comparatively broad beam of the ships and the compact machinery spaces (26 metres – only 26% of length between perpendiculars as compared with 40% in contemporary French destroyers) to provide capacious store rooms between the main machinery bulkheads and the forward and after magazines (see GA plans).

Accommodation was unusually spacious, air-conditioned and lined with insulation throughout. The commanding officer, who had the rank of *Capitaine de frégate* (RN: Commander), and the executive officer (*Capitaine de corvette* – Lt.-Commander) both had day and sleeping cabins in the third deckhouse at the level of the upper deck. The wardroom was at the after end of the third deckhouse (see GA plans), and the senior officers had their own galley, which was located to starboard in the centre deckhouse., The remaining officer cabins, for the Engineering and Supply Officers, the Medical Officer (2nd class) and eight junior officers (lieutenant, sub-lieutenant and ensign ranks) were on the deck below, between the machinery spaces and the stern; the junior officers' wardroom was aft, just forward of the steering compartment, and their galley was to port of the galley for the senior officers.

There was accommodation forward for six senior Petty Officers, 14/16 junior POs and 89 men. The latter figure was found to be inadequate – there were constant complaints from the COs about undermanning – and the regulation peacetime complement was soon increased to 97 ratings. (In wartime, with the addition of new anti-submarine and anti-aircraft weapons, the figure would rise to 139!) The Petty Officer accommodation was at the after end of the forecastle to starboard, with a capacious and well-equipped sick bay to port; the seamen's messes were at the forward end of the forecastle and on the main accommodation deck (*pont des logements*) below.

One of the requirements for these ships was the ability to accommodate an admiral and his staff; this would generally be the officer commanding an overseas maritime station or naval force. While on station, the ships might also be expected to embark the governor of a colony or other local dignitaries. The mission of showing the flag and securing/extending French influence around the world was accorded considerable importance, and the quarters provided for an embarked admiral or a visiting dignitary were unusually spacious and well-appointed. They were located in a semi-independent block at the forward end of the junior officer accommodation aft (see GA plans). There was a large day cabin and an equally large sleeping cabin to starboard for the admiral or chief guest (designated *le passager* – 'the passenger' – on the official plans), and two large combined sleeping/day cabins to port for his aides, who in the case of an admiral would be his Chief of Staff (*Capitaine de corvette*) and a lieutenant; the flag staff would also include two ensigns, who shared an adjacent cabin.

The ship's company was expected to provide a 28-man boarding party or a 57-man landing party when required by circumstances. This placed severe constraints on the operational capabilities of the ship – hence the constant complaints about undermanning.

A 1940 reevaluation of the use of these vessels as troop transports established the following guidelines:

– 50 men plus their equipment in heavy weather
– 200 men in moderate/rough sea conditions
– 500 men for short transit in calm weather

D'Iberville (A7) at Toulon on 24 April 1941. The mainmast has received reinforcing stays, the 37mm Mle 1925 single mountings mounted on the shelter deck have been fitted with protective screens, and new light AA guns have been installed: two 13.2mm Browning MG on platforms extending from the sides of the after ammunition lobby and Hotchkiss 13.2mm MG atop the forward lobby, protected by a prominent screen. Note the canvas 'hangar' for the reconnaissance aircraft and the tricolour recognition markings on the shields of guns nos.2 & 3. (Marius Bar, courtesy of Robert Dumas)

The troops were to have been distributed between the seamen's messes forward (50 men maximum), the 'promenade' decks and the shelter deck.

A Brief Service History

All eight ships of the class were despatched to overseas stations shortly after completion; one (*Amiral Charner*) would never return to metropolitan France.

The first ship to deploy, *Dumont d'Urville* (A2), left for the Far East in May 1932, arriving at Saigon on 25 July. Ships deployed to the Far East generally served with the *Division Navale de l'Extrême Orient* (DNEO), the flagship of which was a cruiser of *Duguay-Trouin* class. There were regular deployments to the China Station at Shanghai, occasional deployments to Djibouti on the Red Sea and the *Station Navale du Pacifique* (SN Pac), and port visits to Korea, Japan, Singapore and Australia.

Bougainville (A1) and *Savorgnan de Brazza* (A3) were the next to deploy, leaving for Casablanca on 14 February 1933. From there they would then go their separate ways. *Bougainville* was to be based at Diego Suarez, at the northern tip of Madagascar; she would serve on the *Station Navale de l'Océan Indien* (SNOI) up to the outbreak of war, patrolling the French territories in that region with occasional deployments to Djibouti. Her sister *Savorgnan de Brazza* was destined for the *Station Navale du Pacifique*; she arrived at Papeete (Tahiti) via Fort-de-France (Martinique) and the Panama Canal on 7 May. She then conducted a tour of the Pacific island territories of New Caledonia and French Polynesia before moving on to Saigon. There were docking and maintenance facilities at Diego Suarez (see accompanying map), Djibouti, Nouméa and Papeete, but major refits for ships attached to the FNEO were generally undertaken by the dockyard at Saigon, which had a large graving dock capable of accommodating a cruiser plus two smaller docks which could take a ship the size of the *Bougainville* class (see map).

The fourth ship of the class, *D'Entrecasteaux* (A4), was assigned to the *Division Navale de l'Atlantique* (DNA) and arrived at Fort-de-France on 27 May 1933 via the Azores. There were port visits to Canada and the eastern seaboard of the USA, Bermuda, St Pierre & Miquelon, the Antilles, and to the west coast of Africa. *D'Entrecasteaux* also joined up with the training cruiser *Jeanne d'Arc* for her annual summer campaigns. For ships deployed to the DNA there were docking and maintenance facilities at Fort-de-France (see map), and also at Dakar and Casablanca on the west coast of Africa, but for major refits the ships generally returned to Lorient.

With the political situation in the Far East fast deteriorating the next two ships, *Rigault de Genouilly* (A5) and *Amiral Charner* (A6), were both assigned to the FNEO, the latter beginning her career with an extensive series of port visits to Australia (see photo caption), while *D'Iberville* (A7), which entered service in September 1935, spent the early part of her career in North Africa

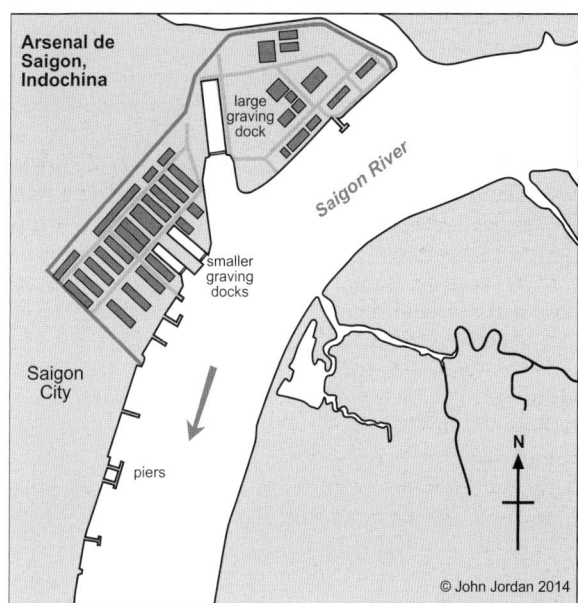

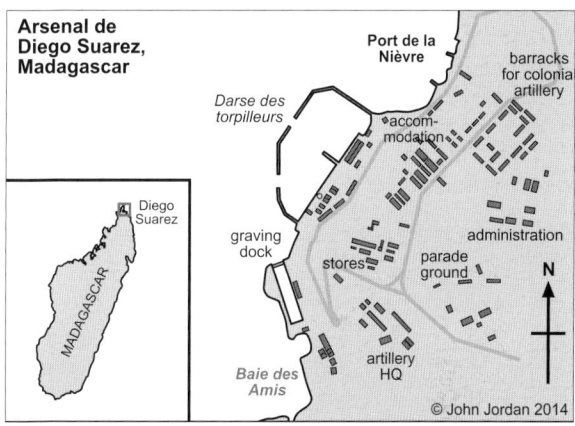

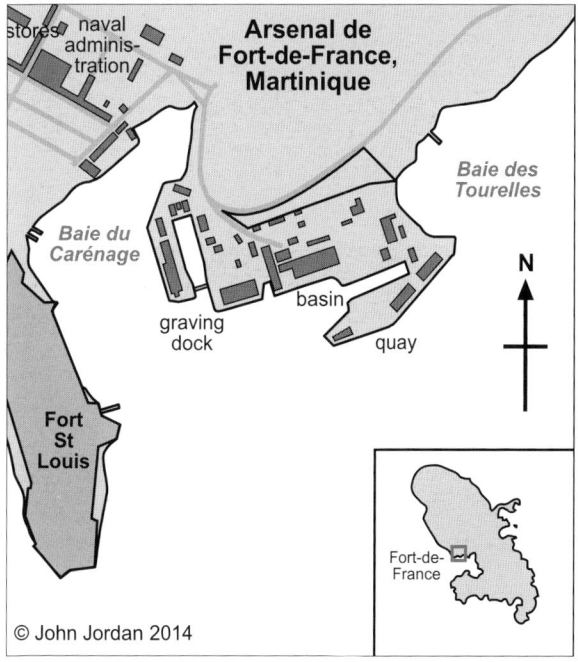

and the Levant (DNL); she was the last ship of the class to be completed before the outbreak of war.

War between France and Germany inevitably meant some adjustments to deployments. *Savorgnan de Brazza* returned to metropolitan France from the Far East for a major refit at La Pallice, while the two ships based in the Atlantic, *D'Entrecasteaux* and *D'Iberville*, were employed on escort duties for transatlantic convoys and for mercantile traffic between West Africa and France.

Savorgnan de Brazza was seized by the British at Portsmouth on 3 July 1940 as part of Operation 'Catapult', and subsequently became one of the two most significant and active ships of the French French Naval Forces (FNFL) – the other was the modern *contre-torpilleur Le Triomphant*. The other seven units in service, which remained loyal to Vichy France, found themselves in the firing line when the French colonies were attacked, first by the British and the FNFL, then by the Siamese and the Japanese. *Rigault de Genouilly* was sunk by HMS *Pandora* off Oran on 4 July following Mers el-Kebir. *D'Entrecasteaux* and *D'Iberville* were present, together with the Free French *Savorgnan de Brazza*, at Dakar during Operation 'Menace' (23–25 September 1940), and the latter ship then proceeded to French Equatorial Africa to rally those countries to De Gaulle; in one of the most notorious incidents in the history of the French Navy, she sank her undermanned and under-prepared sister *Bougainville* at Libreville (Gabon) on 9 November.

A more glorious episode took place on 17 January 1941 when, following border incidents between French Indochina and Siam, the colonial sloops *Dumont d'Urville* and *Amiral Charner* took part in the Battle of Koh-Chang, helping to disable key units of the Siamese Navy in a pre-emptive strike; the conflict was terminated on 28 January after mediation by the Japanese, and a peace treaty followed in May. Meanwhile, on the other side of the world, the British and the Free French continued to chip away at the Vichy French overseas possessions. Operation 'Ironclad', which was launched in May 1942 against Madagascar, saw the loss of the

Right: Two images of *D'Entrecasteaux* at Diego Suarez on 7 May 1942. Damaged by British strafing and shell fire, the ship had been abandoned and sunk in shallow water the previous day. The ensign continues to fly from the mainmast; it would be struck on 11 May. The close-up view shows the damage caused by aerial bombs to the forecastle and bridge. (Courtesy of Conrad Waters)

A superb photograph of *Amiral Charner* (A6) from the Allan C Green collection. Allan C Green was a Melbourne-based photographer, and this photo is one of a series taken when the *Charner* arrived in the city on 4 November 1934 for the Centenary Celebrations. According to the *Sydney Morning Herald* of 3 October, *Amiral Charner* undertook the transit from Toulon to the Dutch East Indies without refuelling. She was due to call in at Darwin, Fremantle, Albany and Adelaide before arriving for the celebrations; from there she would proceed to Sydney before joining the DNEO at Saigon. (Allan C Green collection, State Library of Victoria)

THE COLONIAL SLOOPS OF THE *BOUGAINVILLE* CLASS

D'Entrecasteaux to British naval forces on 6 May at Diego Suarez; damaged by strafing and shell fire, the ship was abandoned and sunk in shallow water.

It was now the turn of the Axis powers to turn the screw. On 27 November 1942 *D'Iberville*, which had been undergoing her first major refit, was scuttled when the German Panzers occupied Toulon. Finally, on 10 March 1945, the unmodified *Amiral Charner*, which had never returned to France from the Far East since her initial deployment in 1934, was scuttled in the Mytho River to avoid capture by the Japanese.

Only three ships of the class survived the war: the Free French *Savorgnan de Brazza*, which had been extensively modernised at La Pallice shortly before her seizure by the British in July 1940 and then in the UK in 1942, and the formerly Vichy French *Dumont d'Urville* and *La Grandière*, both of which had been modernised in the United States during 1944 (see below). All three would see extensive postwar service in their original role, in which they were joined by the ex-Italian colonial sloop *Eritrea*, now re-christened *Francis Garnier* (see Michele Cosentino's article pp.30–41).

Modifications and Modernisations

The outbreak of war in September 1939 saw a change of role for these ships, or at least for those units based in Europe, the Caribbean and West Africa. Their moderate speed and good endurance made them ideally suited to convoy escort duties, and the first priority was to upgrade their anti-submarine capabilities. Guiraud 200kg depth charges were embarked on specially-designed trolleys which used the mine rails on either side of the quarterdeck; each of the rails could accommodate eight charges. The French also developed a conventional depth charge rail, the F28, capable of accommodating twelve of the smaller 100kg depth charges normally fired by the Thornycroft depth charge thrower Mle 1918 (built under licence by the French company Stockes-Brandt). The F28 DC rail, which was mounted on (or just off) the centreline, featured a continuous chain mechanism like the French stern DC chutes but this was manually operated. It was fitted in all except *Rigault de Genouilly* and *Amiral Charner*, which were serving in the Far East.

The intention was to complement the F28 rail with two or four depth charge throwers, mounted on either side of the quarterdeck and angled 50° aft, with stowage for additional charges on adjacent racks. However, as the DCTs were also needed for the destroyers (from which they had been disembarked during the early 1930s) and for a variety of converted auxiliary vessels, this decision could not be fully implemented. Initially only *D'Entrecasteaux*, which was refitted at Lorient July to September 1939, was so fitted, followed by *Savorgnan de Brazza*, which received the full complement of four during her major refit at La Pallice in 1940 (see below).

From early 1940 some ships began to be fitted for or with Asdic 128 (rechristened 'Alpha' in French service) supplied by Britain, but only the one fitted in *Savorgnan de Brazza* was operational before the June Armistice.

La Grandière (A8) at Toulon on 21 July 1942, shortly after her major refit at La Ciotat. The mainmast has been suppressed, and in its place there is a centre-line platform for two 37mm Mle 1925 single mountings, with wing platforms for 13.2mm Browning MG; two additional 13.2mm Browning MG have been mounted in the wings of the upper bridge, and the forward 37mm CAS have been relocated to the forecastle abeam the bridge. An unusual protective housing has been provided for the 3-metre rangefinder atop the bridge, Brest-type life rafts have been fitted forward of the first funnel, and *La Grandière* has received the hull number A 61. Note the aerial spreader fixed to the after AA platform. (Marius Bar, courtesy of Robert Dumas)

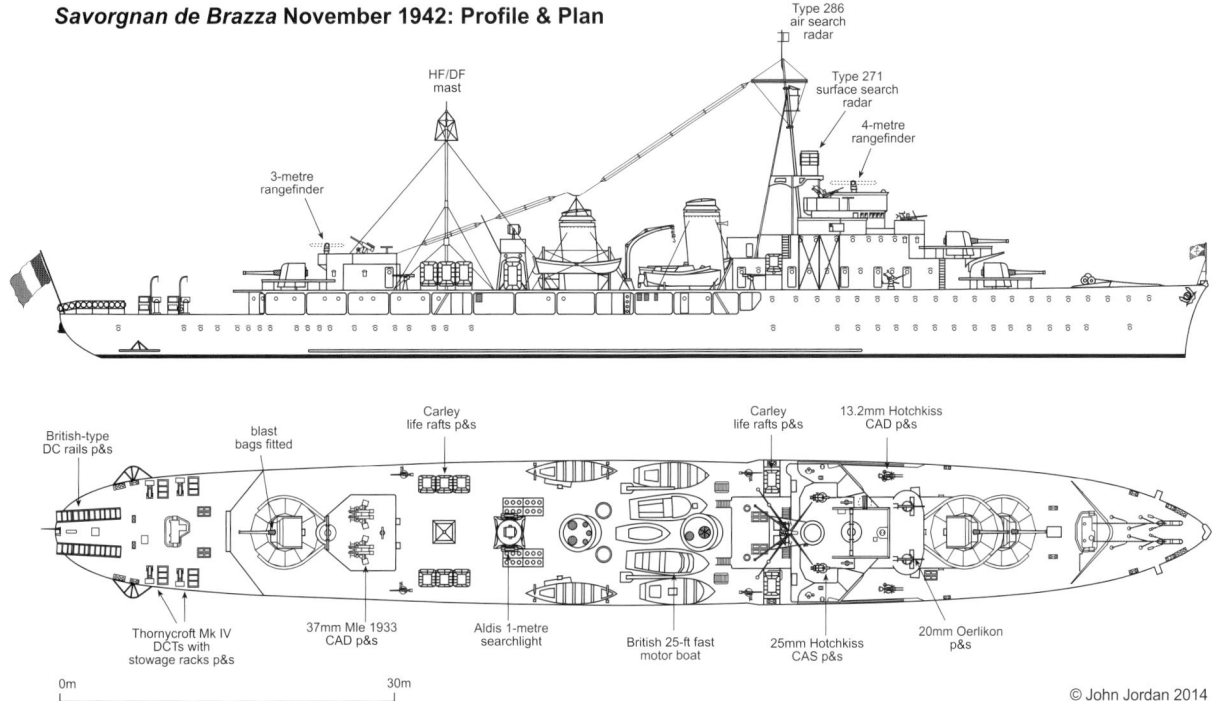

Savorgnan de Brazza November 1942: Profile & Plan

From the spring of 1940 it also became apparent that the light anti-aircraft guns of all French ships in the destroyer and sloop categories were inadequate to deal with the aerial threat when operating off the European coasts. The proposed solution in all cases was the suppression of the tripod mainmast – which in the case of the colonial sloops would also mean the loss of their reconnaissance aircraft – and its replacement by platforms for light AA weapons atop (and to the sides of) the after ammunition lobby. Again the major difficulty in implementing this proposal was a chronic shortage of suitable guns, a shortage which would subsequently be aggravated by the loss of the munitions factories in Belgium and northern France. After June 1940 there were also 'political' difficulties associated with the terms of the Armistice, which effectively prohibited any modifications which increased military capabilities; all proposed modifications, including those undertaken outside the Occupied Zone, had to be authorised by the Italo-German Armistice Commission. The Navy would have liked to fit the sloops and the flotilla craft with the 37mm Mle 1933 twin mounting, but with these in short supply (despite the 'cannibalisation' of ships in Care & Maintenance), those units of the *Bougainville* class which underwent modernisation received a mix of 37mm twin and single (Mle 1925) mountings, and two received the new Hotchkiss 25mm single mounting Mle 1939. When the 13.2mm Browning machine gun became available in numbers in early 1941, two/four of these were fitted in some ships, often in addition to the older-model Hotchkiss 13.2mm and 8mm MG (in single and twin mountings).

Some ships were lost before any of the AA modifications could be implemented, but *Dumont d'Urville*, *La Grandière* and *D'Iberville* all underwent extensive AA modifications at La Ciotat (near Marseille) and Toulon during 1942.

Savorgnan de Brazza

Savorgnan de Brazza served with the DNEO until 19 December 1939, when she left for France via Singapore, Colombo, Djibouti, and the Suez Canal. She then underwent a major refit at La Pallice which was almost certainly intended to serve as a template for all the ships of the class serving in Europe and the Atlantic. The refit took place from 14 February 1940 to 15 May, and the following modifications were made:

– the mainmast was suppressed and a platform constructed atop the after ammunition lobby for two twin 37mm Mle 1933 and two twin 8mm Hotchkiss MG; the original 3-metre rangefinder was relocated from the bridge to the after end of the platform;
– the after 75cm searchlight was moved to a new tower abaft the second funnel;
– two of the latest 25mm Hotchkiss single mountings Mle 1939 were installed in extensions to the after end of the upper bridge and protected by prominent screens;
– a platform was constructed atop the forward ammunition lobby for two twin Hotchkiss 13.2mm MG;
– an F28 depth charge rail was fitted above the stern, and four Thornycroft depth charge throwers Mle 1918 were installed on either side of the quarterdeck; the port mine rail was replaced by a rail for two of the larger 200kg depth charges;

- two twin 8mm Hotchkiss MG were mounted on the quarterdeck forward of the DCTs;
- smoke generators were fitted in place of the after pair of paravanes.

The ship was subsequently moved to Cherbourg (29 May), where she was fitted with:
- a 4-metre rangefinder in an open mounting atop the bridge;
- Brest life rafts (in particular, on the sides of the searchlight tower);
- Asdic 128A (not yet operational).

Following this first modernisation *Savorgnan de Brazza* became flagship of the Pas de Calais Flotilla (CA Landriau), comprising the *contre-torpilleurs Léopard* and *Epervier*, the 2nd/4th/6th *Divisions de torpilleurs* (1500-tonne destroyers), and the 11th/14th DT (600-tonne torpedo boats).

Following the evacuation of Dunkirk, *Savorgnan de Brazza* left for Britain on 18 June 1940, and was seized at Portsmouth on 3 July as part of Operation 'Catapult'. On 17 July she recommissioned with a mixed crew, then from 23 August was transferred to the newly-formed FNFL under the command of CC Roux. Initially assigned to convoy escort duties from Plymouth, she subsequently played an active part in the attempted 'liberation' of French West and Equatorial Africa; she embarked De Gaulle and D'Argenlieu for the abortive operation against Dakar ('Menace'), then sank her sister *Bougainville* at Libreville. In a short refit at Douala in October/November 1940 her Asdic 128 was made operational, and two single 20mm Oerlikons were fitted forward of the bridge, the twin 13.2mm Hotchkiss being moved down to the forecastle; she was also given a new Admiralty disruptive camouflage scheme.

Savorgnan de Brazza spent most of 1941 on active duties in the Indian Ocean, taking part in the blockade of Djibouti. She was now assessed as being in need of a major refit, and following a docking at Durban was despatched to the Swan Hunter shipyard at Wallsend; she remained there until 12 November 1942. In addition to a complete overhaul of her machinery, a number of modifications were made, the primary focus being on bringing her antisubmarine capabilities up to British standards and providing her with the latest small-ship

Left and right: Aerial views of *Dumont d'Urville* following her 1944 modernisation in the USA. Prominent in the bow view are the two Mousetrap launchers on the forecastle; these would later be relocated abeam no.1 gun. The stern view shows the layout of the anti-submarine weapons on the quarterdeck. (For a complete breakdown of the modifications see the labelled drawings.) The ship has been painted in a US Navy Measure 32 disruptive scheme featuring irregular blocks of Navy Blue, Haze Gray, Light Gray and Pale Gray. Note the tubular bars around the AA mountings which limited the training and elevation of the guns to prevent accidental damage to the ship. (US Navy courtesy of Rick E Davis)

THE COLONIAL SLOOPS OF THE *BOUGAINVILLE* CLASS

Dumont d'Urville June 1944: Profile & Plan

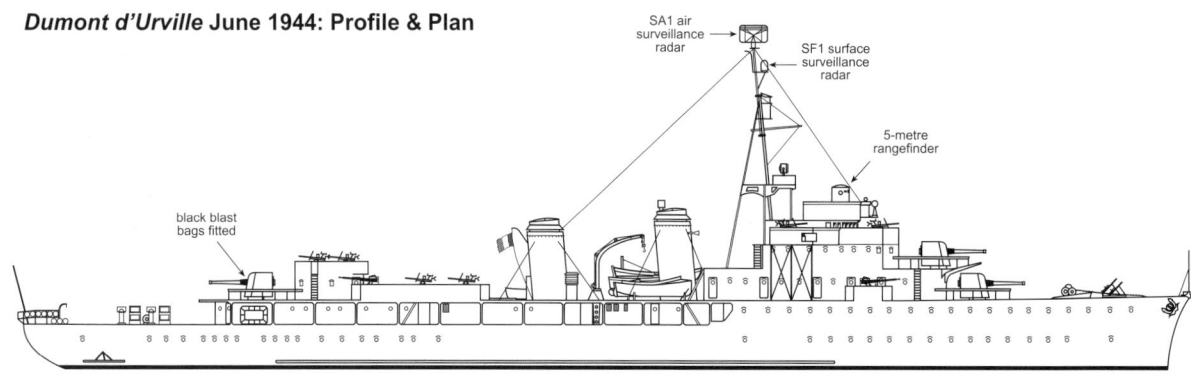
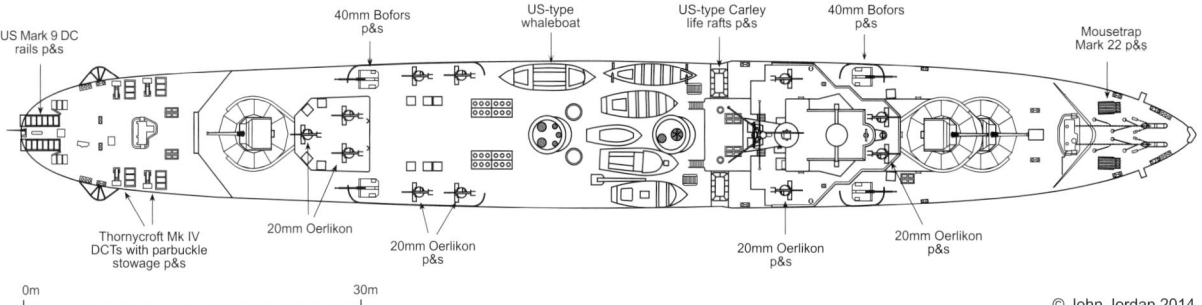

radars and radio intercept gear. The major modifications were as follows:

- blast bags were fitted to the 138mm guns to improve the watertightness of the mountings;
- the forward searchlight was removed and replaced by the distinctive 'dustbin' of the British Type 271 surface search radar (the 4-metre RF was moved one metre farther forward);
- foremast yards in an 'X' configuration were fitted, as in *Le Triomphant*, and the topmast received an aerial for the 'small-ship' Type 286 air search radar;
- a prominent HF/DF mast was stepped amidships, abaft the searchlight tower;
- the after searchlight was replaced by an Aldis 1-metre model; one motor boat and the dinghy were replaced by British models, and Carley floats were fitted outboard of the HF/DF mast;
- the French anti-submarine weaponry on the quarter-deck was removed and replaced by two British-type

The modernisation of *La Grandière*, which took place at the Norfolk Navy Yard, was similar in principle to that of *Dumont d'Urville*, but differed in detail. There appear to be five (unshielded) 40mm Bofors mountings, three aft and two forward of the bridge, and ten 20mm Oerlikons. US-model Mark 9 depth charge rails have been fitted at the stern, but no depth charge throwers are visible. The ship has received the standard US-model SA and SF radars, and is the US Navy's Measure 22: Navy Blue (5-N) hull and Haze Gray (5-H) upperworks. (US Navy courtesy of Rick E Davis)

DC rails, each with 12 depth charges, and four Thornycroft DCT Mk IV with stowage racks, each for three depth charges; in place of the French Guiraud 100/200kg depth charges, the ship was now uniformly equipped with the standard British Mk VII and MkVIIH (heavy) depth charges;
- a new, more powerful capstan was installed aft;
- the Asdic 128A was upgraded to 'C' standard.

The ship would fight in this configuration until a major refit at La Ciotat from January to September 1945.

Dumont d'Urville

Following the Allied invasion of North Africa (Operation 'Torch' of November 1942) the Vichy French ships in North Africa rejoined the Allies. Subsequently, it was agreed with the Americans that the most modern units would undergo modernisation in the United States. Apart from *Amiral Charner*, which remained under Vichy French control in the Far East, and the FNFL *Savorgnan de Brazza*, only two units of the class had survived the first three years of war: *Dumont d'Urville* and *La Grandière*.

On 25 October 1943 *Dumont d'Urville* left Casablanca for Fort-de-France, Martinique, arriving on 7 November; she deammunitioned, then left for Norfolk Navy Yard, from there proceeding to Charleston NY, where she arrived on 7 December. In April 1944 she underwent a major modernisation which was completed on 14 June. The following modifications were made:

- one pair of gangways was removed, the other relocated farther forward, and the accommodation was revised, the former admiral's apartments being converted into additional spaces for the larger crew which the new weaponry would require;
- black blast bags were fitted to the 138mm guns;
- a 5-metre rangefinder in a trainable 'turret', probably from one of the 1500-tonne destroyers of the *L'Adroit* class disabled at Casablanca, replaced the original 3-metre model;
- a completely new AA armament was installed, comprising four 40mm Bofors guns (unshielded) and eleven 20mm Oerlikon;
- US Navy-type SA1 air surveillance and SF1 surface surveillance radars were fitted, the aerials being carried atop the foremast;
- the W/T outfit was upgraded, and now included the US Navy's TBS line-of-sight voice radio for bridge-to-bridge communication;
- two US Navy Mk 9 depth charge rails were fitted at the stern, and four British Thornycroft Mk IV depth charge throwers with the latest parbuckle stowage installed on either side of the quarterdeck;
- two Mousetrap anti-submarine rocket launchers were fitted close to the bow (these were later moved aft to a less exposed position abeam no.2 gun mounting);
- four large Carley-type life rafts were installed on inclined rails abaft bridge, and the complement of boats was revised: some of the original boats were landed, being replaced by a US Navy-type whaler and a plastic Wizard boat;
- the ship was painted in a Measure 32 disruptive camouflage scheme.

Following her modernisation *Dumont d'Urville* was despatched to the Indian Ocean; she arrived at Diego Suarez on 5 September 1944, and remained in the region until the end of the war. Her sister *La Grandière* also arrived in the USA for modernisation in 1944, and was refitted at the Norfolk Navy Yard from 9 March to 30 April. The modifications she received were similar, but by no means identical, to those of *Dumont d'Urville* (see photo and caption for details). Following modernisation she proceeded to the south Pacific via the Panama Canal.

Acknowledgements:

The author wishes to thank Robert Dumas, Jean Moulin, Conrad Waters and Rick E Davis for their help with photographs.

Sources:

Henri Landais, *Les Avisos Coloniaux de 2000tW (1930–1960)*. Tome 1, Lela Presse (Outreau 2012).

Official plans of *Bougainville* and *Amiral Charner* held by the Centre d'Archives de l'Armement, Châtellerault.

THE COLONIAL SLOOP
ERITREA

In order to police their new colonial possessions in the Red Sea, the Italians aimed to create a force of purpose-designed sloops to be employed in the 'presence' and showing-the-flag roles. **Michele Cosentino** writes about the prototype of these ships, *Eritrea*, which incorporated a number of novel design features.

The colonial expansion which occurred from the eighteenth century onwards prompted several European navies to develop a type of warship suited to colonial service. These ships, widely known as 'colonial sloops',[1] were mostly used for showing the flag of their countries to the local population, but they might be involved in real maritime operations in case of regional tensions, crises or conflicts. The growth of colonial empires, especially in the last quarter of the nineteenth century, and the concurrent emergence of new technologies for hull construction, propulsion, and artillery also favoured the development of colonial sloops with characteristics and capabilities not dissimilar from those of traditional warships such as cruisers. In fact, some European navies considered their colonial sloops to be auxiliary cruisers, to be employed for combat duties in 'benign' environments.

The First Steps

The *Regia Marina* (Royal Italian Navy) shared these concepts, although on a lesser scale when compared to the British or French colonial fleets. Italy established its first colonies in Eritrea and Somalia in the late 1880s, and their geographical distance from the motherland obliged the *Regia Marina* to deploy there on a permanent basis a warship specifically designed to show the flag, to enforce Italian sovereignty, to carry out hydrographical and scientific research, and to police maritime areas of interest.[2] The concept of the medium-sized colonial cruiser, well-armed and with good habitability, was reconsidered in 1912. Two ships, *Campania* and *Basilicata*, with a displacement of 2,780 tonnes and a speed of 15.5 knots, were commissioned in 1917 and homeported in Massawa, on the Red Sea, the main port of Eritrea. Other smaller gunboats were built and homeported in China, where they operated in support of Italian political and commercial interests.

At the end of the First World War new foreign policy guidelines were drawn up, and the permanent basing and temporary deployments of Italian warships abroad changed significantly. The 'presence' mission was concentrated in theatres much closer to Italy, namely in the Aegean Sea, North Africa, and the Red Sea, with second-line and auxiliary ships usually deployed on a rotational basis to these stations. However, none of these ships, despite modification to enable them to cope with the tropical climate, could be defined as a true colonial sloop. The modest but growing Italian colonial commitment soon required the construction of a new class of ships specifically intended for overseas service. In 1926–27, the *Regia Marina* commissioned six minelayers, the *Legnano* class, which operated in the Eastern Mediterranean, the Red Sea and the Sea of Japan not only as colonial sloops but also as hydrographical and training ships. In the mid-1920s, Italy was involved in a number of colonial operations, and learned lessons which were used to convert two troop transports, *Cherso* and *Lussino*, as 'transport ships for colonial service'. They had a 3,990-tonne displacement, were equipped with four single 120/45 and two 76/40 guns, and could embark an infantry company and an artillery section.

The real leap forward occurred in late 1932, when the *Regia Marina* submitted to Ansaldo the specifications for a colonial sloop with a legend displacement of 2,000 tons standard and a speed of 20 knots.[3] The vessel was apparently inspired by the French '*avisos coloniaux*' of the *Bougainville* class (the first was commissioned in 1932), but she had some Italian peculiarities. Firstly, she was to be equipped with combined diesel-electric machinery, the result of an experiment carried out by FIAT; there were to be two 4,400bhp diesel engines and two 650hp Marelli electric motors, the latter being driven by a diesel-generator set. The electric motors were intended for cruise speed, while the diesel engines were devoted to high-speed sprints. The availability of an electric motor for cruise speed, already adopted in the 'cruiser' submarines of the *Balilla* class and later used on the long-range submarines of the *Caracciolo* class, would ensure a range of 7,000nm at 18 knots. The armament would include four 120/50 in two twin mountings, two 37/54 AA guns, several 12.7mm machine guns, and 100 mines. This comparatively heavy armament had been conceived following the lessons learned from recent fighting between Italian troops and their supporting ships and Somali tribes operating ashore. In addition the new ship, with a scheduled in-service date of 1935, might become a useful asset in the event of intervention in the Yemen, which Italy planned to occupy swiftly when the local Imam died.

In February 1934 the design prepared by Ansaldo was approved by the *Regia Marina*. However, changes in the Middle East situation compelled the Italian Navy to update its requirements. The risk of a confrontation between Italy and Britain led the *Regia Marina* to enhance its modest forces deployed in the Red Sea with a submarine squadron, and to revisit the design for the colonial sloop. Major General (Naval Engineering Corps[4]) Icilio d'Esposito was tasked with drawing up a new design which would allow the ship to act as a tender for two long-range or four coastal submarines. The requirements for this additional role included a fully-equipped workshop for repairs, the equipment necessary to charge submarine batteries (generators and cables) and to produce compressed air, and accommodation for submarine crews. An additional requirement was added by late 1934: the embarkation of one of the new Ro.43 floatplanes, which was to be lowered and hoisted by a derrick. The ship was also to have a small hospital, equipped with surgical and medical equipment, and tracks for mine laying. The legend displacement increased to 2,200 tonnes, while the adoption of electrical welding helped save some weight. However, the Italian Navy decided to avoid a further increase in displacement, which would have breached treaty restrictions, and chose to downgrade the gun mounting to the older (and lighter) 120/45 model and to embark fewer antiaircraft guns.

Eritrea: Design and Construction

While the plans were being finalised, the ship was officially registered as an Italian military vessel on 20 July 1934 and her construction assigned to the Royal Dockyard of Castellammare di Stabia, near Naples. She was laid down on 25 July 1935, named *Eritrea*, and launched on 28 September 1936. Delivery to the *Regia Marina* took place on 10 June 1937, and she was officially commissioned on 28 June 1937 as a colonial sloop.

Eritrea had a raked yacht-like bow and a rounded stern with a semi-balanced rudder. Bilge keels were fitted for stability along the central section of the hull. The forecastle ran aft as far as the quarterdeck; the after section housing the officer accommodation. The superstructure, two decks high, incorporated the bridge, which had prominent wings and housed the operations rooms and accommodation for the flag staff. A tall pole foremast was located abaft the bridge, equipped with a lookout position and twin platforms for a pair of searchlights. There was a broad single funnel amidships housing uptakes for the diesel engines and generators, while the pole mainmast supported a heavy-lift derrick. A seaplane with folded wings was to have been stowed on the after part of the forecastle deck; however, it was never embarked. Service boats were stowed at the sides of the deck, and there were twin paravanes forward of the superstructure.

Eritrea's full load displacement was more than 3,110 tonnes. Length overall was 96.9 metres, with a beam of 13.3 metres and a maximum draft of 4.7 metres. Internally, the hull was divided horizontally into three decks, with a double bottom running for 80 metres. There were 145 frames and twelve watertight transverse bulkheads from the keel to the upper deck (*ponte di coperta*), which was the strength deck. Forward of the officers' accommodation were the galleys, including one for the native Africans who were usually part of the ship's crew,[5] and several workshops; the forward section housed the hospital and petty officer accommodation.

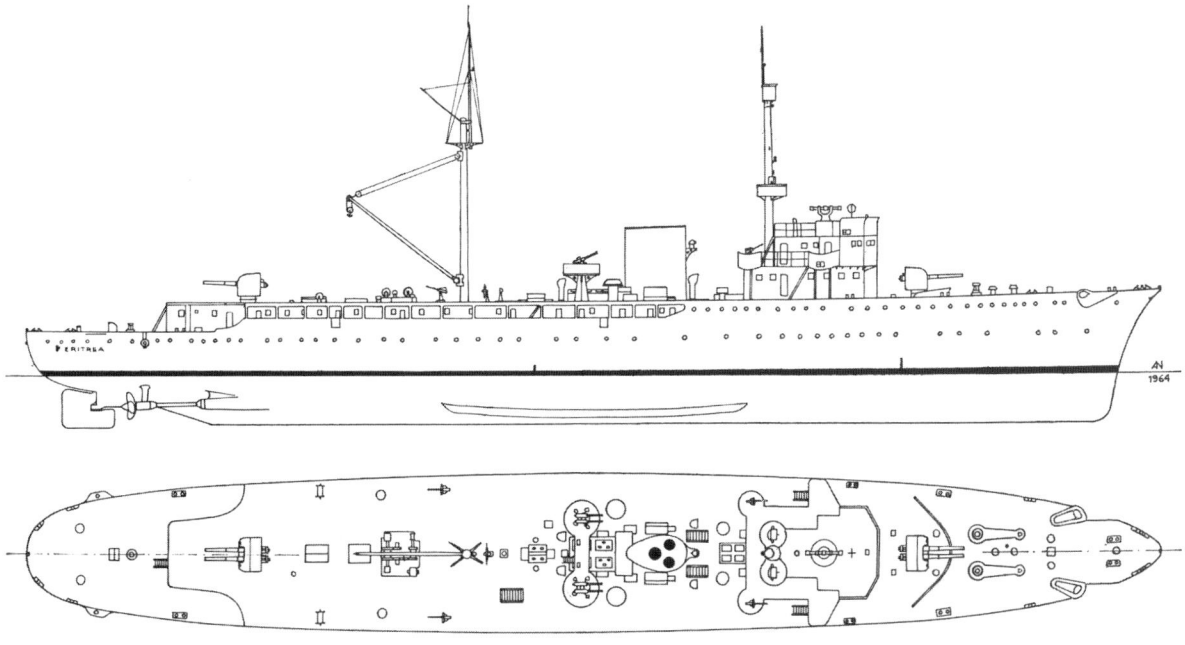

Profile and plan of *Eritrea*. (A Nani, USSM)

The launch of the colonial sloop *Eritrea* at the Royal Dockyard of Castellammare di Stabia (Naples) on 28 September 1936. (Collection Erminio Bagnasco)

The upper platform deck (*ponte di corridoio*) was the main accommodation deck for junior petty officers (aft), seamen (forward), and transported troops. This deck also housed the upper part of the diesel room, a workshop, and the auxiliary boilers and distiller plant room. The machinery spaces occupied the lower platform deck (*copertino*) and the hold (*stiva*): from forward to aft were the main diesel-generator room, a pump room, the main diesel engine room, the auxiliary diesel-generator room, and the electric motor rooms. Fuel, fresh water and oil tanks were located outboard of these spaces on two decks. The layout of the machinery spaces allowed for

This photo, taken a few days after the official commissioning of *Eritrea*, dates from late June 1937; the hull and superstructures are painted white. (Collection Erminio Bagnasco)

THE COLONIAL SLOOP *ERITREA*

A Ro 43 reconnaissance floatplane on the forward catapult of an Italian cruiser. *Eritrea* was to have a Ro 43 aircraft, stowed on the after part of the forecastle deck and handled by the mainmast derrick. (Collection Giorgio Parodi)

Table: *Eritrea*
Characteristics

Built:	Cantiere di Castellammare di Stabia
Building dates:	Laid down 25 Jul 1935, launched 28 Sep 1936, completed 28 Jun 1937
Displacement:	2,200 tonnes standard, 3,117 tonnes deep load
Dimensions:	87m pp, 96.9m oa x 13.3m x 4.7m (max.)
Machinery:	two-shaft diesel-electric; two FIAT 6-cyl diesels, each 3,500bhp; two Marelli electric motors, each 650hp; 8,300hp = 20 knots (designed)
Endurance:	8,000nm at 10kts
Armament:	4 – 120/45, 2 – 40/39, 4 – 13.2mm MG 100 Bollo-type mines
Complement:	13 officers and 221 men

two adjacent compartments to be flooded while retaining a margin of buoyancy and operational capability.

The two FIAT 6-cylinder diesel engines each had a sustained power of 3,500bhp and a maximum power rating of 3,900bhp. They drove two shafts directly via couplings; there was no gearing. Two Marelli 650hp electric motors driven by the main FIAT main diesel-generator sets were also coupled to the two shafts. The machinery arrangement permitted four operational modes:

– electric motors only, driven by the two main diesel-generators sets and providing 1,300hp, for a corresponding cruise speed of 11 knots – the auxiliary diesel-generator sets provided power for the 'hotel services';
– electric motors only, driven by a single diesel-generator set which provided power also for the hotel services; in this configuration the cruise speed was 8 knots;
– main diesel engines only, providing a total of 7,000bhp for 18 knots and with the diesel-generator sets providing power for the hotel services;
– diesel engines and electric motors combined, for a total of 8,300hp and 20 knots. (Note, however, that this top speed was not achievable in practice because the diesel engines and the electric motors ran at different revolutions.)

All the speeds listed above assumed a clean hull and engines/motors in good working condition; in practice the sustained speed in modes c) and d) was 15 and 18 knots respectively. The normal oil stowage was 283 tonnes, while the radius of action varied from 5,000nm

A view of the port of Massawa taken during a ceremony (probably the Italian Navy Day) in June 1940. *Eritrea* is in the left foreground, with Italian destroyers at anchor. (Collection Erminio Bagnasco)

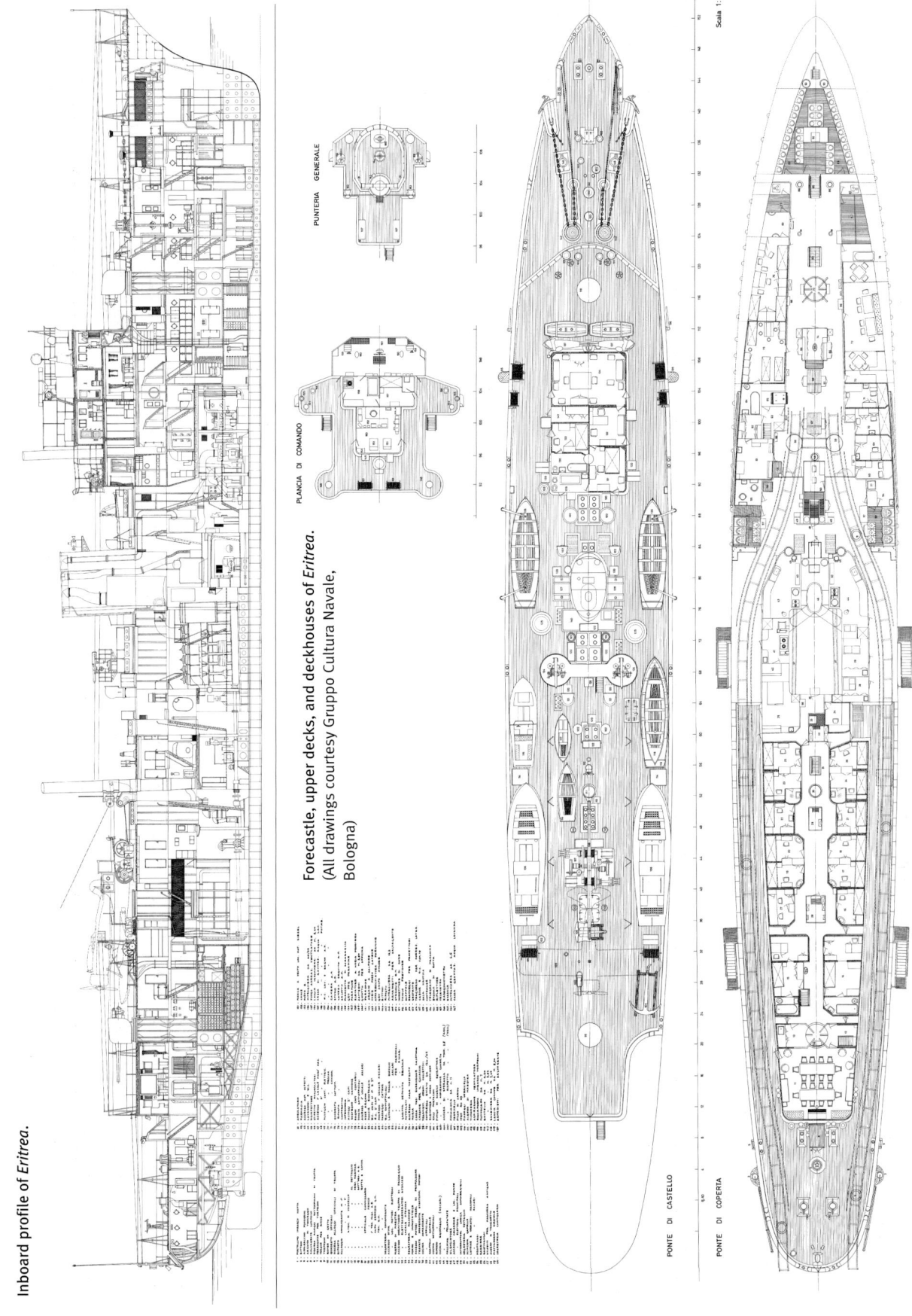

Inboard profile of *Eritrea*.

Forecastle, upper decks, and deckhouses of *Eritrea*.
(All drawings courtesy Gruppo Cultura Navale, Bologna)

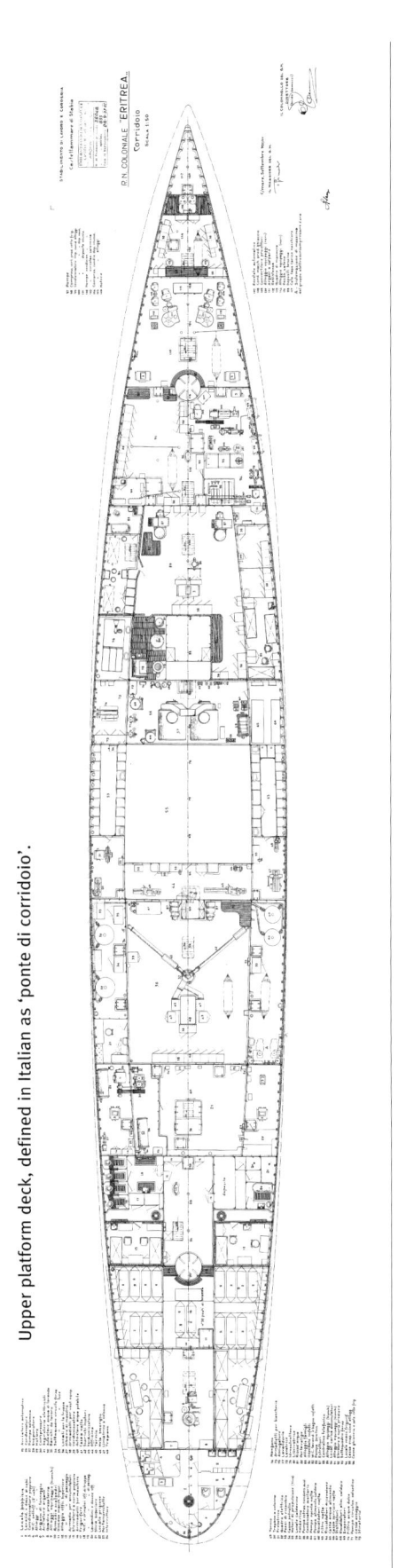

THE COLONIAL SLOOP *ERITREA*

Upper platform deck, defined in Italian as 'ponte di corridoio'.

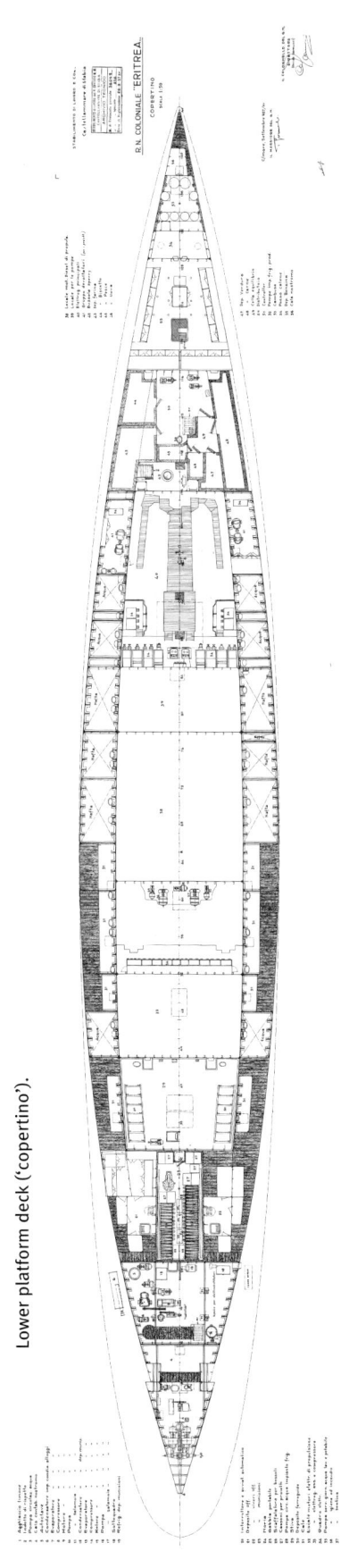

Lower platform deck ('copertino').

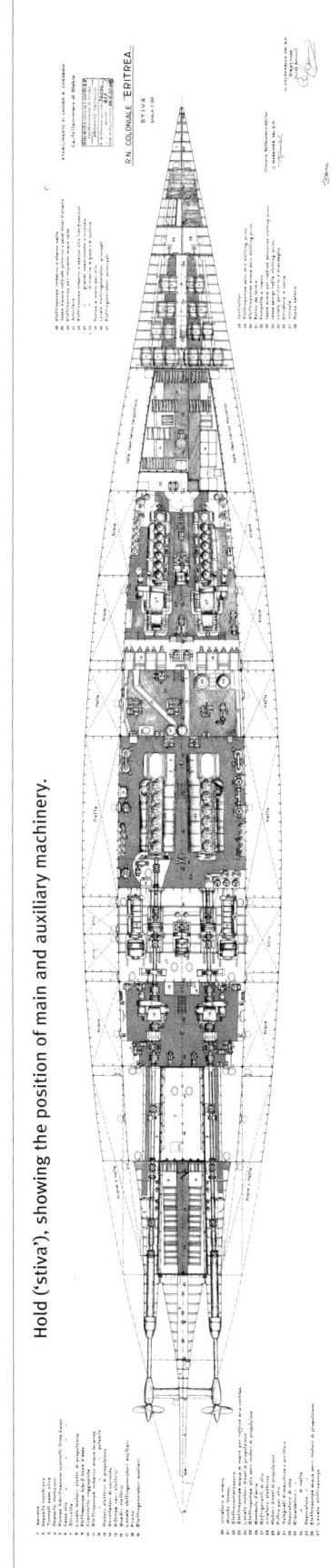

Hold ('stiva'), showing the position of main and auxiliary machinery.

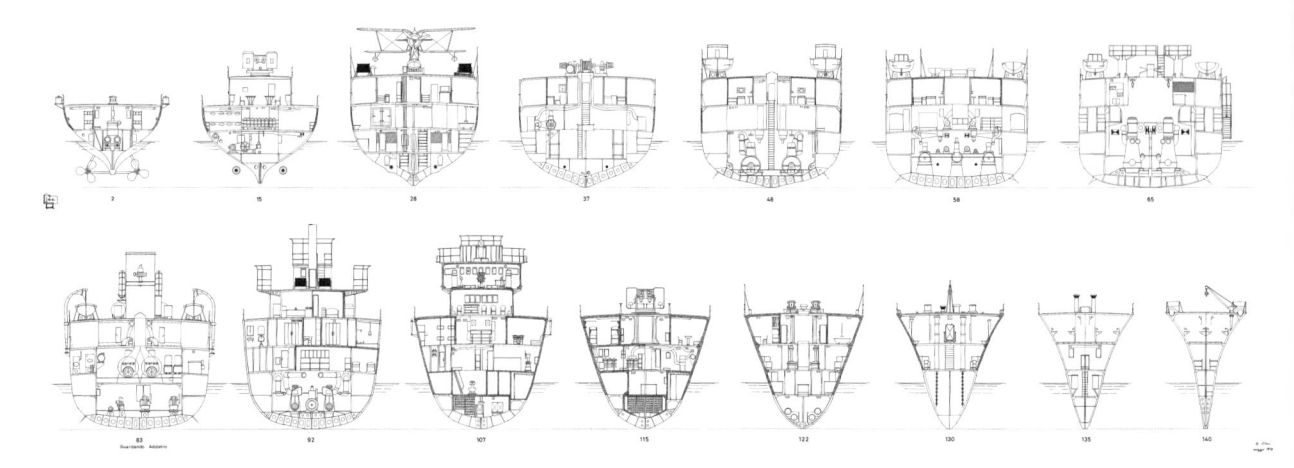

Transverse sections of *Eritrea*, looking forward. (Courtesy Gruppo Cultura Navale, Bologna)

to 8,000nm, depending on speed and machinery configuration. The most economical configuration was 10 knots with the diesel-electric combination, with the diesel engines running at minimum power.

Apart from the provision of large spaces for accommodation, habitability was improved by the installation of large fibreglass panels which insulated the internal space from the heat radiated by the hull plating. Food, wine, and other provisions were stored in large rooms most of which were located forward. Two auxiliary boilers provided steam for the production of fresh water, using a set of evaporators and distillers. The planned complement was 13 officers and 221 men, with accommodation available for an additional 100 officers and men.

Eritrea's armament included two Canet-Schneider-Armstrong 120/45 mm Mod. 1918–19 guns in twin shielded mounts, located forward of the superstructure and at the after end of the forecastle deck. These guns were of an older type generally fitted on auxiliary cruisers and troop transports; their maximum range was 12,600 metres, with a rate of fire of six rounds per minute.[6] Rudimentary fire control for these guns was provided in the form of a Galileo rangefinder atop the bridge. Anti-air armament included two Vickers-Terni 40/39 Mod. 1917 guns in single mountings located abaft the funnel on two raised platforms. These guns were water-cooled and their production had ended in 1935. Two single 13.2mm machine guns (later twins), mounted at the after end of the bridge, and two 6.5mm machine guns, atop the bridge, provided short-range protection. *Eritrea* was also designed to carry 100 Bollo-type mines,[7] to be stowed on tracks on the upper deck, which were fitted in 1940.

When completed, *Eritrea*'s hull was painted white but later, due to a scarcity of resources, the ship was compelled to adopt a light grey livery.

Etiopia and the Planned Conversion of the Cruiser *Ancona*

In 1939–40, the Italian Navy planned to build a second colonial sloop, to be named *Etiopia*.[8] She would have been an improved *Eritrea*, with similar dimensions, armament, and appearance. The machinery would be based instead on the new Tosi diesel engines, providing a motive power greater than the FIAT engines in *Eritrea* for a top speed of 23 knots. The outbreak of the Second World War hampered the construction of *Etiopia*, which

A close-up of *Eritrea*'s quarterdeck, with sailors manning the rails. The aft twin gun mount is visible in the upper part of the photo. (Italian Navy Historical Office)

A picture of *Eritrea* ready to set sail, probably during the cruise carried out in the Persian Gulf in March 1938. (Italian Navy Historical Office)

THE COLONIAL SLOOP ERITREA

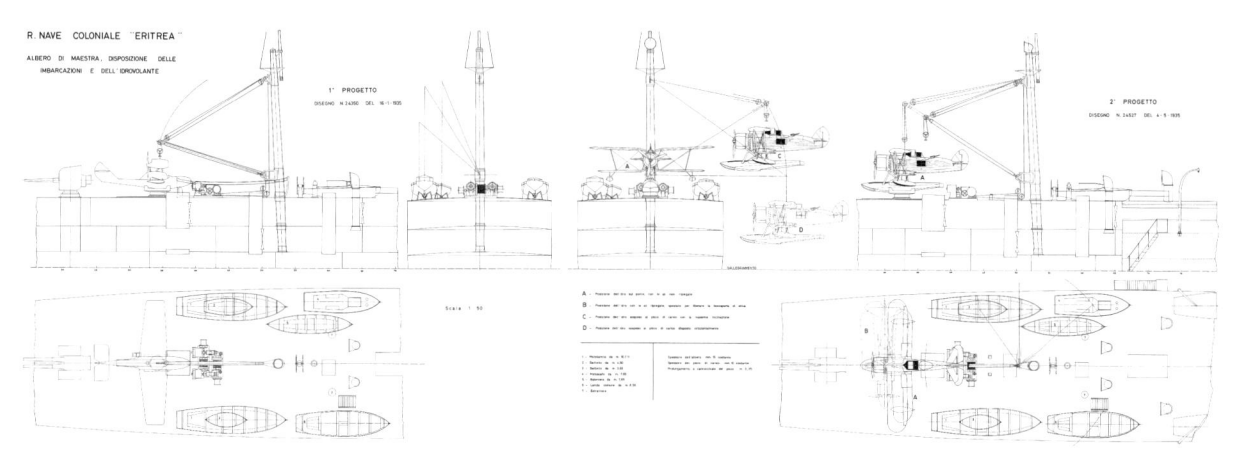

Handling arrangements for the embarked seaplane and the service boats. (Courtesy Gruppo Cultura Navale, Bologna)

was designed for superior sea-going performance to *Eritrea* at the expense of the submarine tender role. Not only would she have more powerful engines capable of greater speeds, but the simplicity of the machinery linkage would make for greater reliability; however, the armament remained inadequate for commerce raiding.

Another attempt to operate a colonial sloop was made through the planned conversion of *Ancona*, an older cruiser dating back to the Great War.[9] The preliminary design work for her reconstruction had been carried out in the second half of 1936 by the Royal Dockyard of Taranto. Her original armament, comprising seven 15cm SK L/45 guns in single mountings, was to be supplemented by three 76/40 single mounts intended for the engagement of small vessels. *Ancona*, which had a full load displacement of about 6,400 tonnes, had already had her original machinery upgraded with new oil-fired boilers, and the original three funnels were to be replaced by a single funnel. The new machinery and the ample space available on board were features which could be utilised to increase *Ancona*'s range and to transform her into a useful flagship for colonial duties. However, her inherent obsolescence adversely affected her combat value, while a lack of funding forced the *Regia Marina* to abandon the reconstruction, even though at the time Italian colonial commitments were increasing.

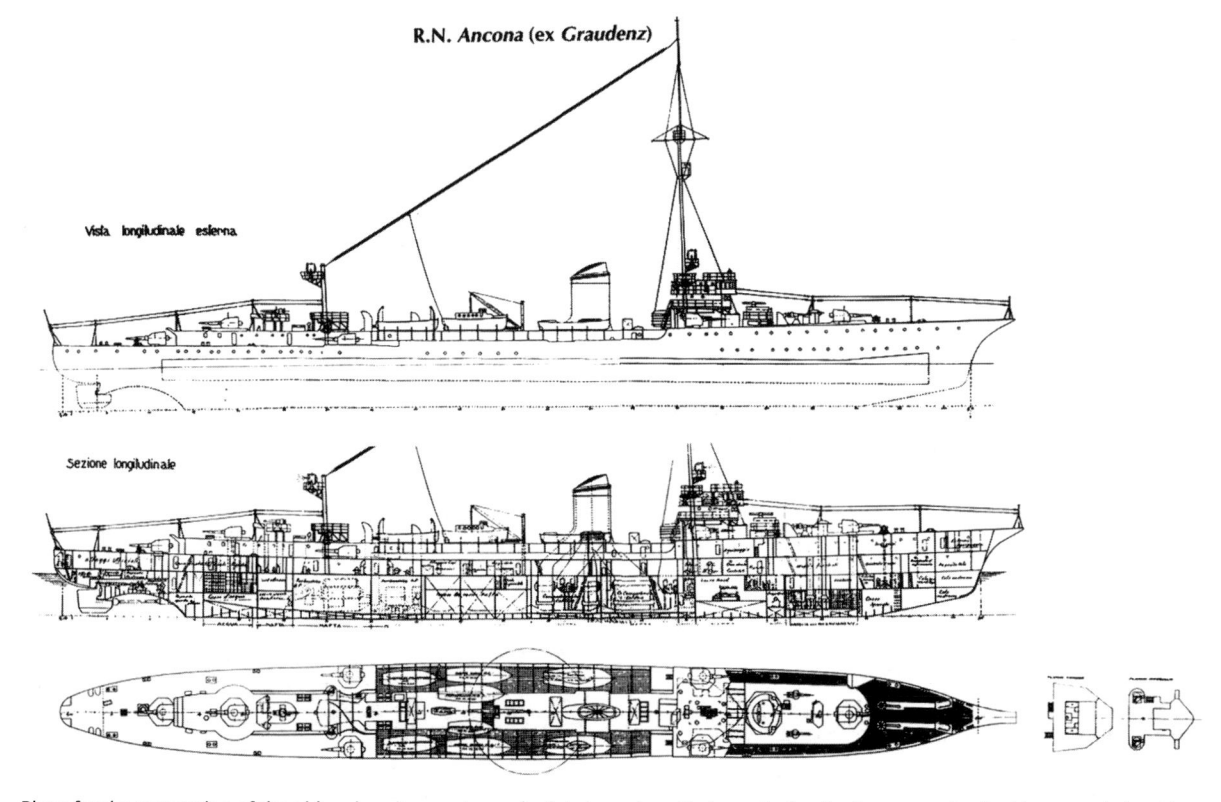

Plans for the conversion of the old cruiser *Ancona* to a colonial sloop. A preliminary design for the conversion had been carried out in the second half of 1936 by the Royal Dockyard of Taranto, but it was later cancelled. (*Storia Militare*, via Maurizio Brescia)

Eritrea underway at sea, conducting trials of the paravanes. This photo shows the general arrangement of the ship. (Italian Navy Historical Office)

A further attempt to build a colonial sloop dates back to 1933–34, when Ansaldo drafted a sketch design of a 93m-long ship with two funnels and simpler machinery based on two diesel engines with direct drive. The legend displacement was 1,940 tonnes; horsepower and speed were 8,800bhp and 20 knots respectively. The armament was to include two 152/50 twin mountings, two 37/54 mm twin mountings, a number of 13.2mm machine guns, and mines. However, it appears that this effort was overtaken by the development of *Eritrea*'s design.

Sailing on Two Oceans

In principle, *Eritrea* was an excellent sea boat and performed well as a colonial sloop; for the most part she was moored alongside, occasionally sailing to show the flag. In service, her machinery proved unsuited to long deployments; it was difficult to operate and prone to breakdown. Before being deployed in the Red Sea, in late 1937 *Eritrea* carried out three missions in the western Mediterranean during the Spanish Civil War. Later, she was based at Pola, which she departed bound for Massawa, arriving there on 8 May 1938. During the late 1930s, *Eritrea* paid visits to many ports, including Aden and Hodeida (Yemen), as the flagship of the Italian naval command in East Africa. When Italy entered the war on 10 June 1940, the East Africa command comprised *Eritrea*, two flotillas of elderly destroyers (7–8 ships) and a single submarine squadron (6–7 boats). In early June 1940, *Eritrea* laid a defensive minefield at the entrance of the port of Massawa. During the following months, her officers and men were used as a reserve to replace personnel on Italian destroyers, while her machine guns contributed to the air defence of Massawa.

In early 1941, the Italian Navy was aware that British forces were to occupy Eritrea and decided to send *Eritrea* to Japan.[10] The ship was prepared for a long ocean crossing, including work aimed at facilitating her operation in distant waters and concealing her presence in maritime areas controlled by the Royal Navy. The height of both masts was reduced, while the mine-laying equipment, the searchlights and the cables for charging submarine batteries were all removed. The major concern was the machinery, which had already given problems and had not been operated for two months. In early February, the ship had her bottom scraped, an operation which in tropical waters was often essential. Another major issue to resolve was fuel stowage, which was deemed insufficient for a long oceanic voyage. *Eritrea* therefore embarked 780 fuel barrels: 240 stowed on the lower decks and the others placed on the upper deck – an arrangement which could prove dangerous in the event of air attacks.[11] In order to provide sufficient space for the fuel barrels and improve arcs of fire for the after guns, several of the service boats were landed. The captain of *Eritrea*, Commander Marino Iannucci, also decided to leave ashore some 100 men deemed not essential for the planned voyage. The crew would consist of 13 officers and 188 men, 19 of them *ascari*.

On 16 February 1941, *Eritrea* carried out trials at sea to work up and calibrate her machinery, perform practice gunnery shoots, and verify the condition of the ship, her crew and her equipment. *Eritrea* left Massawa at dusk on 19 February 1941.[12] After a relatively uneventful voyage, she arrived in Kobe (Japan) on 22 March, having safely negotiated the naval blockade established by the Royal Navy off Aden and in the Indian Ocean, and the surveillance of the Royal Netherlands Navy in the region of the Dutch East Indies.[13] During the crossing of the Sunda Straits (7–11 March), *Eritrea* was camouflaged to look like the *Pedro Nuñes*, a small Portuguese escort ship, in order to deceive Dutch surveillance in that area. At the end of the voyage, Commander Iannucci simply wrote in the ship's log: '9,555 miles; 754 hours of navigation; consumption of fuel 440 tonnes'.

In Kobe *Eritrea* was visited by a Japanese Navy's technical-military commission to assess what kind of repairs she needed, with a particular focus on her machinery. Discussions were prolonged because the Japanese authorities were not keen to provide support to a belligerent nation such as Italy. However, in April the Japanese authorised the company Mitsubishi to perform the work needed to ensure the ship could sail safely. *Eritrea* was ready again to set sail in July, but she left Kobe only on 24 August and four days later arrived in Shanghai. The international situation then changed dramatically, with Japan attacking the United States fleet in Pearl Harbor and expanding the conflict into the Pacific Ocean. *Eritrea* left Shanghai on 8 July 1942 and two days later arrived in Tsingtao, where she remained until 27 September.

Another chapter of *Eritrea*'s deployment to the Far East opened when the *Regia Marina* decided to use the colonial sloop as a support ship for the Italian submarines operating between France and Japan for the transfer of special materiel. Thus, *Eritrea* operated on a regular basis between Singapore, Penang and Sabang, sailing across the Malacca Strait. This activity ended on 9 September 1943, when the Italian government signed an armistice with the Allies. At that moment, *Eritrea* was at sea in the Malacca Strait and Commander Iannucci ordered to set course for Colombo, where the ship arrived on 14 September 1943.

After Italy declared war on Germany (13 October 1943), *Eritrea* remained at Colombo and served as a support and training ship for Allied submarines. On 11 September 1944, the ship set sail from Colombo, called at Aden, Massawa and Port Said, and arrived at Taranto on 23 October. After months of repair work, *Eritrea* again deployed to the Ocean Indian, operating as a training ship between Colombo and Trincomalee and

Eritrea in the Red Sea in early 1941. She was the flagship of the Italian Naval Command in Eastern Africa. In February 1941 *Eritrea* set sail for Japan, arriving at Kobe on 22 March. (Collection Erminio Bagnasco)

returning to Taranto on 7 February 1946. Two months later, the ship was back in the Far East, calling at Port Said, Aden, Colombo, Singapore, Hong Kong, and Shanghai. This new deployment to the Far East ended on 16 October 1946, when *Eritrea* arrived in Taranto.

Another Life

The peace Treaty imposed on Italy by the Allied powers stipulated the transfer of *Eritrea* to France. The ship was sent to the shipyard of Castellammare di Stabia for some preparatory work, officially decommissioned from the Italian Navy on 27 January 1948, and renamed F1. She sailed to Toulon with a civilian crew and the French Navy took her under its control on 12 February. Due to the scale of France's own colonial commitments, the former *Eritrea* was commissioned into the French Navy

Eritrea at anchor off Massawa; the searchlight is missing from the starboard platform. (Collection Giuseppe Garufi)

Eritrea was transferred to France in 1948, commissioned as *Francis Garnier* in January 1950, and reclassified as 'aviso', with the pennant number A04. (Collection Erminio Bagnasco)

on 12 January 1950, renamed *Francis Garnier* after the prominent 19th century explorer, and classified as '*aviso*' with the pennant number A04. She conducted a number of hydrographical campaigns in the Pacific Ocean, and showed the flag in French Polynesia; she was also docked in Auckland, New Zealand, for hull scraping.

Francis Garnier underwent a much-needed refit at Toulon from October 1951 to September 1952, when her original main armament was replaced by guns taken from the former Italian destroyers transferred to France following the peace Treaty.[14] The ship embarked a single 120/50 Mod. Ansaldo 1940 forward and a twin 120/50 Mod. Ansaldo 1937 aft. The armament also included four 40mm Bofors and four 20mm Oerlikon guns for air defence, together with antisubmarine weapons. The original fire control director was refurbished, while a small surface surveillance radar was installed atop the foremast; communications equipment was also upgraded. *Francis Garnier* was reclassified as an escort ship (*aviso-escorteur*) with the pennant number F730, and participated in the war in Indochina. She operated mainly in the South China Sea, participating in the evacuation of refugees from the Tonkin Gulf and leaving Saigon for the last time in August 1955. In the following years, *Francis Garnier* operated mostly in the Pacific Ocean as a support ship during the construction of the facilities for the nuclear tests to be conducted in that maritime theatre by the French Navy. On 1 January 1966 she was placed in reserve at Papeete, having totalled 234,000 miles under the French flag. Finally *Francis Garnier* was sunk

From October 1951 to September 1952 *Francis Garnier* was refitted in Toulon. Her original main armament was replaced with guns taken from ex-Italian destroyers already transferred to France. The ship was reclassified as 'aviso-escorteur', with the pennant number F730. (Collection Maurizio Brescia)

THE COLONIAL SLOOP ERITREA

A photo of *Francis Garnier* mooring alongside, probably at Papeete. The ship was sunk as a target by ships and aircraft belonging to 'Force Alpha' on 29 October 1966. (Collection Robert Dumas, via Maurizio Brescia)

as a target by ships and aircraft belonging to 'Force Alpha' on 29 October 1966;[15] the hull now rests on the bottom at a depth of 1,300 metres.

Acknowledgements:
The Author is very grateful to Gino Chiesi, head of the 'Gruppo Cultura Navale', Bologna, for the provision of *Eritrea*'s large-scale plans.

Footnotes:
1. The term 'sloop' and its derivative 'sloop of war' come from the classification introduced by the Royal Navy during the era of sailing vessels and referred to ships equipped with fewer than 18 guns. It was later generically adopted for steam-powered ships which were neither capital ships, cruisers, nor destroyers.
2. For these roles, the *Regia Marina* developed a new type of protected cruiser, the *Regioni* class, of which six were built.
3. This was the maximum displacement and speed permitted for the 'treaty exempt' category defined by the London Treaty of 1930 (see the Editor's article on the French colonial sloops of the *Bougainville* class pp.8–29).
4. In the Italian Navy, the Naval Engineering Corps is roughly equivalent to the Royal Corps of Naval Constructors.
5. The native Africans, called *ascari*, had their own separate accommodation, toilets, and messes.
6. The adoption of such heavy main guns for a colonial sloop is due to the fact that *Eritrea* was also designed to serve as a commerce raider, particularly in the Indian Ocean. However, the older-model guns and her comparatively low speed made her unsuited to this role.
7. These moored contact mines were spherical and contained 125kg of explosive; they were best suited to defensive minefields with a maximum anchoring depth of 350 metres.
8. Ethiopia had been conquered in 1936, leading to the proclamation of the Italian empire.
9. *Ancona* was the former German cruiser *Graudenz*, built in the Imperial Dockyard of Wilhelmshaven and transferred to the *Regia Marina* in 1920.
10. Although not yet involved in the Second World War, Japan had a political agreement with Italy and Germany ('Tripartite Pact').
11. Fuel stowage was thereby increased by about 180 tonnes. After use the empty barrels were jettisoned at sea.
12. The operational order stated: 'Sail to Japan. Prepare to scuttle the ship and destroy the documents in case of necessity. Avoid at all cost any engagement with the enemy. Keep absolute radio silence'.
13. During the voyage, numerous problems were experienced with the diesel engines and *Eritrea* had to run on her electric motors. She maintained an average speed of 12 knots, which proved sufficient to reach Japan unscathed.
14. They were *Oriani* (later *D'Estaing*), *Legionario* (later *Duchaffault*), and *Mitragliere* (later *Jurien de la Gravière*).
15. The fleet escorts *La Bourdonnais*, *Forbin* and *Jauréguiberry* (6e DEE) fired HE piercing rounds against the *Francis Garnier* to little effect. She then survived 100mm rounds fired by the carrier *Foch* and 400kg and 250kg bombs dropped by the latter's Etendard strike aircraft, before finally succumbing to depth charges dropped by *Foch*'s Alizé anti-submarine aircraft.

41

THE JAPANESE DESTROYERS OF THE *ASASHIO* CLASS

With the expiry of the Washington Treaty in prospect, the *Asashio* class marked a return to the large destroyers which the Imperial Japanese Navy had favoured since the late 1920s, when it had embarked on the construction of the 'Special Type' (*Fubuki* class). **Hans Lengerer** looks at the design characteristics of these ships and, in particular, the problems they experienced with their turbines.

The *Hatsuharu* class was designed under the new constraints of the London Naval Arms Limitation Treaty of 1930 as a 'reduced *Fubuki*'. The aim was to provide the maximum fighting power in a smaller, lighter hull. However, the ships of this class had insufficient stability, highlighted by the *Tomozuru* Incident (see the author's article in *Warship 2011*, pp.148–164), which required urgent remedial action, and insufficient hull strength, highlighted by the Fourth Fleet Incident (see *Warship 2013*, pp.30–45), which led to reinforcement of the hull members. These successive reconstructions resulted in a marked reduction in fighting power, in particular the number of torpedoes which could be embarked, and also in maximum speed, both of which were key characteristics of this type of warship.

The ships of the *Shiratsuyu* class were designed to restore some of the capabilities lost when the *Hatsuharu* class had to be rebuilt, with a consequent reduction in armament and speed (see the author's article in *Warship 2007*, pp.91–112). Offensive capability was boosted by the adoption of the newly-developed Type 92 61cm quad torpedo mounting, thereby increasing the number of tubes from six to eight, but it proved impossible to boost maximum speed beyond 34 knots using the same propulsion machinery as the *Hatsuharu* class. Moreover, the smaller hull dimensions compared to the *Fubuki* class meant that seaworthiness was also impaired, and it was difficult to sustain high speed in heavy weather. Like their immediate predecessors, the *Shiratsuyu* class also had to be reconstructed after the Fourth Fleet Incident, and only

Asashio, the name ship of the class, running at high speed in July 1937. Built at Sasebo NY she was delivered and commissioned on 31 August of the same year, so this photo must have been taken earlier during trials. Despite her apparently high speed the absence of smoke is remarkable, and it is possible the photo has been retouched. *Asashio* was the second ship bearing this name; the first was the British Thornycroft-built four-funnelled sister of the destroyer *Shirakumo* (I), a 400-ton vessel, and belonged to the second and final batch of Japanese destroyers ordered from that shipyard. (Fukui Shizuo collection)

Table 1: **Decision-making Steps in the Authorisation Process**

	NGS Requirement	Finance Ministry Decision	Audit Dept. Decision	NGS Conference	Planning Revised	Cabinet Decision & Diet Authorisation
Destroyers	14	8 (amended to 14)	14	14	14	14
Total no. warships	22	14 (amended to 22)	22	22	22	22
Total no. auxiliaries	65	5 (amended to 12)	22	38	32 (amended to 35)	26
Grand Total	87	19 (amended to 34)	44	60	54 (amended to 57)	48

Notes:
1. Four of the 14 DDs to be modified *Shiratsuyu* class (*Umikaze* etc.) and ten to be *Asashio* class.
2. In summary of the IJN's revised DD building policy after the *Tomozuru* Incident, the main items were to:
   – to complete *Ariake* and *Yugure* as reconstructed *Hatsuharu* class;
   – to complete the next ten ships whose materials were prepared as improved *Hatsuharu* class (*Shiratsuyu* class), and to form a destroyer squadron comprising four divisions (each of four ships) with the six ships of *Hatsuharu* class and the ten of the *Shirayuki* class;
   – to build the remaining ten ships (*Asashio* class) to a larger design comparable to the *Fubuki* ('Tokkei') class, the notification of the treaty termination being expected at the end of 1934.

the last four ships of the class (*Umikaze, Yamakaze, Kawakaze, Suzukaze*) had uniform hull construction, the earlier ships having a mix of riveting and electric welding to ensure sufficient strength.

With the end of the arms limitation treaties in prospect – Japan intended to notify the other powers in late 1934 of its intention to withdraw from the Washington Treaty – it was decided that the next class of destroyer revert to the size of the 'Special Type' with a view to improving seaworthiness and stability, and would mount a comparable gun and torpedo armament. In addition, speed and range were to be increased. In brief: the new ships were to be an enlargement of the *Shiratsuyu* class to meet the standards set by the *Fubuki* class.[1]

The Second Naval Replenishment Program of 1934[2]

Discussion of the new naval programme took place in the context of a worsening of Japan's foreign relations following the Manchurian Incident and her withdrawal from the League of Nations. Other key considerations were the transfer of the American battle force from the Atlantic to the Pacific Fleet, US attempts to secure air bases in central and southeast China for use in time of war – seen as a revision of US Pacific strategy from defence to offence, and a strong demonstration of power in opposition to Japan's China policy – and a steady increase in the strength of American naval forces (culminating later with the first Vinson programme of March 1934). Given the rapidly-deteriorating political and military situation, the Naval General Staff (NGS) considered it essential to build up to existing treaty limits at the earliest possible moment.

The formal process was set in motion by the submission of NGS secret document no.154 dated 6 May 1933 to the Navy Minister. It related to the naval programmes to be adopted from fiscal year (FY) 1934 and, citing a previous submission (NGS secret document no.215 dated 17 August 1932) insisted that the primary aim should be to compensate for numerical inferiority by focusing on delivering significant qualitative improvements in individual units. It was quickly followed by NGS secret document no.199 dated 14 June 1933, the appendix of which (dated 12 June) requested the construction of 87 warships and auxiliary vessels displacing c.159,000 tonnes (Yen 664,793,079) over three years (FY1934–1936). Among the 22 warships were 14 destroyers.

This programme was closely related to the 1929 plans, which were unfulfilled due to the conclusion of the London Treaty, and the related First Naval Replenishment Program of 1931. It was an attempt to build the ships of that programme which had not been authorised, and to expand the IJN to the maximum tonnage permitted by the London Treaty of 1930 before the opening of the Second London Conference scheduled for 1935. Because of its relationship to previous planning there was no official request from the Navy Minister to the Prime Minister, but there were repeated meetings between the Navy Ministry and the NGS, and between the former and the Finance Ministry concerning budgetary appropriations. Consent was difficult to obtain, and the programme was revised repeatedly before the cabinet meeting finally accepted the 22 combatants (which included the 14 destroyers), but only 26 of the original 65 auxiliaries[3] (the latter were not subject to treaty limits) with a total displacement of 137,350 tons (see Table 1). Moreover, instead of the three-year programme requested, the NGS also had to accept that it would be spread over four years (FY1934–1937).

In his explanation to the Diet the Navy Minister complained that:

- even the construction of 22 warships within current treaty limits was insufficient to ensure national security;
- supplementing the currently inadequate force strengths by building many ships outside the treaty limits, despite its importance, had had to be restricted to 26 ships due to the parlous situation of the national finances; this highlighted the absolute urgency for increased expenditure on the armed forces.

The budget of Yen 431,688,000 (divided into Warship Building Expenses [*Kantei Seizô Hi*] of Yen 273,948,000 and Auxiliary Warship Building Expenses [*Hôjû Kantei Seizô Hi*] of Yen 157,740,000) was later increased to Yen 432,707,739 owing to a change in the exchange rate. Together with the budget of Yen 959,279,219 already passed, the grand total rose to Yen 1,391,986,958, calculated for the 48 ships authorised in the 65th session of the Diet (opened on 26 December 1933); this was promulgated as law on 20 March 1934 following authorisation by the emperor.

The Design of the *Asashio*

The destroyer design team of the Navy Technical Department (NTD), Fourth Division, designated the design project for the new destroyer F 48. That of the *Hatsuharu* class was F 45, and that of the *Shiratsuyu* class, including the improved ships (*Umikaze* sub-group), was F 45D. The F 48 designation therefore suggests major changes compared to these earlier classes.

Drawn up in the aftermath of the *Tomozuru* Incident, stability was a key consideration for the design. Lowering the centre of gravity (CG), increasing draught and reducing the lateral wind pressure area were priorities. The CG was brought down closer to the waterline and efforts were made to obtain a favourable rolling period.

The increase in draught was at the expense of freeboard; high freeboard was important for seaworthiness and for the maintenance of high speed in rough seas, but the problem was resolved by increasing freeboard at the bow (see Table 2).

In place of the high-tensile and super-high-tensile steel used in the construction of the earlier ships, the new ships employed Ducol steel for the shell/inner bottom plating and for the longitudinal bulkheads, and the thickness of the bottom plates was increased for added hull strength. Frames and other strength members such as deck beams were to be electrically welded. The extensive use of welding improved watertightness and reduced overall hull weight, thereby lowering the centre of gravity.

The keel of the name ship, *Asashio*, was laid down at Sasebo NY on 7 September 1935. On the 26th of the same month, the IJN was again shaken to its foundations by the Fourth Fleet Incident. Construction was temporarily halted, and strength calculations were checked and revised if found to be unsatisfactory. Some material thicknesses were reinforced, and riveting replaced welding for key structural members; these measures resulted in an increase in the trial displacement of 95 tonnes (see Table 2).

The key features of the new design were as follows:

- the main guns were as in the *Fubuki* class, with three twin mountings replacing the two twin and one single of the *Hatsuharu*/*Shiratsuyu* classes;
- the torpedo armament was the same as for the *Shiratsuyu* class;
- a new propulsion system was adopted which incorporated an intermediate pressure (IP) turbine for the first time in a destroyer;
- total horsepower was increased from 42,000shp in the *Hatsuharu*/*Shiratsuyu* classes to 50,000shp (as in the *Fubuki* class) for a higher designed speed of 35 knots;

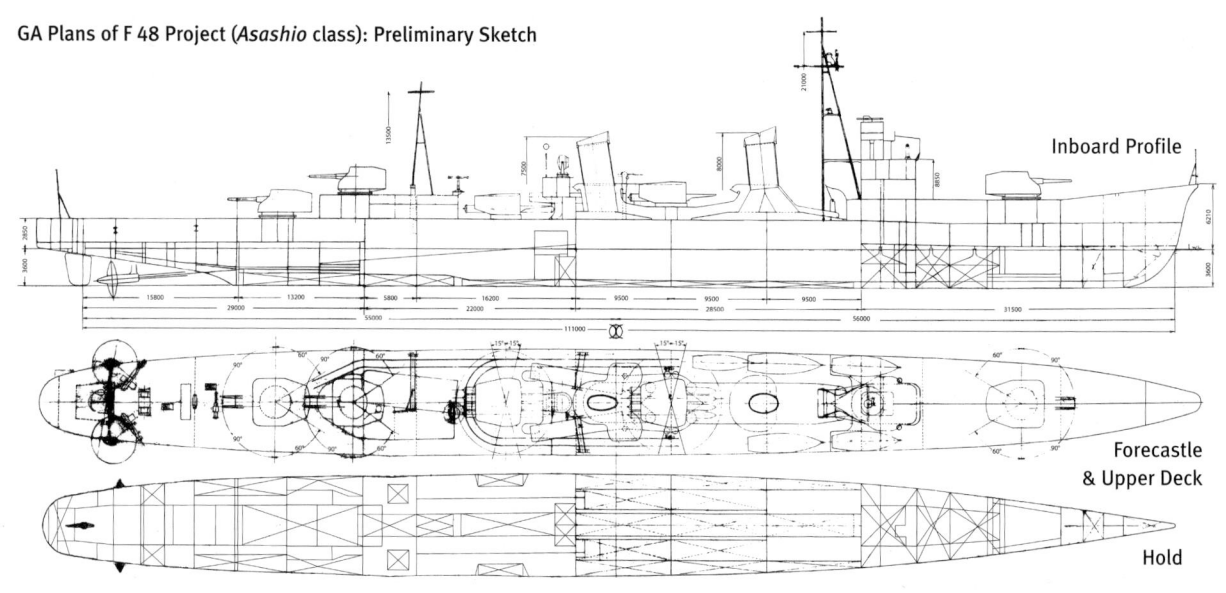

GA Plans of F 48 Project (*Asashio* class): Preliminary Sketch

Inboard Profile

Forecastle & Upper Deck

Hold

Note: The original plan was subject to later revisions, notably the layout of the bridge. (Official plan with dimensions added)

Table 2: Characteristics

	Original Design	Revised Design	*Kasumi* as Completed
Length oa	118.00m	118.00m	–
Length wl	115.00m	115.50m	–
Length pp	111.00m	111.00m	–
Beam (wl)	10.35m	10.35m	10.39m
Depth	6.30m	6.30m	–
Draught	3.60m	3.69m	3.71 m
Freeboard (fwd; midship; aft)	6.21m; 2.61m; 2.85m	6.30m; 2.61m; 2.76m	–
Trial displacement	2,275mt	2,370mt	2,403mt
Speed	35.2kts	35.0kts	35.98kts
Shaft horsepower	50,000shp	50,000shp	51,026shp
Shaft revolutions	380rpm	360rpm	–
Fuel	580mt	580mt	591.9mt
Range	4,000nm @ 18kts	3,800nm @ 18kts	5,190nm @ 18kts
Armament (all variants)	six 12.7cm LA (3 x II)		
	four 25mm MG (2 x II)		
	eight Type 92 61cm TT (2 x IV) for 16 Type 90 torpedoes		
	one Type 92 90cm searchlight		
	one Type 94 S/L controller		
	one 2kW signal lamp		
Complement	12 officers, 217 petty officers and ratings		

Note:
The data given for *Kasumi* as completed is from Tech V-Ad Niwata Shôzô.

Table 3: Weight Distribution

	Fubuki class tonnes (%)	*Hatsuharu* class tonnes (%)	*Shiratsuyu* class tonnes (%)	*Asashio* class tonnes (%)	*Kagero* class tonnes (%)
Hull	585.0 (26.5)	518.5 (26.7)	518.5 (28.0)	732.8 (30.5)	750.0 (29.7)
Fittings	77.6 (3.5)	72.8 (3.7)	73.1 (3.5)	74.5 (3.1)	79.4 (3.2)
Equipment (fixed)	32.6 (1.5)	42.6 (2.2)	39.7 (1.9)	42.5 (1.8)	48.6 (1.9)
Equipment (consumable)	60.6 (2.8)	64.7 (3.3)	67.7 (3.3)	60.8 (2.5)	76.9 (3.1)
Guns	137.6 (6.2)	143.5 (7.4)	139.5 (6.7)	151.3 (6.3)	154.8 (6.1)
Torpedoes	129.0 (5.8)	120.2 (6.2)	116.8 (5.6)	118.4 (4.9)	137.4 (5.4)
Electric	35.6 (1.6)	53.6 (2.8)	54.0 (2.6)	48.1 (2.0)	49.5 (2.0)
Navigation	–	6.0 (0.3)	3.1 (0.2)	3.4 (0.2)	4.2 (0.2)
Machinery	793.6 (35.6)	571.4 (29.4)	589.9 (28.4)	734.7 (30.6)	755.3 (29.9)
Fuel	334.2 (15.1)	312.1 (16.0)	362.2 (17.4)	386.0 (16.1)	410.1 (16.2)
Light oil	–	1.3 (0.1)	1.6 (0.1)	3.4 (0.2)	0.9 (–)
Lubrication oil	–	6.7 (0.3)	8.0 (0.4)	8.5 (0.4)	5.0 (0.2)
Reserve F.W.	23.7 (1.1)	27.5 (1.4)	29.3 (1.4)	33.4 (1.4)	35.1 (1.4)
Margin	–	3.3 (0.2)	11.0 (0.5)	3.0 (0.1)	9.4 (0.4)
Uncertain	–	–	–	–	7.2 (0.3)
Trial Displacement	2,208.8 (100)	1,944.2 (100)	2,077.4 (100)	2,400.5 (100)	2,523.8 (100)
Characteristics					
Length pp	112.00m	103.50m	103.50m	111.00m	111.00m
Beam max.	10.36m	10.00m	10.00m	10.35m	10.80m
Depth	6.25m	6.00m	6.05m	6.30m	6.46m
Frame spacing	0.95m	0.87m	0.87m	0.81m	?

Note in particular the development of hull, armament, engine and fuel weights.

Source: 'Weight and Centre of Gravity for Miscellaneous Ships' by Preliminary Design Group Tech Lt. Kôyama, Eng. Imai, Asst. Eng. Takahashi & Ogino, second revised edition, October 1941, pp.30–45; *Kaigun Zôsen Gijutsu Gaiyô*, Vol.II; *Shôwa Zôsen-shi*, Vol.I; Fukuda *Gunkan Kihon...*, *Niwata Kenkan hiwa* etc.

Table 4: Building Dates & Fate

Name	Builder	Laid down	Launched	Completed	Fate
Asashio	Sasebo NY	7 Sep 1935	16 Dec 1936	31 Aug 1937	Sunk 3/4 Mar 1943 by US & RAAF aircraft SE of Finschhafen (Battle of the Bismarck Sea)
Michishio	Fujinagata	5 Nov 1935	15 Mar 1936	31 Oct 1937	Sunk 25 Oct 1944 by USS *Hutchins* (DD 476) at 10°25 N/123°25 E (Battle for Leyte Gulf)
Oshio	Maizuru NY	5 Aug 1936	19 Apr 1936	31 Oct 1937	Sunk 20 Feb 1943 by USS *Albacore* (SS 218) at 0°50 S/146°06 E c.140nm NNW of Manus, Admiralty Is.
Arashio	Kawasaki	1 Oct 1935	26 May 1936	20 Dec 1937	Sunk 3 Mar 1943 (see *Asashio* for details)
Yamagumo	Fujinagata	4 Nov 1936	24 Jul 1937	15 Jan 1938	Sunk 25 Oct 1944 by USS *McDermut* (DD 677) at 0°25 N/125°20 E (Battle for Leyte Gulf)
Natsugumo	Sasebo NY	1 Jul 1936	26 May 1937	10 Feb 1938	Sunk 12 Oct 1942 by US aircraft off Savo Is. (Solomon Is.)
Asagumo	Kawasaki	23 Dec 1936	5 Nov 1937	31 Mar 1938	Sunk 25 Oct 1944 by USS *Denver* (CL 58) at 10°04 N/125°21 (Battle for Leyte Gulf)
Minegumo	Fujinagata	22 Mar 1937	4 Nov 1937	30 Apr 1938	Sunk 6 Mar 1943 by US CLs and DDs (TF 68) at 08°05 S/157°15 E (Kula Gulf, Solomon Is.)
Arare	Maizuru NY	5 Mar 1937	16 Nov 1937	15 Apr 1939	Sunk 5 Jul 1942 by USS *Growler* (SS 215) at 52°00 N/177°19 E (off Kiska)
Kasumi	Uraga Dock	1 Dec 1936	18 Nov 1937	28 Jun 1939	Sunk 7 Apr 1945 by US aircraft (TF 58) at 30°57 N/127°57 E (East China Sea) as escort of '*Yamato* force'

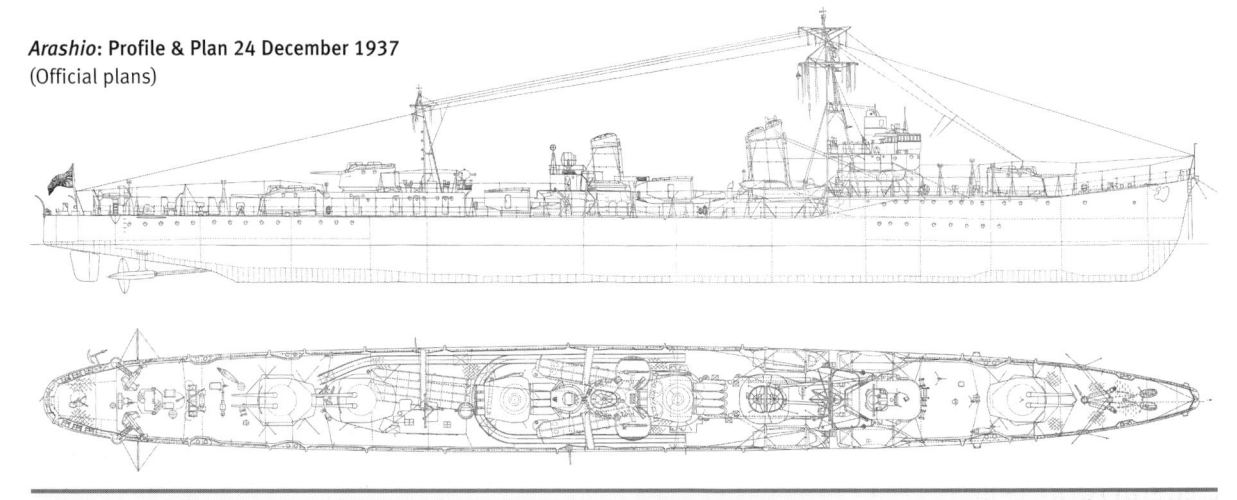

Arashio: Profile & Plan 24 December 1937
(Official plans)

Table 5: Endurance Trials (*Asashio*)

	Designed	After revision	Trials	Oshio	Asagumo
Displacement	2,275t	2,408t	2,403t	2,452t	2,408t
Fuel	580t	578.9t	578.9t	?	574.8t
Speed	18kts	18.18kts	18.18kts	18kts	18kts
Range	4,000nm	4,975nm	4,824nm	4,760nm	4,771nm
Horsepower	4,130shp	4,265shp	4,160shp	4,075shp	4,160shp
PC	48.9t	51.7t	51.7t	50t	?
Fuel: shp/h	0.615t	0.471t	0.496t	0.584t	0.495t

Note:
Due to the use of high-pressure, high temperature (HPHT) steam, theoretical fuel consumption was reduced by 11% compared to the *Fubuki* class. However, the Fifth Division of the NTD claimed that 580 tonnes of heavy oil was necessary for a range of 4,000nm, and this figure was responsible for the increase in displacement to 2,400 tonnes trial and 2,600 tonnes full load. The table shows that the range achieved on trials exceeded the requirement by c.20% – a similar miscalculation occurred with the *Yamato* class. The maximum range recorded was 5,190nm at 18 knots with 591.9 tonnes of fuel (*History of Steam Turbine Research of the IJN* by Amatsusho Yashiyuki & Yasagi Shôichi; also Niwata, above).

- an alternating current (AC) electrical supply replaced the direct current (DC) supply of the earlier ships – the first adoption of AC in an IJN destroyer;
- endurance was to be 4000nm at 18 knots.

The unplanned increase in the displacement resulted in some reductions in speed and range compared with the initial requirements (see Table 2); armament remained unchanged. In the end, however, all ships attained the planned standard in terms of stability, strength, speed and firepower, and they were much better balanced ships than the *Hatsuharu* and *Shiratsuyu* classes.

Seaworthiness

After the *Tomozuru* and Fourth Fleet Incidents the design and construction techniques of light warships received considerable attention, and strict regulations were drawn up. Official trials had to be carried out under their strict observance and the results had to be recorded. It was only after checking all records that a ship could be declared operational. In his destroyer monograph, Hori Motoyoshi describes the seaworthiness and turning trials of the *Oshio*. The trials were carried out off Sasebo in September 1937. The sky was cloudy with occasional rain, wind speed 18–20m/s (gale force), wave height 2.5m, wave length 30–40m. *Oshio* steamed at 6/10 full power (30,000shp). During the trial the wind gusted up to 20–22m/s (strong gale), occasionally 25m/s. The conditions limited speed to 6/10 full power, corresponding to 30 knots in calm waters; in the 25m/s winds only 28 knots were attained, which at times had to be reduced to 26 knots to avoid damage to fittings and equipment. When steering straight ahead the angle of heel was generally in excess of 10°, and pitching was 5–6°. The heeling and pitching moments did not restrict the handling and firing of the guns and torpedoes, and speed was sufficient to escort a fleet. In the end seaworthiness and hull strength were considered adequate both on the crest and in the trough of waves.

General Arrangement

The *Asashio* had a long forecastle with the bridge structure, the size of which was reduced compared to her predecessors, at its after end. A light tripod foremast was stepped around the after end of the bridge, and no.1 12.7cm twin LA mounting was located at the mid-point of the forecastle, just abaft the beginning of the marked deck sheer and well clear of the bridge (see profile drawing). Abaft the foremast, at the level of the forecastle deck were the two funnels: the broad first funnel with its large ventilation cowls on either side, and the slimmer second funnel with ventilation intakes and a platform above. The forward part of the platform was used to mount two twin 25mm MG, the after part for a Type 92 90cm searchlight and the MF/DF loop antenna. The first of the two Type 92 61cm quad torpedo mountings was between the two funnels; the second was abaft the second funnel; both were mounted on the centreline. The reserve torpedo containers for rapid reloading of no.1 torpedo mounting, each holding two torpedoes, were angled inboard on either side of the second funnel, while the reload container for no.2 mounting held four torpedoes and was located obliquely abaft its mounting and integrated into the port side of the after deckhouse, upon which were located, from fore to aft: a Type 14 2-metre rangefinder, a short tripod mainmast, and no.2 12.7cm twin mounting, which was superimposed above no.3 12.7cm twin mounting on the main deck. The quarter-deck accommodated a single Type 94 depth charge thrower (similar to the US Navy's 'Y-gun') together with its associated loading platform on the centreline, two minesweeping paravanes, and three chutes each for a single depth charge on either side of the stern.

Machinery

Greater displacement, the need for higher speed, and the general trend towards the improvement of warship characteristics following the *Tomozuru* and Fourth Fleet Incidents naturally impacted on the machinery installation. The design of the main engines was revised, and the *Asashio* class mounted a three-pinion system (the so-called *Asashio*-type turbine) with the additional installation of an intermediate pressure (IP) turbine – the first such application in a Japanese destroyer (see accompanying drawing). High-pressure, high-temperature (HPHT) steam conditions were adopted: pressure was 22kg/cm^2 and temperature 300°C. The main engines were designed for an output of 50,000shp.

The dimensions of the engine room were 22.0m(L) x 10.2m(B) x 5.6m(H), making for a surface area of 226m^2 and a power/area ratio of 225.8shp/m^2. The three boiler

Table 6: **Speed Trials**

Name	Displacement	Speed(kts)	SHP	RPM	Effective SHP	Propulsion Efficiency
Asashio	2,408t	35.29	50,280	355	–	–
	2,389t	35.87	53,030	365	–	–
	2,419t	34.82	50,260	356	24,530	48.8
Michishio	2,424t	35.10	50,195	361	–	50.0
Oshio	2,410t	35.44	50.324	361	25,580	50.8
	2,385t	36.38	53,018	371	27,031	51.0
Arashio	2,392t	34.69	50,498	357	–	48.6
Yamagumo	2,417t	35.09	?	370	–	48.3

rooms were 28.5m(L) x 7.2m(B) x 5.65m(H) for a surface area of 203m² and a power/area ratio of 244shp/m². The total weight of the machinery was 731.1 tonnes, giving a ratio of 68.4shp/tonne.

The name ship *Asashio* was completed on 31 August 1937, followed by *Michishio*, *Oshio* and *Arashio* before the end of that year. On 29 December *Asashio*'s turbines were inspected at Sasebo NY. The result was a shock: four of the moving blades in the second stage of the newly-introduced IP turbine were found to be damaged. A check on the turbines of the other ships on trials revealed numerous cracked and damaged blades in the second and third stages of the IP turbine. The cracking and shearing always occurred at a position about one-third from the top of the blades, but there was also damage to the shrouding rings and joints, and some had even sheared. Inspection of the HP and LP turbines did not reveal any problems; these turbines were completely intact. As an emergency measure the blade rows of the second and third stages of the IP turbines were removed, and a reduction in the turbine output was imposed, the maximum permitted speed being c.25 knots.

A turbine has several thousand blades. Damage to a few of these blades is generally not a serious problem. But if the problem continues unchecked, all the ships fitted with a particular type of turbine could develop failures one after another, thereby bringing about the temporary immobility of the greater part of the fleet. The IJN had experienced a similar incident almost 20 years previously, when frequent turbine blade damage occurred with the all-geared turbines introduced in 1918–19. This problem had been resolved in 1922 when the Kampon turbine was introduced. The IJN was proud of the Kampon turbine, which had been developed without outside assistance, and was confident about its reliability. The shock caused by the latest incident was therefore that much greater than in the case of the 'First Turbine Blade Incident', which had occurred when Japanese turbine design and manufacture were still in their infancy.

When the problem was first discovered, nobody seriously believed that the cracks and shearing of the IP turbine blades would limit the activities of all naval ships mounting this type of turbine, or that a period of five years and six months would be required before the cause was ascertained with a high degree of certainty. The studies, tests and experiments conducted in relation to this incident were on a large scale. A real turbine was used to monitor the blade and wheel vibrations, and the Japanese naval engineers involved believed this to be the first experiment of its kind in the world. Investigating the turbine blade and wheel vibrations and their interdependence was one of the most complex technical problems of the day, and experiments with real turbines on ships, on land and on a testing apparatus developed particularly for this purpose were carried out in parallel to determine the necessary remedial measures.

A Committee of Investigation is Appointed

On 19 January 1938 the Ad Hoc Committee of Investigation of Engines (*Kaigun Kantei Rinji Kikan Chôsa Iinkai*; abbreviated *Rin-ki-chô*) was established, with deputy navy minister Yamamoto Isoroku in the chair. Its brief was (i) to ascertain the cause of the problem, (ii) to propose remedial action, (iii) to investigate the characteristics and performance of the main engines, and (iv) to carry out its investigations in conditions of extreme secrecy. (The committee's secondary brief was to establish a project for improving the reliability of the Kampon DA diesel engine, which had experienced repeated technical problems, leading to cancella-

A starboard broadside view of the Maizuru-built *Michishio* mooring at a buoy in the naval port of Maizuru. This photo was taken on 31 October 1937, her commissioning day. Note the 'S'-like shape of the bow, the marked sheer of the forecastle, the rounded connection between the sheer strake and outermost strake of the forecastle (see *Warship 2013*, p.39) and the large shields of the Type 92 61cm quadruple torpedo tube mountings. Allied investigator teams were impressed with the torpedo reloading system and by the outstanding performance characteristics of the Type 93 oxygen-propelled torpedo. Note the significantly higher position of the first torpedo mounting, which was on a raised deckhouse between the funnels; it would be lowered in the following *Kagero* class due to its adverse effect on stability. (Fukui Shizuo collection)

Asashio Class: Turbine Layout (Port Shaft)

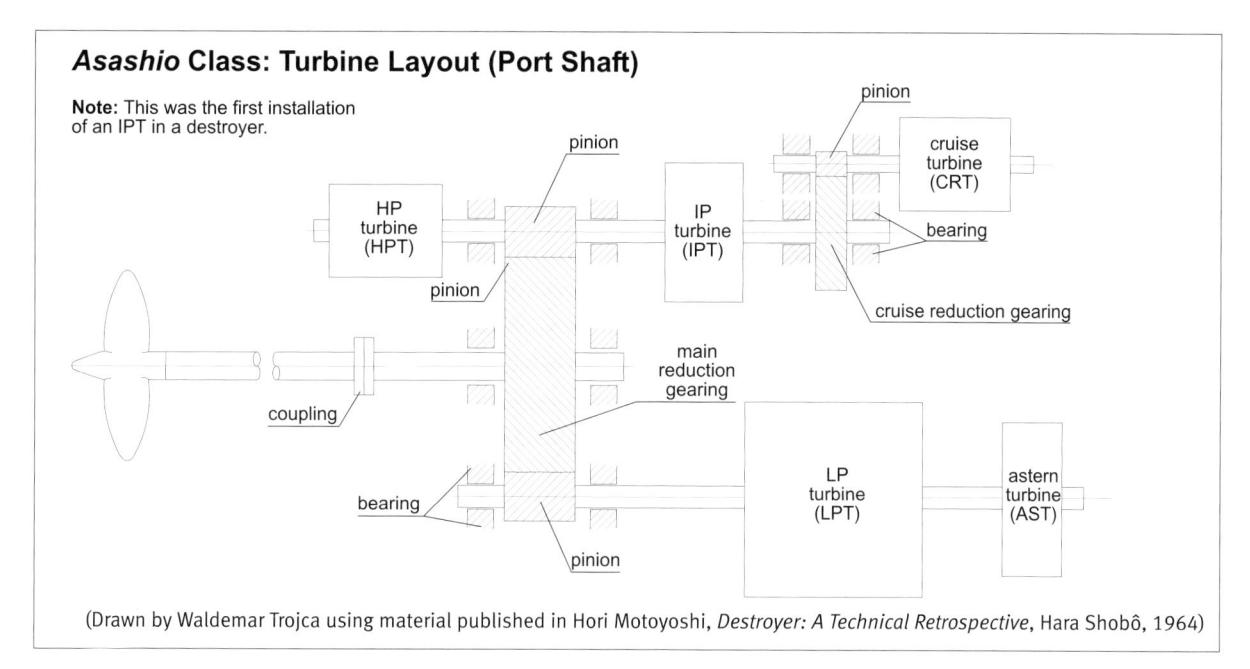

Note: This was the first installation of an IPT in a destroyer.

(Drawn by Waldemar Trojca using material published in Hori Motoyoshi, *Destroyer: A Technical Retrospective*, Hara Shobô, 1964)

tion of its proposed installation as the diesel main engine on the battleships of the *Yamato* class, but this falls outside the scope of the present article.)

The committee was made up of 34 members from the Naval Affairs Division of the Navy Ministry, the Fourth, Fifth and Sixth Divisions of the NTD, the NGS, the Navy University (*Kaigun Daigakkô*) and the Navigation School (*Kôkigakkô*). It held 13 plenary sessions and 53 sub-committee meetings to determine the causes of the turbine problems and to recommend the necessary remedial measures. All the main turbines of warships designed from the heavy cruisers of the *Myôkô* class and the destroyers of the *Fubuki* class were investigated.

On 2 November 1938 Yamamoto submitted his report to the Navy Minister. The committee considered the main cause to be the resonant vibration of the first order of the discs;[4] this assumption was based on the test results of 1928 – trials in the Hiro Engine Laboratory using the method established by the American General Electric Co. which, however, stopped short of investigating the problems of vibration resistance – and on foreign as well as domestic technical literature. It was thought that the cause of the resonant vibration was the configuration of the turbine blade and the inadequate vibration number of the wheel (or disc); faulty design and the method of fixing the blades to the wheel were thought to be responsible. If correct, all IP turbines of the same design would have to be rebuilt. This gave cause for alarm because a period of about three years would be necessary for the reconstruction of all ships, and during this period a substantial part of the fleet would be out of service at a time of increasing political tensions between Japan and the USA. Moreover, the execution of this work would have huge financial and labour costs and, given the limited capacity of the Japanese shipyards, would inevitably delay the scheduled completion dates of new ships at a time when a second naval arms race was already underway.

Meanwhile, on 3 February 1938, the Navy Minister had ordered the C-in-C of the Kure Naval Station to investigate and analyse the problems experienced with the *Asashio* turbine at the Hiro NY Engine Experimental Division, and to propose solutions. Based on the result of these tests the intermediate pressure turbines of the *Asashio* were rebuilt,[5] and subsequent trials at every power up to 10.5/10 full power were completed without incident. This appeared to confirm the success of the reconstruction, and the conclusions of the committee were based largely upon this result. However, some members of the committee, and in particular the turbine experts of the NTD, Fifth Division, dissented from the conclusions; there was still concern that similar incidents could occur with every turbine of every capital ship, cruiser, and destroyer, and that the true cause of the problem had not yet been fully determined.

The need for further investigation was recognised to ascertain at which speed of revolution (rpm) the turbine blades broke. In the interim the ships were to be maintained in operational status with restrictions on maximum speed until their turbines could be rebuilt. Based on a proposal by R-Ad. Kubota Yoshio, secret order no.1973 was issued on 1 April 1939 under which the light cruiser *Mikuma*,[6] the destroyers *Hatsuharu*, *Fubuki* and *Sagiri*, and the torpedo-boats *Otori* and *Chidori* were selected to conduct long-range cruising trials corresponding to a ten-year service cycle. An average range was decided for each fiscal year; the figure was multiplied by ten, and the turbines run for a corresponding period without interruption as part of the investigation; complaints about the consumption of heavy oil were put to one side. The test results confirmed the reliability of the IP turbines of the cruisers and the HP and LP turbines mounted in the destroyers, and suggested that remedial action was necessary only for the IP turbines mounted in the *Asashio* class, which had already

Turbine Load Vibration Experiment Equipment

(Drawn by Waldemar Trojca using material from the 'Destroyer Monograph' of Hori Motoyoshi supplied by the author)

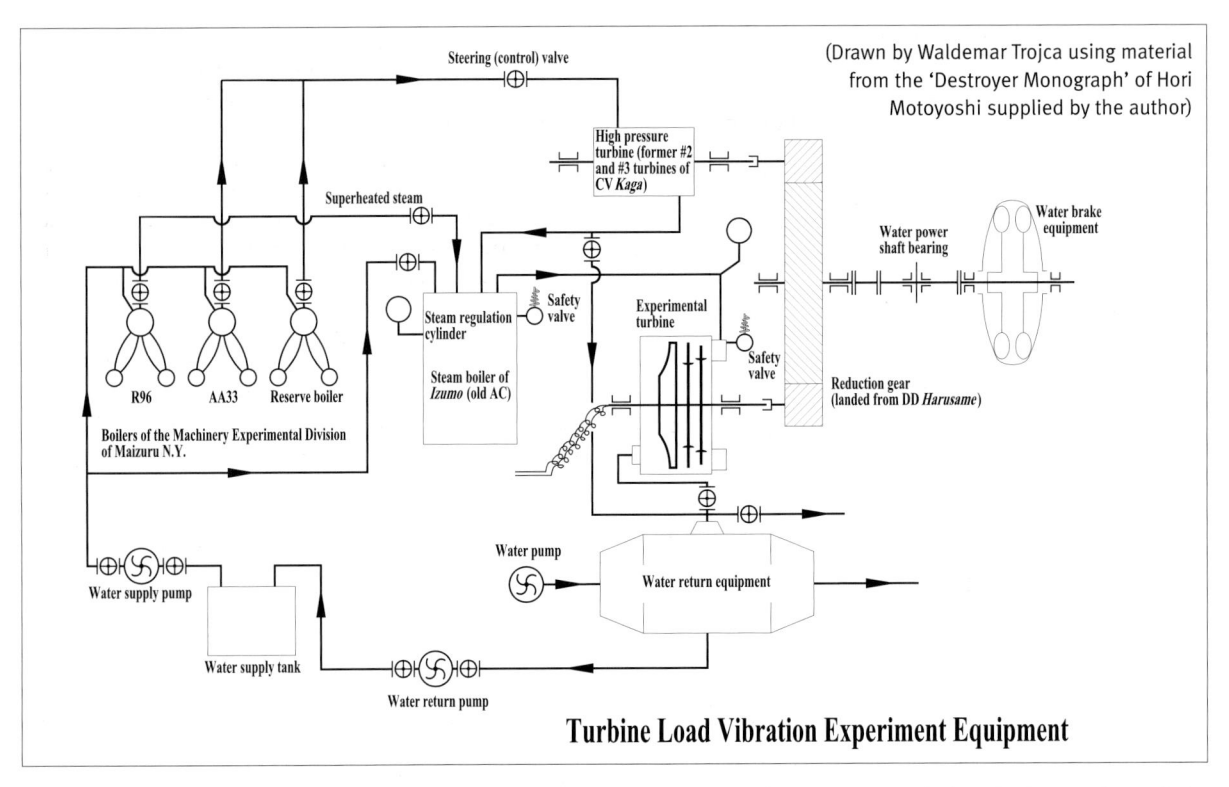

been rebuilt. The trials suggested that the conclusions of the committee regarding the probable cause were in error; consequently modifications to the turbines of earlier ships were postponed pending testing on land.

Despite the success of the above trials, the relationship between the vibrations of the discs and the shape, strength and method of attachment of the turbine blades remained unclear. Meanwhile, in December 1939, the land experiment at the Hiro branch of Kure NY to investigate the dynamic vibrations of discs using special nozzles with small, circular holes recorded a phenomenon of resonant vibration at the point where vibrations of the second order (bi-nodal vibrations) in a circular (tangential) direction coincided with the product of the total number of nozzles (p) and the number of revolutions (N) – ie p x N. Findings were as follows:

1. when the maximum stress of the blade exceeds its fatigue limit, cracking is to be expected;
2. singularity occurs in the case of a group of blades when in the state of resonant vibration of the second order;
3. there are symmetric and asymmetric forms, hence many frequencies of resonant vibration of the second order;
4. shrouding also influences the natural frequencies of a group of turbine blades, but it is difficult to calculate the natural frequencies using theoretical methods; they can only be obtained by experimentation.[7]

Arrangement of the Wheel Room of the Turbine Load Vibration Experiment

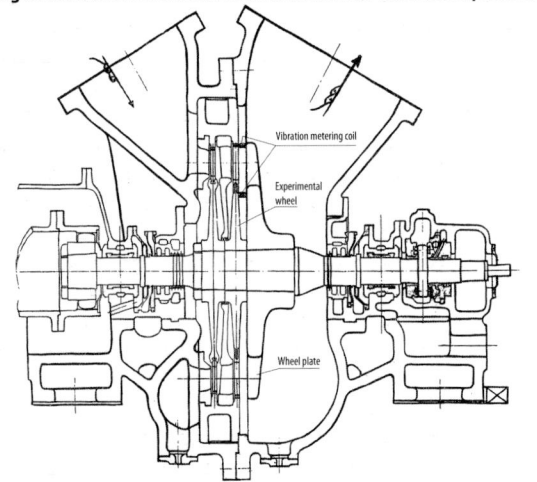

(Drawn by Waldemar Trojca using material from Technical Material of the Former Navy – see Footnote 7)

Turbine Blades: 3rd Stage of IP Turbines

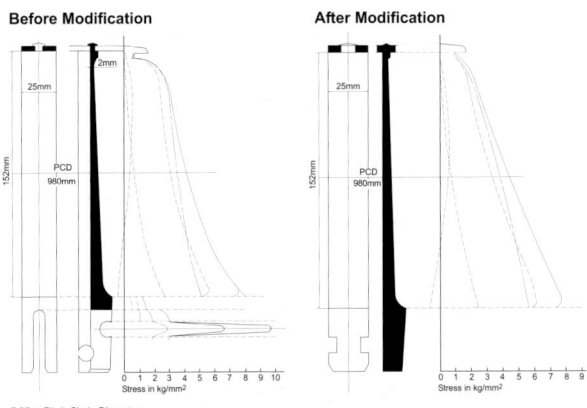

Before Modification / After Modification

PCD = Pitch Circle Diameter

Note the different thicknesses and, in particular, the change to the blade foot, which was now secured to the rim of the wheel. Note also the different stress distribution.

Source: Rinchiko Top Secret Report No.1 dated 18 February 1938.

THE JAPANESE DESTROYERS OF THE ASASHIO CLASS

Group (Rows) of turbine blades		Single turbine blade	Terms	
Symmetric vibration	Asymmetric vibration			
	Actual blade — Easily generated	Actual blade	Resonant vibrations of the first order (first node vibrations)	Shapes of Turbine Blade Group (Rows) Vibrations — Resonant vibrations to the direction of the tangent of the wheel shaft
Actual blade — Easily generated	Actual blade — Generated with difficulty	Actual blade	Resonant vibrations of the second order (second node vibrations)	
This shape is only generated with the two blade model	This shape is only generated with the two blade model	Model	Resonant vibrations of the third order (third node vibrations)	
Easily generated	Actual blade — Easily generated			Resonant vibration to the direction of the wheel shaft
Easily generated	Generated with difficulty			

(Drawn by Waldemar Trojca using material from the 'Destroyer Monograph' of Hori Motoyoshi supplied by the author)

To confirm the result of this land experiment and determine the relationship between resonant vibration and vibration resistance, secret order no.1122 dated 12 February 1940 was issued, stipulating an experiment involving a real ship (Hiro had no suitable boiler capable of generating the necessary amount of steam). The destroyer *Yamagumo* of the *Asashio* class was fitted with an IP turbine of the original design and a modified turbine with the new wheels and blades to investigate the speed at which the original designed blades suffered damage. The trials were carried out from March to May 1940. Initial speed was 12 knots, and the turbines were run for a similar period of time to that which first revealed the damage in the ships of the *Asashio* class ships; they were then dismantled and inspected. Speed was then gradually increased by increments of two knots. Cracked and broken blades were found at 22 knots. After renewal of the blades the turbines were then run at decreasing speeds, beginning with 10/10 power, then 8/10, then 6/10 power. Again, damage was sustained at around 22 knots, representing c.1/20 of full power. The turbine blades in the modified turbine (port shaft) sustained no damage at this speed.

The shearing of 30 moving blades was restricted to the second stage of the IP turbine corresponding to the rotation in the condition 'cruising full power'; the moving blades of the third stage, corresponding to 'cruising super full power', revealed no damage. The cause of the bi-nodal vibration resonance in the tangential direction of the disk was therefore uncertain.

Secret order no.1122 dated 12 February 1940 stipulated that the land testing be moved to Maizuru NY, where high-powered boilers were available, until a new boiler could be provided for the Hiro NY Engine Experimental Division (which again conducted tests from December 1941 onwards).

On 20 June 1941 secret order no.5389 was issued: the experiments were to be concluded on 31 March 1943,[8] and the proposed rebuilding of the turbines postponed except for the IP turbines of the M type (see Table 7, Note 2). Depending on the results of the land testing, the new investigations were to be completed by the end of June 1943.

In parallel, the basic research conducted by the Navy Technical Research Institute (NTRI) and the studies of manufacturing techniques undertaken at Yokosuka NY began to bear fruit. The land studies and the trials with *Yamagumo* demonstrated that the cause of the damage to the turbine blades was the resonant vibration between the steam pulsation, the product of p x N and the inherent bi-nodal frequency of the wheel.

The final report was submitted in April 1943. It proposed the following remedial action:

1. modification of the turbine blade section;
2. a change in the self-vibration number of the blade;
3. the use of very stable blades.

The 'Engine Committee' was a large-scale project which lasted more than five years and involved many experts;[9] the outcomes were remarkably successful, despite the tremendous cost in manpower and resources.

The Adoption of AC instead of DC

Ever since the mounting of a searchlight on the cruiser *Tsukushi* (ex-*Arturo Prat*) in 1883 the IJN had used direct current (DC) for electrical power supply on warships. Over the years, however, power demands increased, requiring the installation of more powerful generators. On the other hand, treaty restrictions impacting on weight and volume had important consequences for the expansion of electrical power. In this situation the IJN concluded that the installation of an AC power plant, which was in general use on land, was also advantageous for naval vessels. Given the higher voltage implicit in AC plant, weight and volume of the generators could be reduced. Electrical engineers also believed that reduction in size, allied to improvements in cabling, meant fitting and maintenance would be simplified and service life increased.

Based upon these conclusions, the design of a 230V three-phase AC power system and an in-line switchboard distribution system for destroyers was studied from November 1933 to October 1935. The system was to be

51

Table 7: Turbine Modifications to IJN Warships

Type	Names	Remarks
BB	Fusô, Yamashiro, Ise, Hyûga	I type turbines; reconstruction scheduled in 1940, IP turbines in 1939.
BB	Kongô, Hiei, Haruna, Kirishima	M type turbines; same schedule as above; Hiei was undergoing modernisation, and this opportunity was used for the turbine work.
CV	Kaga, Sôryû, Hiryu, Shôkaku class, Taiyô, Zuihô, Ryûhô, Kaiyô	Kaga also had M type turbines in her forward ERs; the other carriers were not retro-fitted, but had their design revised to incorporate M type turbines.
CA	Mogami class, Tone class, Aoba	M type except Aoba (see text).
CL	Agano class, Oyodo	Turbine design revised.
DD	Ten Asashio class, 16 Hatsuharu and Shiratsuyu classes, (23 Fubuki class), 18 Kagero class, Shimakaze, 12 Akizuki class	Some units (eg Hatsuharu & Nenohi) had H type turbines; the Asashio class and Hiei were refitted as the first ships; in the Kagero and Akizuki classes and Shimakaze the design of the IP turbines was revised.
TB	Four Chidori class, eight Otori class	The former had C type, the latter K type turbines.
AV	Chitose class, Mizuho	These ships were also the object of an investigation into their DA diesel engines. When Chitose and Chiyoda were reconstructed as light carriers they received steam turbine machinery.
CM	Okinoshima	

Notes:
1. For M type turbines, refitting was to be carried out in the order: BBs, CVs, CAs, CLs; the Nagato class had a different turbine type which gave no problems. For Yamato and Musashi the mounting of DA diesel engines was cancelled; they had revised Hatsuharu-type turbines and the design was rechecked.
2. The destroyers of the Asashio class and the battleships of the Kongô class had their original turbines replaced. In other ships modifications to the turbine design resolved the problem.

According to secret order no.417 dated 12 December 1938, the time schedule for upgrading the turbines of the ships listed in the table was the end of June 1942; in the event the work was actually finished in the autumn of 1941, and the IJN was confident that these ships were ready for service at any time and that their turbine problems had been resolved.

The same order, based on the conclusions of the (first) report, stated that a decision to refit the heavy cruisers of the Myôkô and Takao classes, the light carrier Ryûjô and the destroyers of the Fubuki class would depend on the results of further trials.

Revisions to the design meant that the turbine and its foundations had to be lengthened, sometimes resulting in revisions to the layout of the engine room.

---

trialled in the *Asashio* class, which became the first destroyers with a 230V AC power supply. The trials were a complete success and led to this system being adopted as standard for IJN destroyers.

The IJN considered the principal advantages of the AC systems to be:

- lighter weight and volume;[10]
- reduced vulnerability to damage;
- lower maintenance requirements;
- easy adaptation of commercial equipment for naval installation.

Disadvantages included the difficulties involved in the use of batteries and speed control of motors – the motor commutator problems were subsequently resolved and the cost of electric motors was reduced by almost 50%. The IJN also had to develop a suitable AC circuit breaker for shipboard use.

Main Armament

Torpedoes

Two quadruple torpedo mountings, designated Type 92 Model 2 Mod.1, were mounted on the centreline, between the two funnels and abaft the second funnel. This torpedo mounting was developed in response to an urgent requirement to reinforce torpedo armament at minimal cost in weight when the destroyers of the *Shiratsuyu* class were designed. The Type 92 was to become the standard torpedo mounting of the IJN for more than ten years, was mounted on most surface ships (including cruisers), and was manufactured in large numbers. Allied to the latest torpedo fire control system (FCS), excellent reloading equipment and the outstanding qualities of the Type 90 torpedo, the Type 92 was recognised as the best torpedo mounting produced for the IJN.

Table 8 lists the principal particulars of the Type 92 quad mounting Model 2, and provides a comparison with the Type 90 triple mounting mounted in the *Hatsuharu* class.

The following description of the torpedo mounting and its rapid-reloading gear is based largely on Report 0-01-3 *Japanese Torpedoes and Tubes, Article 3, Above-Water Tubes* of the US Naval Technical Mission to Japan (USNTMtJ), from which the figures are taken. Four variants of the original type of mounting were manufactured, but the changes were of a minor nature and involved only modifications required to adapt the mounting to the type of ship in which it was to be fitted. The basic structure and operation were essentially the same, so the following description applies to all variants.

The mounting combined four torpedo tubes, together with the handling and firing devices, on a platform which used roller bearings. It could be trained either manually by means of two large hand-wheels (see accompanying drawing and caption), with an operator for each, but more usually by a 10hp motor using compressed-air, operating at 16kg/cm² pressure and 600rpm. The angle of train was generally transmitted from a torpedo director, from which the torpedoes were also fired remotely. For local (back-up) control there was a Type 14 torpedo sight and the torpedoes were fired manually using an on-mount firing mechanism. The firing orders were shown by an indicator panel located above this mechanism. When the torpedo left the tube, it tripped a lever, completing an electrical circuit which indicated at the fire control director and at the mounting that the torpedo had been fired. The usual firing method was by compressed air, each tube having its own air bottle, to avoid easily visible flashes, but for back-up a black powder charge of 600g could be used. The maximum tube pressure was 3.5–4.6kg/cm², torpedo muzzle velocity 11–12m/s. Depth and gyro angle setting in the torpedo was made in the usual way inside the tube, but there was no device for speed setting: in general the high-speed setting was used, and changes to torpedo speed were rarely made.

The torpedo tube was made of 5mm mild steel and had internal steel guides 50mm wide running for the entire length of the tube on each side and along the bottom. The tube door was of cast steel and was hinged to the tube. When the door was closed, a steel ring was rotated by a hand-wheel and gear rack, locking the door in position. The air bottles for torpedo launch, each of which contained 160 litres of compressed air with a pressure 20–25kg/cm², were made of special HT steel, while nickel-bronze was used for the firing valves and pistons. All the torpedo mountings fitted in IJN destroyers were fitted with shields.

According to *Kaigun Suirai-shi* (p.208), the time

Table 8: Comparison between Type 90 triple and Type 92 quad TT

Type 90 triple Model 2: *Hatsuharu* class
Type 92 quad Model 2: *Shiratsuyu* and *Asashio* classes

	Type 90	Type 92
Length (all measurements mm)	8,870	8,870
Height of the centre of TT	855	955
Maximum width of the shield	4,465	4,600
Diameter of the training platform	3,000	3,000
Diameter of the (training) axis	1,160	1,160
Height of the shield	3,105	2,470
Position of the CG (unloaded)	774	783
Position of the CG (loaded)	803	809
Length of the shield	4,100	4,650
Distance from centre of axis:		
to fore end of spoon	4,650	4,650
to the rear end of tube	4,220	4,220
to the left side	2,200	2,300
to the right side	2,265	2,300
Distance between the tube axes	780	780
Breadth of the tubes	2,230	3,110
Training gear	Motor & manual	
Janney electro-hydraulic motor:		
power	5hp	10hp
rpm	800rpm	700rpm
voltage	100V	100V
training time (motor)	25 secs.	25.2 secs
training time (manual)	120 secs.	112 secs.
Time to load torpedoes	23 secs.	32 secs.
Weights:		
torpedo tubes	12.10 tonnes	16.20 tonnes
shield	1.70 tonnes	2.00 tonnes
related parts	0.17 tonnes	0.16 tonnes
Total weight of torpedo mounting	14.14 tonnes	18.36 tonnes

Source: *History of Naval Underwater Weapons* (*Kaigun Suirai-shi*), pp.208–09

A starboard bow view of *Arashio* taken on 21 December 1937, the day after her builder, Fujinagata shipyard, delivered her to the IJN. The arrangement of the main guns closely resembled the *Fubuki* class but the shape of the gunhouses was different (Type C instead of Types A and B). The 1.5-metre rangefinder in front of the mainmast was for navigation purposes, while the 3-metre base-length type located atop the bridge structure supplied target range for the main guns, and was part of the fire control system. (Fukui Shizuo collection)

between torpedo launchings following reloading was about the same as with the earlier system. The following breakdown is given:

- training of the torpedo mounting to align with the reload containers: 30–35 seconds;
- reloading from the containers: 17–20 seconds
- closing the tube door(s): c.10 seconds
- re-training of the torpedo mounting to the centreline: 30–35 seconds.

This meant that the tubes could in theory be reloaded in 90–100 seconds provided that everything functioned correctly.

Sixteen Type 90 torpedoes were carried; eight in the tubes and eight in the reload containers. The performance data for this torpedo can be found in *Warship 2007*, pp.103–04. For charging the air vessels a Y 6 model air compressor was fitted. Developed from 1916 to 1917 and manufactured by Kobe Seikoshô, this diesel-driven compressor – the diesel operated at 350rpm – generated 720 litres per hour with a pressure of 250kg/cm², and weighed 1980kg. The lead ship of the class, *Asashio*, trialled a lightweight German Junkers type air compressor which was very easy to operate and generated less noise and vibration. The German VDI (Society of German Engineers) magazine published a report on this model in February 1935; a single unit was purchased by the IJN in 1936 following inspection and trialled from 1937. On a weight of only 850kg it produced 350 litres of compressed air at 300kg/cm² per hour.

The Torpedo FCS

The torpedo FCS or Torpedo Battle Command Equipment (*gyôraisen shiki sôchi*) comprised two Type

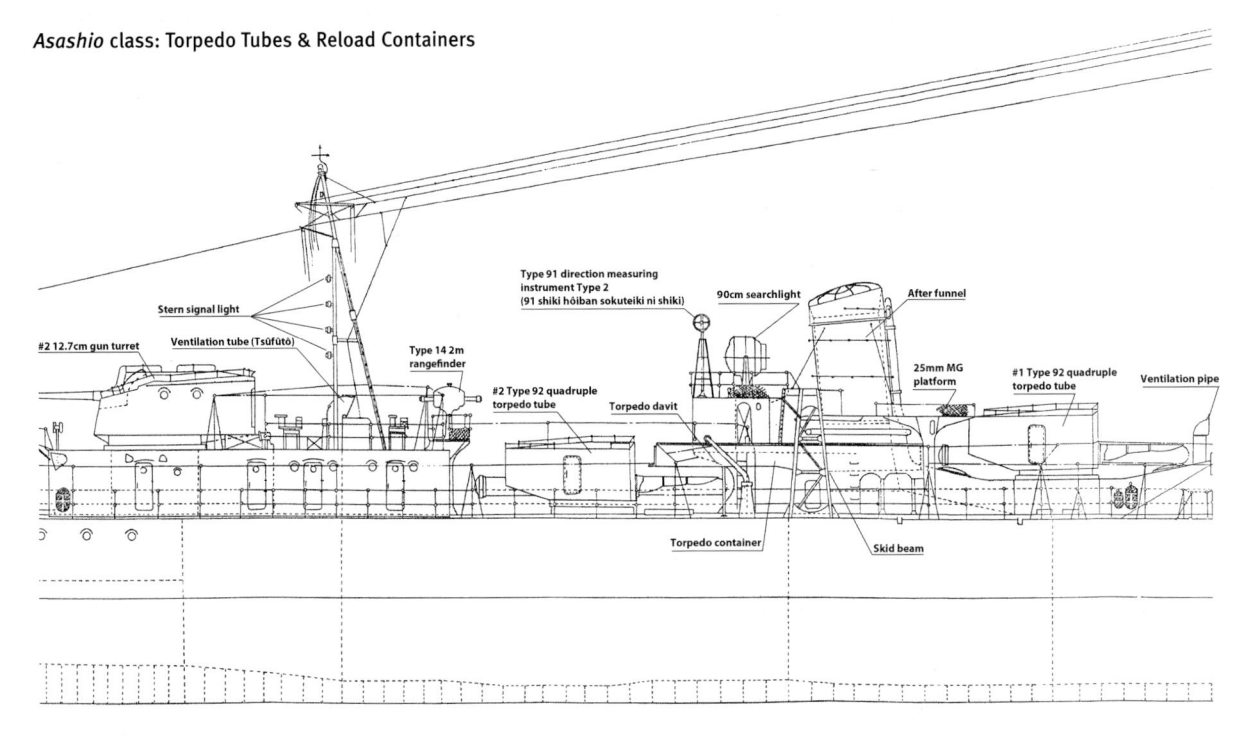

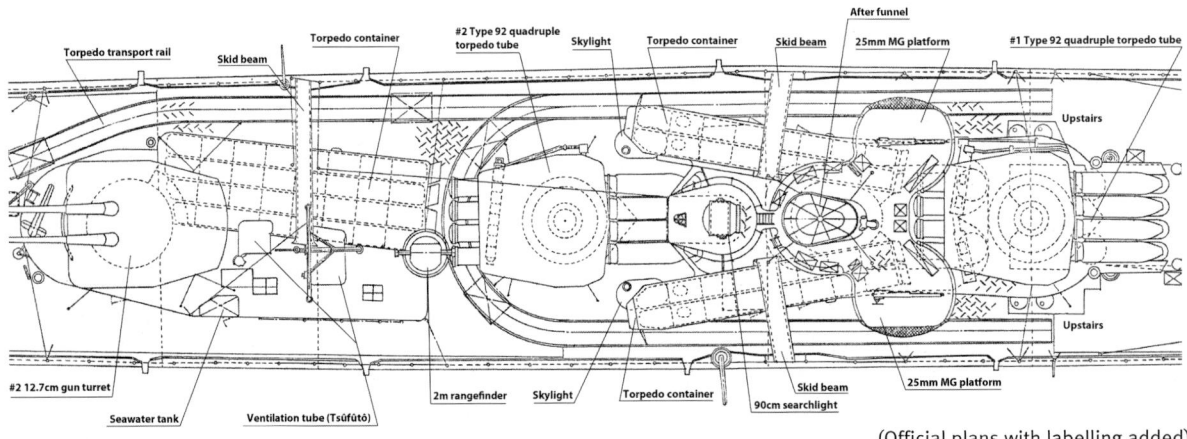

Asashio class: Torpedo Tubes & Reload Containers

(Official plans with labelling added)

THE JAPANESE DESTROYERS OF THE ASASHIO CLASS

Torpedo Launch Procedure

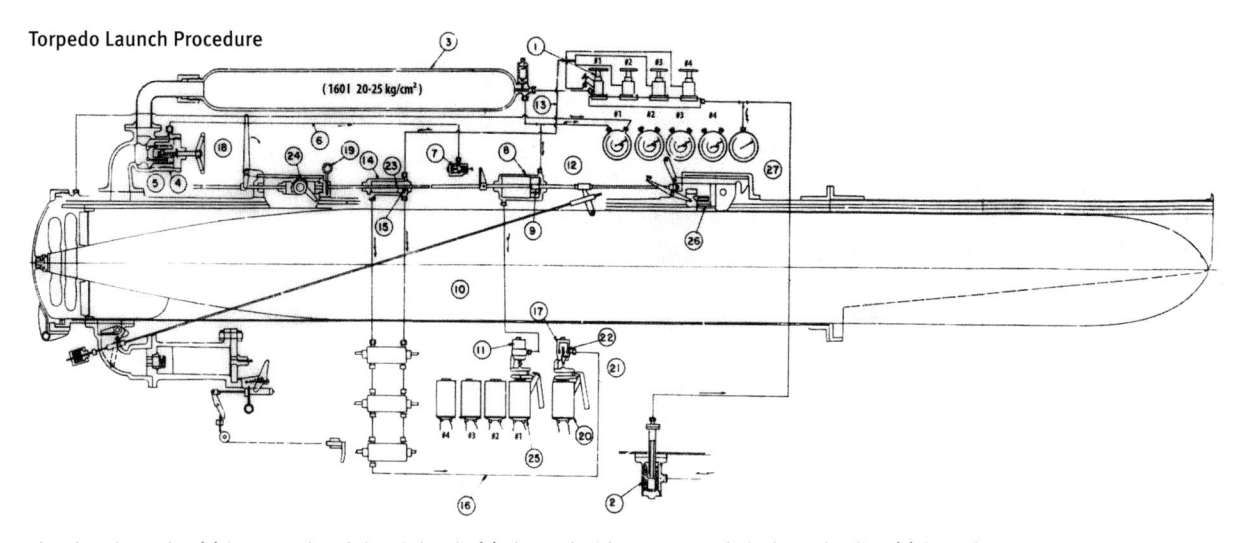

The charging valve (1) is opened and the air bottle (3) charged with compressed air through a line (2) from the main compressors. Compressed air fills two cylinders (8,14) and three releasing valves (7,11,17). Pressure is indicated by a gauge (12). When the firing button on the director is pushed, the first firing circuit is switched in and the electro-magnet (20) is actuated, opening the first relief valve (17) and discharging the compressed air from one cylinder (14). The piston (23) is moved to the left, the starting lever of the torpedo is tripped by a lever (24), and the torpedo gyro is started. A second firing circuit is automatically switched in and the electro-magnet (25) is activated, opening the second relief valve (11) and discharging the compressed air from the cylinder (8). The piston (9) is moved to the left and the forward stop (27) is lifted, freeing the torpedo. Simultaneously, the lug (7) pushes against the third relief valve (7) and releases the compressed air holding the firing valve on its seat. The firing valve opens and compressed air is forced into the tube, launching the torpedo. (USNTMtJ: Report 0-01-3)

91 Model 3 torpedo fire directors (*hassha hôiban*) and two Type 92 torpedo fire control panels (*hassha shikiban*).

As with IJN gun fire control systems very little firm data about torpedo FCS can be found in the literature. The following description of its operation is therefore necessarily rudimentary and incomplete, but will hopefully serve to explain the key features.

The early types of torpedo FCS were designed by the Second Division of the NTD and mostly produced in the naval dockyards, but early in the Shôwa era (supposedly late in 1927) the Second Division asked the Japan Optical Co. (Nippon Kogaku K.K.) to develop a prototype fire director with more powerful optical sights and other improvements. From 1929 to 1930 Lt-Cdr. Tamura and Eng. Kazumori Toshirô oversaw the detailed design and production of the next generation of fire directors which included Type 90, Type 91 Model 3 and also Type 97 Models 1 and 2.[11]

In a torpedo battle the prime consideration is to attain the most advantageous position of the ship for torpedo launch; in addition to determining the direction of launch (ie angle of train of the mounting), it is necessary to calculate the oblique (gyro) angle and the lead angle for spread in order to maximise the probability of hitting the target. However, it was only with the Type 91 Model 3 fire director that the gyro angle was introduced. Even then the fundamental principle of the *shahoban hôiban* remained, and the torpedo officer had to observe and monitor the situation within the minutes which preceded launch before transmitting the launch angle and applying the necessary corrections to the theoretical solution.

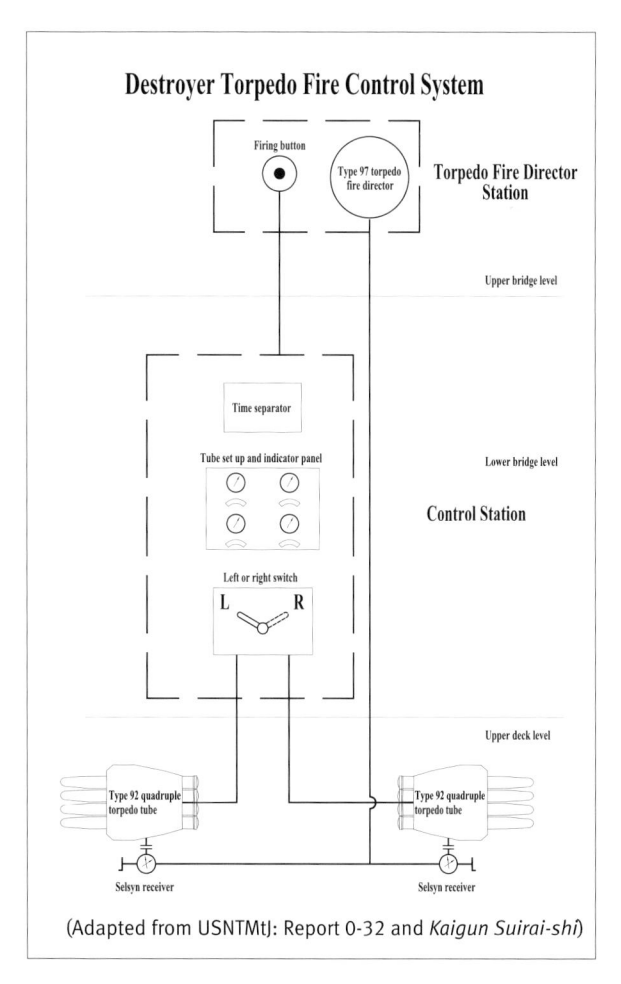

Destroyer Torpedo Fire Control System

(Adapted from USNTMtJ: Report 0-32 and *Kaigun Suirai-shi*)

Torpedo Reloading Gear

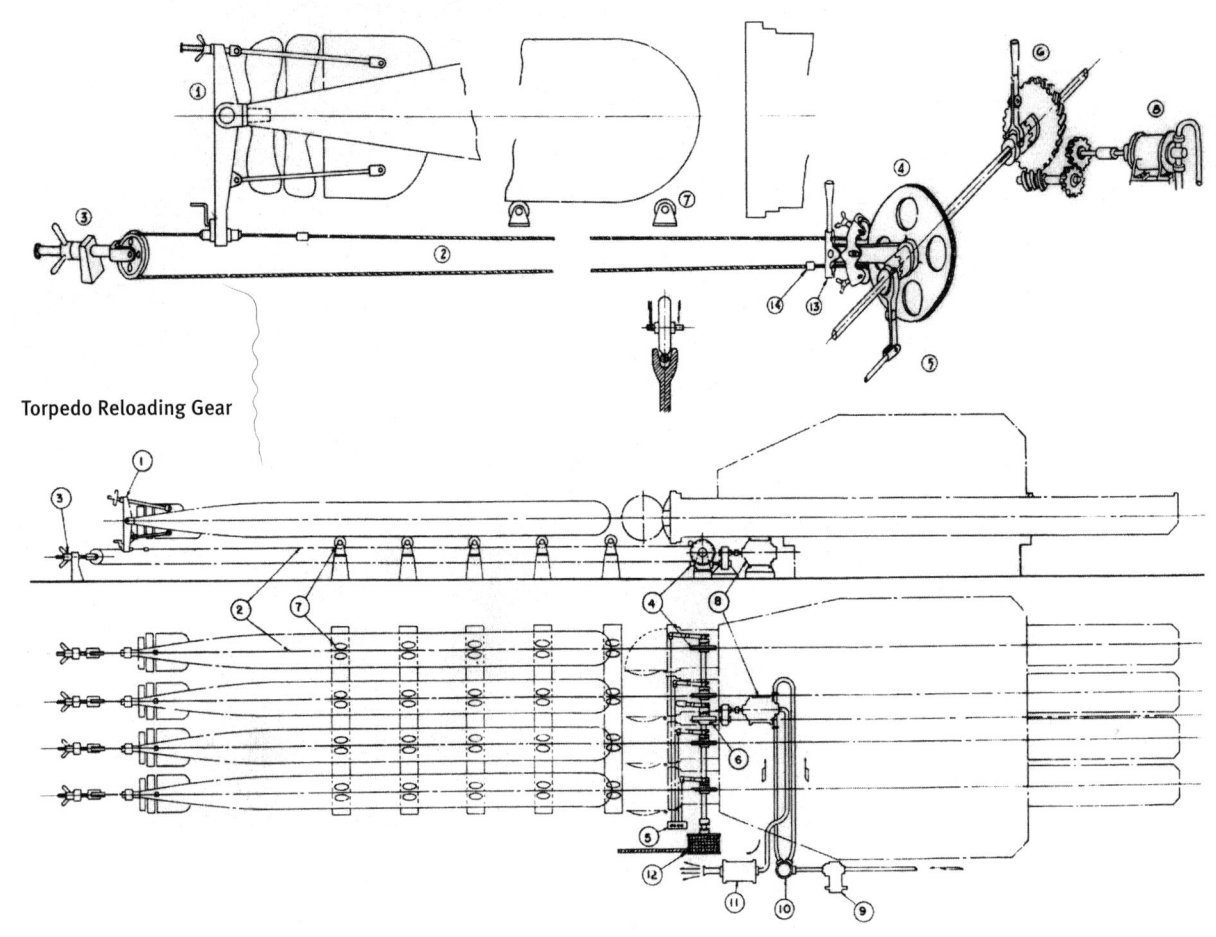

While the American investigators could not discover any special feature in the torpedo mounting, the reloading gear was judged to be exceptional. Instead of the former bicycle chain system, an endless wire rope (2) and friction ring (14) system driven by a 10hp compressed air motor was used. The rpm of the motor (8) was controlled by a valve (10) and the torpedoes could be loaded either separately or all four together with a speed of c.0.4m/sec.[1] The revolutions of the compressed air motor shaft were transmitted to a pulley (4) via 1:45 reduction gears (6). Special attachments (1) connected to the torpedo and the endless wire rope (2) were provided for moving the torpedo forward using rollers (7). An idling pulley (13) secured positive transmission between the pulley (4) and the endless wire rope. Just before the torpedoes were all the way in or out of the tubes, friction rings (14) automatically released the idling pulleys (13) and prevented further movement. Manual loading was also possible using a rope wound to a drum (12), but this method required ten men and took much longer. Consequently, it was used only in the event of failure of the motor-driven system. (USNTMtJ: Report 0-01-3)

Note:
1. Report 0-01-3, p.17, gives a figure of 0.2m/sec, but this could not be confirmed by Japanese sources.

Table 9: Torpedo Fire Control Development

	mechanism/function	development & construction 2nd Div. NTD	Japan Optical Co.	year of production	units built
Type 91 Mod. 3 torpedo FC director	Fitted with 8cm or 12cm binocular type sight. Calculation & transmission of the gyro angle to the TT.	Tamura, Eng. Kazumori	Eng. Murata, Yamagashira	1931	c.120
Type 92 torpedo fire control panel	Fitted with 12cm or 15cm binocular type sight. Equipped with sub-compass. Delivered correction values of the gyro and launch angles.	as above	Eng. Ninomiya, Ishiwara, Nakayama	1932	c.215

Source: *Kaigun Suirai-shi*, p.268ff.

Asashio class: Quarterdeck Arrangements

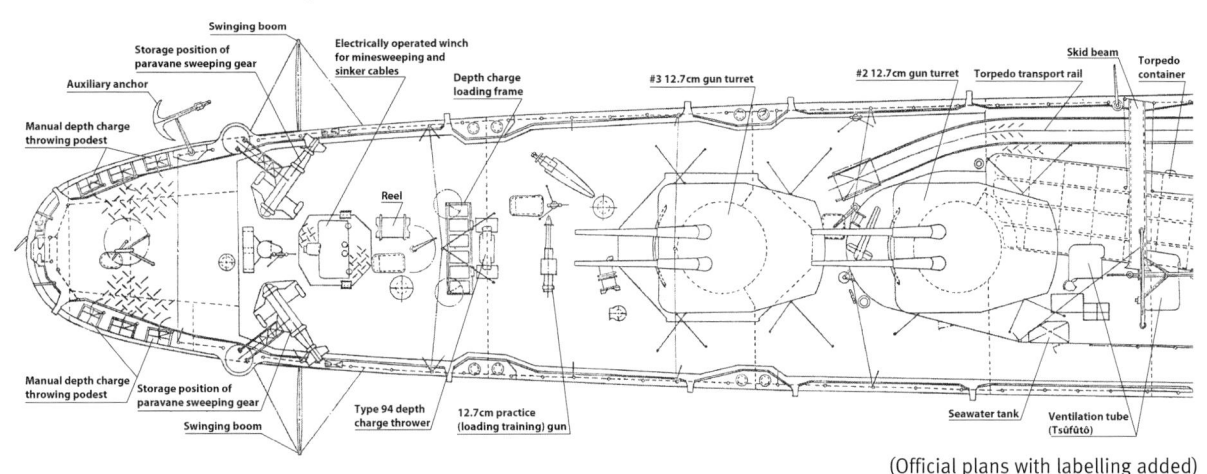

(Official plans with labelling added)

From 1930 onwards the development of torpedo fire control was accelerated, and to replace the earlier directors[12] the Type 92 fire control panel (*hassha shikiban*) and, in succession, the Type 93 target course and speed metering panel (*sokutekiban*) and Type 93 Model 1 computer (*shahoban*) were developed and produced in close co-operation between the NTD and the Japan Optical Co. At the same time there were improvements in the communications between these components, the director (torpedo officer) and the weapons; at the end of this process the IJN was satisfied that torpedo fire control had been taken as far as was possible using current technology.

The principle of the Type 91 Model 3 torpedo fire director was essentially the same as for the Type 97 Model 2 which equipped later classes of destroyer, and there was little difference in their respective appearances. The following outline description of the Type 97 Model 2 fire director, taken from *Kaigun Suirai-shi*, p.263ff and the USNTMtJ-Report 0-32 *Japanese Torpedo Fire Control*, p.17ff, is therefore equally applicable to the Type 91 Model 3.

The director comprised a binocular type sight on a supporting pedestal, with adjustable annular rings scales to provide corrections for deflection. The principal method for using the director on board cruisers was to track the target continuously with the pin engaged in the hole provided. Target bearing was transmitted to the computer and the target course and speed metering panels (i e the director acted merely as a target data transmitter). However, if the functions of the computer or target metering panel failed, or damage occurred to them, the function of the director was much expanded and it showed its true capability. The target was observed and various inputs made to the annular scale rings; these included gyro angle, angle of train of the torpedo mounting, deflection angle (dead time, parallax, etc.), gyro correction angle to standard firing line (lead angle for spread), and range. The marker which indicated firing bearing (ie the bearing of the target ship when the torpedo should be fired) could be matched with the

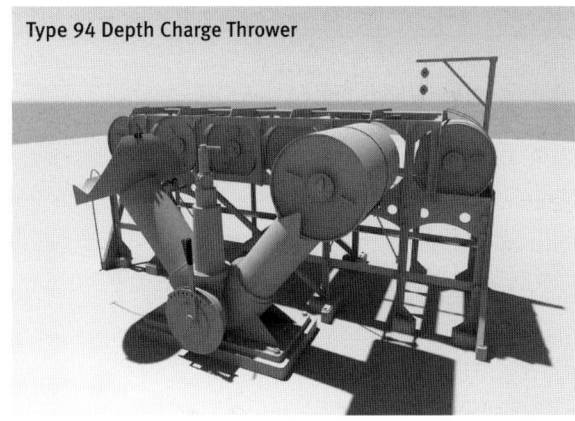

A recreation of the IJN's Type 94 depth charge thrower by J Ed Low. The Type 91 charges in the rack were aligned with one of the two arms of the thrower and shunted via steel pinch rollers onto the arbor, which was expendable. The thrower used a single powder charge; range was 105m for a single charge and 75m when both charges were fired simultaneously. The rack for the depth charges was reloaded using a davit. (© 2015 J Ed Low IJNWarship.com)

marker for line of sight (LOS). When the target once again came into view in the sights, the signal was given for the launching of the torpedoes. There were two ways of acquiring the target: one was to wait until the relative movement of the launch and target ships brought the latter into the sights; the other was to manoeuvre the ship into the required position for earlier firing, which was the method generally favoured by the IJN.

The Type 92 torpedo firing control panel had the appearance of a director. A casing with a sub-compass was mounted on an anti-vibration plate at the bottom. The lower part of the pedestal for the large binocular-type sight – initially a 12cm model but this was soon upgraded to 15cm with up to 18.8 times magnification – was inside the casing, the upper panel projected out of the roof, and around this various scale rings, input handles, interconnections, and dials were arranged in a similar fashion to the Type 97 Model 2 fire director. After computing gyro angle and launch angle, the panel provided the necessary corrections.

This panel was in production for more than ten years before being superseded by the modified Type 1, which had enhanced computation capabilities, in 1942. With about 215 units built it was the most widely-produced item among all torpedo FCS components both for submarine and surface ship use, and was mounted on every class of ship equipped with torpedo tubes completed during the period of manufacture.

Guns

The main gunnery armament of the *Asashio* class consisted of six 50-cal Third Year (1928) Type 12.7cm guns in three C-type twin mountings (weight 32 tonnes; angle of elevation +55° to -7°; muzzle velocity 915m/s; 4.4 rounds/min; 16 men) and four Type 96 Model 2 25mm MG in two twin mountings. Fire control for the main guns was provided by a Type 94 combined fire director and computer (*shageki hôiban*) with a separate 3-metre rangefinder. The machine guns were fired in local control using the French Le Prieur sight.

The 12.7cm main gun is described in *Warship 2007*, pp.104–06, where there is also some data for the standard 25mm close-range AA gun.

Anti-submarine Weapons & Sensors

A Type 94 'Y gun' (depth charge thrower) with its associated loading platform for six DCs was located on the centreline abaft no.3 gun mounting. The Type 94 DCT weighed 680kg and threw two Type 91 depth charges from arbors mounted at an angle of 50° to either side of the ship; using a single no.4 powder charge of 180g maximum range was 105m for a single DC and 75m when both DCs were fired simultaneously. The Type 91 was a cylindrical depth charge with a length of 775mm and a diameter of 450mm; total weight was 160kg, of which the bursting charge accounted for 100kg; sinking speed was 2m/s. The Type 3 fuse was activated by water pressure but there were only two depth settings, of 25m and 50m, which were selected manually prior to throwing. The effective distance at which damage would be caused to a submarine was c.33m. In addition to the DCT there were individual chutes for three depth charges per side at the stern; these were released manually.

For submarine detection *Kaigun Suirai-shi* (p.368) states that *Kasumi* and *Arare* were fitted with a Type 93 passive hydrophone and a Type 93 active ranging sonar. It is possible that the other ships were likewise equipped. The principal technical data for these two systems are listed in Tables 10 and 11. The performance of both was markedly inferior to the underwater sensors in service with the Allies. Japanese experts were well aware of this, but in the absence of anything better the crews of the ships placed considerable reliance upon them. The hydrophone was generally operated continuously at sea, while the sonar was used at intervals of about 15 minutes to render passive detection more difficult.

Hydrophone

Following the import of hydrophones from France, USA and Britain, serious research was begun in 1928. However, lacking good microphones and other components it was decided to revert to importing American technology; in 1929 Lt. Hisayama visited the Submarine Signal Co., and in the following year some MV-type hydrophones were purchased. A little later Capt. (later Admiral) Nomura Naokuni, stationed in Germany, reported on a new type of hydrophone developed by the Elektroakustik Co. at Kiel. Two units, designated Type Ho in the IJN, were purchased in 1932 and trialled. Following the acquisition of further technical know-how from Germany[13] this Type Ho was used as the model for the IJN's domestically-developed Type 93 hydrophone. *Kaigun Suirai-shi* (p.343) states that this type of hydrophone employed 16 microphones, and the requirement to match the characteristics of these created a bottleneck in production. The fitting of hydrophones also caused many problems ranging from the optimal position of installation to the suppression of self-noise.[14] It was not until 1939 that the hydrophones entered service in numbers. In that same year regulations regarding wireless telegraphy and sounding equipment were issued, classifying it by ship type. In addition to saving labour and time, these regulations meant that hydrophones and sonar were now recognised as standard practical systems.

Sonar

In 1923 Lt. (later Tech. V-Ad.) Nawa Takeshi, stationed in Paris as superintendent of the NTD, reported on the Langevin-type crystal emitter and proposed that it be purchased by the IJN. However, the Second Division of the NTD decided to use it for depth sounding. Two or three years later, two ultrasonic devices were imported from the French SCAM Co. and with them the domestic

Table 10: Passive Underwater Detection Array

Arrangement:	16 microphones in two elliptical arrays with a main axis of 3m mounted on the hull sides: c.5° angle accuracy and 50% background noise.
Microphones:	Moving coil type with membrane 145mm in diameter, sensitivity 35dB, frequency range 500–2500Hz, weight 18kg.
Listening:	Comprised compensator, amplifier, filter and electrical source. The compensator with contact bar grating control system was manually operated; bearing was indicated by a bearing repeater on bridge.
Switching:	Two complex connection (switching) boxes, one switch.
Performance:	Detection depended on various factors; range data considered unreliable.

Table 11: Sonar

Frequency:	17.5kHz.
Projector:	One quartz, retractable within a circular trunk sealed off with a sluice valve. Electro-hydraulic raise and lower system, also used for training. Remotely controlled from the Sonar Room.
Transmitter:	Hartley oscillator circuit with 1.7kW output, automatic pulse repetition with adjustable intervals in seconds or milliseconds. The sound wave was non-decreasing.
Receiver:	Heterodyne amplifier tuneable over 16–31kHz, and with a gain of 120dB.
Range:	A pointer traversed a circular scale in conjunction with echo responses in the receiver.
Indicator:	Headphones; ranges were 1,500m, 3,000m and 6,000m. Accuracy was ±100m.
Bearing:	Directivity 12°, accuracy ±3°, discrimination 10°.
Electrical source:	Motor generator 100V DC
Performance:	Range up to 1,300m at 12 knots.

development of active sonar was begun. In 1929 naval Eng. Saigô of the NTRI conducted a study for an underwater sound emitting and receiving system which was independent of the Pierre Langevin system. Trials of a prototype were begun, but when Saigô died research was halted.

Given the circumstances it was decided in 1931 to utilise the Langevin system, and the IJN formally adopted the Type 93 sonar, which was based on it, in 1933. Like the Type 93 hydrophone it was to be the principal IJN sonar until the end of the Pacific War.

Modified Stern Configuration

A large turning circle is problematic in a destroyer because the effect of evasive manoeuvres in the event of attack is markedly reduced, and counter-attacks against submarines are rendered less effective. In addition, steerability and stability are closely related. During trials with *Asashio* the ship was found to have a larger turning circle than expected; the cause was immediately investigated and remedial measures proposed.

The hull-form of the *Asashio* class was very different to that of the *Fubuki* class. The draught of the latter was unusually shallow, and this was also a feature of the *Hatsuharu* and *Shiratsuyu* classes. In the *Asashio* class the designers aimed to expand the underwater volume, thereby lowering the centre of gravity. However, the cut-up above the propellers was at a comparatively shallow depth. The vertical hull sides were modified so that the bottom formed a sharp angle at this point; the bilge radius was very small and the shape of the stern was different.

When a ship is making high speed and the rudder is activated the stern moves outwards. At that moment it generates a lot of foam on the surface, with multiple air bubbles which descend along the hull sides to below the stern. In the case of the *Asashio* class the air bubbles

A clear port broadside view of *Asagumo* steaming at high speed 18 months after her commissioning. The trials of the first ships completed revealed defects, and initiated modifications resulting in an improved shape of the stern and rudder which affected both speed (increased) and turning circle (reduced). Despite the high speed indicated by the bow wave, no stern wave can be seen. A censor has apparently retouched the background and, perhaps, also smoke from the funnels. Note the reduced size and clear lines of the bridge structure carrying the fire control director and rangefinder tower, the two light tripod masts and the angle of the funnels, with the oblique top contributing to the elegant appearance. The Type 94 DC thrower, its loading frame, and a minesweeping paravane are clearly visible on the quarterdeck. (Fukui Shizuo collection)

Asashio Class: Rudder & Stern Configuration

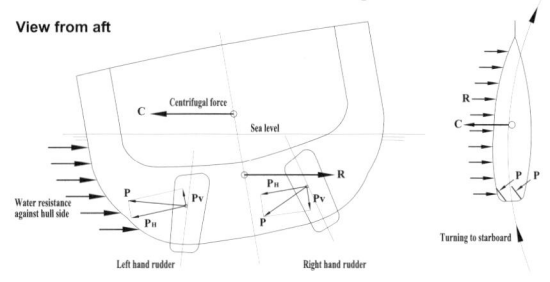

View from aft

Note: Twin inclined rudders were used to resist heel when turning.

Shiratsuyu Class (also Asashio as completed): Rounded Stern

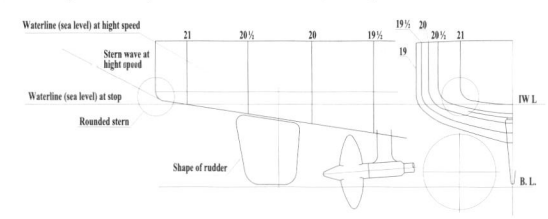

Kagero Class (also Asashio as rebuilt): Knuckle Stern

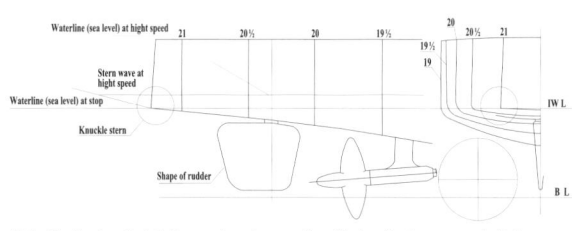

Note: The drawings illustrate the superior water separation of the knuckle stern compared with the earlier type, and show the markedly flatter stern wave.

(Drawn by Waldemar Trojca using material from the 'Destroyer Monograph' of Hori Motoyoshi supplied by the author)

Table 12: Armament Summer 1944

Michishio	20 Aug 1944	25mm MG: 4 triple, 1 twin, 12 single
		13mm MG: 2 twin
		two DC rails; no DCT
		one Type 22, one Type 13 radars
		two infra-red IFF lamps
Asagumo	30 Jun 1944	25mm MG: 4 triple, 1 twin, 8 single
		no DC rails; no DCT
		one Type 22, one Type 13 radars
Yamagumo	20 Aug 1944	25mm MG: 4 triple, 1 twin, 12 single
		two DC rails; no DCT
		one Type 22, one Type 13 radars
		two infra-red IFF lamps
Kasumi	02 Sep 1944	25mm MG: 4 triple, 1 twin, 8–10 single
		13mm MG: 4 single
		two DC rails; one DCT
		one Type 22, one Type 13 radars
		two infra-red IFF lamps

Source: Official data after *A Gô Sakusen* (Marianas) for *Shô Gô Sakusen* (Leyte); this may be considered reliable.

continued down to the rudder but remained behind its after edge, thereby lowering the pressure on the main body of the rudder and markedly reducing its effectiveness. In order to counteract this effect and change the flow, a new type of stern with a sharp knuckle was developed for the next type of destroyer, the *Kagero*; the research and trials were conducted while the later units of the *Asashio* class were building, it is possible that the last two units, *Arare* and *Kasumi*, were completed with the revised stern and rudder. The modified stern reduced the turning circle and also resulted in a slight increase of 0.4 knots in maximum speed.

Wartime Modifications

During the war the anti-submarine weapons were reinforced: the minesweeping gear was landed and replaced by DC rails, and there was a significant increase in the number of depth charges carried. The type of depth charge was reportedly changed to a more modern type (Type 95/Type 2?) with a modified configuration to increase sinking speed. Some ships had the DC thrower removed. Data about the installation of more modern underwater detection equipment is lacking.

The standard twin-horn antenna for the No.22 surface search radar was fitted atop a modified tripod foremast, the crow's nest being relocated at the midpoint of the pole topmast with an antenna for No.27 ECM near the top. The distinctive ladder-type antenna for the No.13 air search radar was located on the forward side of the mainmast. The radar room was in an extension to the lower bridge structure.

No.2 main gun mounting (deckhouse aft) and the after 2-metre RF were removed, and two superimposed 25mm triple MGs mounted on platforms in their place. A twin 25mm MG was located on a platform which extended from the front of the bridge, and single 25mm MGs were located on both sides on the forecastle, abreast the after deckhouse, and on the quarterdeck. The twin MG on the platform around the after funnel were replaced by triples. Numbers are generally uncertain and differed from ship to ship, despite attempts by the IJN to standardise the AA outfit (eg in July 1944).

The increase in ASW weaponry and the reinforcement of the AA armament began in the autumn of 1943. No.2 main gun was landed in 1944, and in that same year the radars were fitted; however, some sources state that the no.13 radar was installed in late 1943.

Footnotes:

[1.] *Outline of Warship Construction Technique* (Kaigun Zôsen Gijutsu Gaiyô), Vol. V, p.1283.
[2.] *Outline History of Naval Armament* (Kaigun Gunbi Enkaku), pp.622–645, *Naval Armaments and War Preparations*, Vol. 1 (Kaigun Gunsembi, ichi... = Senshi sôsho Vol. 31) pp.420-437; *Navy* (Kaigun), Vol. IV, pp.99–102, *History of Shipbuilding in Shôwa Era* (Shôwa Zôsen-shi), Vol. 1, pp.456–57, 505. The reinforcement of the naval air force was also part of the programme, but discussion of this is omitted.

This photo of *Yamagumo* was taken on 15 September 1939, one day after the photo of *Asagumo* in the same location, at almost the same angle and apparently also at similar speed. In contrast to the photo of *Asagumo*, the state of the sea is different (slight to moderate compared to smooth), the bow wave higher and longer and she has a distinct stern wave. The number 41 painted on the bow of *Asagumo* and *Yamagumo* indicates that both were assigned to the same destroyer division. (Fukui Shizuo collection)

3. Among them were ships designed to be quickly converted into aircraft carriers when the situation required, and also tankers, a repair ship and other auxiliary types recognised as essential for ocean warfare. Appendices to the document provided comparative strength data for the IJN, the US Navy, and the Royal Navy, and even a comparison of building expenses in these countries.
4. On this assumption the maximum stress would occur at the clamped end or root of the blade. In fact the breaks frequently occurred near the middle of the blade, between one third of the length from the top and the mid-point.
5. The turbine blades were clipped at the centre of the wheel and then riveted. They were usually situated in a recess of the wheel but this was dispensed with to ease production and save weight. Assuming resonance vibrations of blade and wheel, the shape of the latter was improved, the recess again adopted and the thickness of the blades and their radius at the fore and rear ends increased. Due to these revisions the weight of the turbine increased, and the efficiency ratio was slightly reduced.
6. One turbine blade of the after starboard LP turbine broke, but the cause was put down to a material or production defect.
7. As well as *Technical Material of the Former Navy* (Nihon Kyu Kaigun Gijutsu Shiryô), and *History of Shipbuilding in Sôwa Era* (Shôwa Zôsen-shi), there is a fine summarised report of a Japanese engineer (Tohara) as enclosure (C) in Target Report S-01-11 *Main and Auxiliary Machinery*.
8. The land tests are described by Admiral Shibuya Ryûtarô in *Technical Material...*, but cover almost 30 pages and are so specialised that they are useful for only turbine experts.
9. R-Ad. Kondo Ichirô highlights the achievements of naval engineer Mori Shigeru and Tech. Capt. Yasugi Shôichi, and regrets that the designers of the IP turbines were disciplined despite the fact that the true cause of this phenomenon was not known at that time in the world.
10. According to the authors of *History of Electrical Technology in Japanese War Vessels and Merchantmen* (*Nippon no Kantei, Shôsen no Denki Gijutsu-shi*) the reduction was about 25 per cent; no figures are given for volume.
11. Those in charge at the NTD were Cdr. Ooe Ranji and Lt-Cdr. Imazato Kazuo.
12. In 1931 the IJN purchased a torpedo fire director from British Vickers Co. and gave it to Japan Optical Co. for testing. At the same time the test production of a submarine torpedo fire director was ordered from the same company, but it is unclear what influence this may have had on the development of surface warship FC systems.
13. Japanese technicians were surprised to find that permanent magnetic steel, invented by Dr Mishima in Japan, was already employed in this hydrophone.
14. Noise was generated by the main engines, auxiliary machinery, propellers, water stream along the hull sides, bubbles, etc. and masked the signal from the target. The rapid training of operators and the maintenance of the hydrophones proved inadequate; a shortage of spare parts and a lack of regular inspection and maintenance were responsible for further problems.

THE NAVAL WAR IN THE ADRIATIC PART 2: 1917–18

In the concluding part of their article on the naval war in the Adriatic, **Enrico Cernuschi** and **Vincent P O'Hara** cover the years 1917–18, which saw the triumph of the Entente forces and the defeat of the Imperial Austro-Hungarian Navy.

The first thirty months of war in the Adriatic had seen the Austro-Hungarian k.u.k. Kriegsmarine preserve its battle fleet in Pola while using its light units to conduct hit and run raids. It had bombarded many lightly-defended targets along the Italian coast, but had disrupted traffic between Italy and the Balkans only briefly. Italy's Regia Marina, meanwhile, patrolled vigorously, worked at ways of attacking enemy units in their harbours and, most importantly, established an effective army support force with specialised units and doctrine. The British and French navies supported Italian efforts in the lower Adriatic, but poor coordination marred the occasions when the three navies acted in concert against enemy raids.

The year 1917 began with major changes in the high command of both the Italian and Austro-Hungarian fleets. On 8 February 1917 Admiral Haus died of pneumonia after returning from a meeting in Germany in an unheated railway carriage. This, however, did not change Imperial naval strategy as the new *Flottenkommandant*, Admiral Maximilian Njegovan, held a similar appreciation of the naval situation. The battle force stayed at Pola while the modern light cruisers and destroyers, supported by the armoured cruisers *Sankt Georg* and *Kaiser Karl VI*, pursued their guerrilla campaign in the south, increasingly feeling the strain of two and a half years of war.

On the Italian side Admiral Thaon di Revel, who had been naval Chief of Staff from 1913 to October 1915, when he was demoted to commander of the Upper Adriatic regional command, returned to his old job on 16 February 1917, while the Duke of Abruzzi retired as fleet

The Italian battleship *Cavour* returning to Taranto from Corfu on 20 January 1917. (Enrico Cernuschi collection)

commander – causing no sorrow to his cousin the king. The Duke had fallen victim to a storm of criticism that followed the loss of the pre-dreadnought *Regina Margherita* on a German mine on 11 December 1916. Upon his return to power Thaon di Revel promoted his own supporters to head the battle fleet in Taranto, the navy ministry in Rome, and the Brindisi and Valona regional commands while retaining his Upper Adriatic command.

The Battle of Otranto

During the early months of 1917 the Otranto Barrage was an important target for the Habsburg fleet. On the night of 11/12 March four Austro-Hungarian destroyers swept the strait, but their attack on a French steamer failed. In the next raid, on the night of 21/22 April, Austro-Hungarian torpedo boats sank the Italian steamer *Japigia*, which was sailing independently. On 15 May the Austro-Hungarians mounted their largest action against the Otranto Barrage in the war to date. This involved three light cruisers and two destroyers, while the Entente forces that put to sea in response included two British light cruisers, one Italian scout cruiser, three Italian flotilla leaders, and eight Italian and seven French destroyers. In the subsequent surface action an additional k.u.k. force of one coastal battleship, one armoured cruiser, two destroyers and seven torpedo boats sortied from Cattaro to support the raiders on their return to base.

The Austro-Hungarian armoured cruiser *Sankt Georg*, 15 May 1917. (Rivista Marittima)

An aerial view of the three Austro-Hungarian *Novara*s manoeuvring during the Battle of Otranto Strait on 15 May 1917. (Rivista Marittima)

Splinter damage on the Austro-Hungarian scout *Novara* after the 15 May 1917 action. (Rivista Marittima)

The Austro-Hungarian plan called for the Austro-Hungarian destroyers *Csepel* and *Balaton* to raid along the Albanian coast as a diversion while the three light cruisers attacked the barrage itself. The destroyers ran into a three-ship convoy escorted by a single Italian destroyer. Surprise and good shooting allowed them to sink the destroyer *Borea* and the steamer *Carroccio* (1,657 GRT) and to damage the merchant vessels *Verità* and *Bersagliere*. Meanwhile, the light cruisers sank fourteen and damaged three of the forty-seven British drifters present that night.

After this successful attack the two k.u.k. forces turned for home. However, an Entente patrol consisting of the flotilla leader *Mirabello* and three French destroyers had arrived off Durazzo at dawn and they turned south to intercept. Meanwhile, the light cruisers *Bristol* and *Dartmouth*, the flotilla leader *Aquila* and four Italian destroyers put out from Brindisi. The scout cruiser *Marsala* and three more destroyers were raising steam to follow. A series of high speed engagements resulted. The *Mirabello* group made first contact with the k.u.k. cruisers, but superior Austro-Hungarian firepower kept them at bay. The *Dartmouth* group, meanwhile, intercepted the two k.u.k. destroyers. A high-speed chase ensued but a lucky shot delivered by *Csepel* hit *Aquila* in the boiler room and disabled the Italian vessel. The Austro-Hungarian ships ultimately escaped to the protection of Durazzo's batteries. After detaching two destroyers to protect *Aquila*, the rest of the *Dartmouth* group encountered the *Novara* force. In the ensuing pursuit *Bristol* (with a foul bottom) and *Mirabello* (with contaminated fuel) fell back; then *Commandant Rivière* broke down and the other two French vessels stayed behind with her. This left *Dartmouth* and the destroyers *Acebri* and *Mosto* to continue the chase with *Bristol* coming up close behind. *Dartmouth* punished *Novara* from 0940, and at 1035 a hit in her after engine room forced half the k.u.k. cruiser's boilers to go off line. *Dartmouth* and *Helgoland* were also hit during this period. At 1045 *Dartmouth* turned away to allow *Bristol* to close. *Saïda*, which was lagging due to machinery problems, became the main target of the British cruisers and was hit twice.

By 1115 *Novara* was dead in the water and, under heavy fire, *Saïda* attempted to take her under tow. *Acebri*, which misread a signal, attacked independently but the combined fire of all three AH cruisers drove her away. Nonetheless, it seemed the Austro-Hungarians were finally trapped. Reinforced by the *Marsala* group the Entente flotilla turned again to the attack as the Austro-Hungarians laid smoke. At 1205, however, the Entente force spotted smoke from the armoured cruiser *Sankt Georg*, two destroyers and four torpedo boats which were rushing down from the north. The Italian admiral in command, thinking he would shortly be facing two armoured cruisers or possibly battleships, ordered his force to turn away. The nearness of the enemy base and his tail of crippled ships were also factors in this decision. On the way back the German submarine UC 25 torpedoed and seriously damaged *Dartmouth*, while the French destroyer *Boutefeu* hit a mine laid by that same submarine and sank. The coordination of the various groups of Entente warships was poor, and the finger-pointing between the Italians and the British that followed the action added salt to the wounds.

The effects of the Battle of Otranto were mixed. The Austro-Hungarians obtained a clear success, and the barrage was not patrolled at night for a month until the Entente navies had augmented their escorts. However, the k.u.k. Kriegsmarine concluded that the danger to its valuable scouting squadron of irreplaceable light cruisers (necessary for strategic scouting for the battle force) was too great. It did not order another raid into the Otranto Strait until 20/21 September 1917, and the light cruisers were not involved before 18/19 October when *Helgoland* sailed alone, without result, as did *Novara* the next month.

The Northern Adriatic

In the Northern Adriatic MAS 18 under Lt. Luigi Rizzo penetrated Trieste harbour on 15 March. His mission was to scout the defences for a future attack. Minor surface encounters occurred on 11 May, 25 May, 3 June, 23 September, and 24 September. The most significant action to occur in northern waters during the year took

WARSHIP 2016

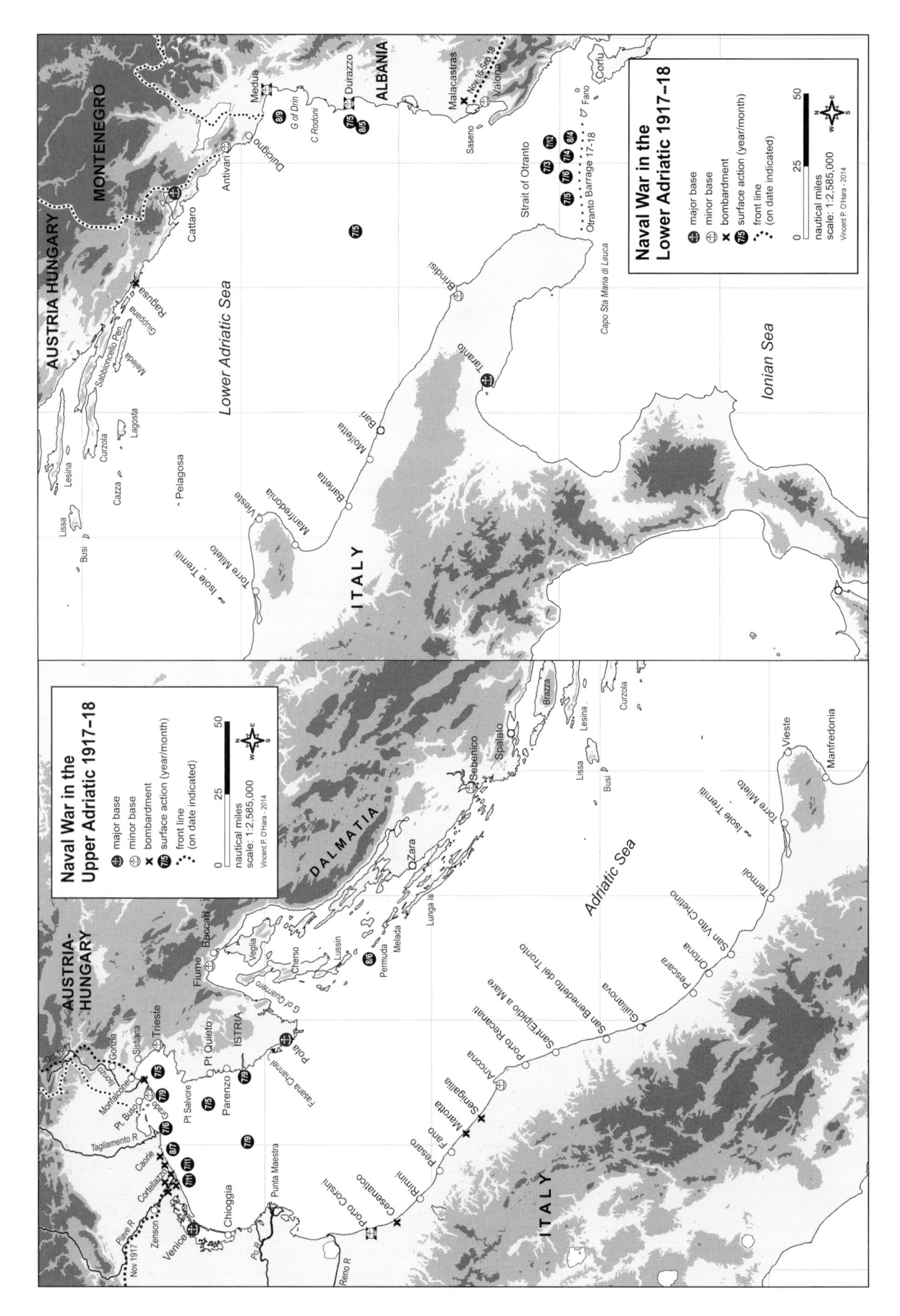

THE NAVAL WAR IN THE ADRIATIC PART 2: 1917–18

The British monitor *Earl of Peterborough*, nicknamed *Il Conte* (the Earl) by the Italians. (Enrico Cernuschi collection)

An Austro-Hungarian flying boat which was seized and towed to Venice. The town was the largest European seaplane base (about 90 aircraft) of the First World War. (Enrico Cernuschi collection)

place on the night of 29/30 September. Four Austro-Hungarian destroyers and four torpedo boats were at sea supporting an air raid against Ferrara. They encountered an Italian flotilla leader and seven destroyers off Punta Maestra at the mouth of the River Po. The squadrons fought a savage high-speed action in the darkness with periodic gunnery exchanges at 1,000–2,000m all the way across the Adriatic to the extensive Austro-Hungarian minefield off Parenzo. Both sides suffered significant damage but no ships were sunk.

Italy continued to emphasise naval support of its army's maritime flank. On 24 May, after a slow transfer from the Aegean Sea, the British monitors *Earl of Peterborough* and *Sir Thomas Picton* ventured into the Gulf of Trieste and shelled Austro-Hungarian lines. Their advantages, armour and mobility, were offset as the Royal Navy asked for a defence system around them formed by destroyers, torpedo boats, MAS, minesweepers, auxiliary vessels, and dozens of flying boats on a scale that paralysed routine operations from Venice and Grado. The Austro-Hungarians reacted with air raids, landing one bomb on *Earl of Peterborough*, and a fruitless night search by a dozen destroyers and torpedo boats.

Between 18 and 24 August the Regia Marina renewed naval bombardment of the Austro-Hungarian lines around Mount Hermada during the new offensive along the Isonzo River. The British monitors were flanked by all the Italian naval artillery on the front line with the addition of the newly-commissioned Italian self-propelled pontoons *Cappellini* and *Faà di Bruno*. The latter were each armed with a pair of 15in guns originally intended for the *Caracciolo* class battleships, the construction of which had been suspended in the spring of 1916 after the Jutland experience of plunging fire revealed that their horizontal armour was too thin and that the required modifications would make their draught too deep. Booms, minefields and the light forces of Venice and Grado, which included fourteen destroyers and fifteen torpedo boats, flanked the bombardment force, while six of the torpedo boats shelled targets of opportunity along the coast.

Fifteen MAS boats joined the party with eleven submarines and minesweepers, auxiliary vessels, and aircraft. The old battleships *Saint Bon* and *Emanuele Filiberto* and their escort were ready to sail from Venice if enemy heavy forces should be sighted. Ordnance expended included 63 15in shells, 243 12in (175 fired by the British monitors), 5,923 8in, 7.5in, 6in, and 4.7in projectiles from the Italian pontoons, and 1,926 3in rounds. The Austro-Hungarian reaction was limited to one 150mm shore battery and some ineffectual air attacks.

Complaints from the Imperial Army about enemy warships sailing by day in the Gulf of Trieste, firing on their troops at will, finally compelled the k.u.k. Kriegsmarine to transfer the old coastal defence ships *Wien* and *Budapest* from Cattaro to Muggia near Trieste. Italian aircraft attacked the squadron on 26 August, the same day it arrived. Over the next 18 days the Italians made 183 bomber sorties at the cost of two aircraft. On 5 September a near miss damaged *Wien* and caused her to withdraw to Pola. By 13 September *Budapest* had likewise retreated, as the enemy's air superiority made the squadron's forward base too dangerous.

On 22 August an Italian flotilla attacked Ragusa in the southern Adriatic as a diversion, while other units laid a minefield off the Sabbioncello peninsula. On 10 September an Italian flotilla, covered by an Entente squadron,

The Italian self-propelled 15-inch pontoon *Cappellini*. (Enrico Cernuschi collection)

The k.u.k. torpedo boat Tb 11 in Venice in October 1917 after the mutiny of her crew. (Rivista Marittima)

bombarded the villages of Poiani and Stula on the Albanian front with minor results. On 5 October there was an unusual episode when the Austro-Hungarian torpedo boat *Tb 11* defected to Ancona after a mutiny by Czech and Italian sailors. The Regia Marina commissioned the vessel as *Francesco Rismondo*.

On the night of 13/14 October Lt. Luigi Rizzo returned to Trieste in *MAS 20*. He moored his boat at the mole to investigate the enemy defences and returned undetected. His report confirmed the feasibility of an attack inside that harbour. Admiral Thaon di Revel, following his calculated risk strategy, was always ready to use expendable warships if the target was worthwhile. At the same time, however, he was careful about the lives of his handpicked personnel, and so two nights later he personally inspected the Austro-Hungarian booms in a motor boat equipped with a silent auxiliary electric engine. He returned to Grado persuaded that the action was feasible. The assault, however, was delayed by the launching of the unexpected massive German and Austro-Hungarian offensive of Caporetto on 24 October.

The Crisis Point

On the night of 27 October, two days after the Italians began a general retreat on the Isonzo front, the Austro-Hungarian torpedo boats *T 4*, *6* and *9* bombarded the Italian coastal batteries of Punta Sdobba. This marked the beginning of new offensive phase by the k.u.k. Kriegsmarine. On 30 October *Wien* and *Budapest* returned to Trieste ready to intervene along the fast-moving coastal front, but the Italian retreat from Grado was largely unhampered by the Habsburg navy.

By 2 November the Italian Navy's coastal artillery forces were deployed from the mouth of the River Piave to Venice, which the German and Austro-Hungarian advance was now approaching. The naval floating batteries conducted their first heavy barrage (almost 300 medium- and large-calibre shells) on 11 November against Habsburg bridgeheads at Zenson and Grisolera, upstream from Cortellazzo. The fire, at a range of 22,000m, lasted eight hours, destroying pontoon bridges and ferries and delaying the advance. The next day the small pontoons, placed in the channels along the front line, stopped an Austro-Hungarian attack between San Donà and Musile. On 13 November the big gun pontoons again hammered the Zenson area for eight hours, halting the advance towards Venice from that point.

That same evening four Italian destroyers fired 800 4in rounds against an enemy column near the mouth of the River Piave. The Imperial Army complained bitterly about the Navy's failure to respond. At dawn, as a result of the Italian attack and signals intelligence, there followed a brief clash between five Austro-Hungarian torpedo boats and four Italian destroyers. A Regia Marina coastal battery of four 6in guns joined the action, and the Austro-Hungarian warships retreated behind a minefield. That battery was the only one present in the area. The pontoons, which were without armour or watertight compartments, could not be used because the barren beach had no channels or locations to camouflage vessels. The battery further distinguished itself on 13 November, bombarding the Austro-Hungarian bridgehead of Fornaci di Brazà, the most advanced point reached by the Imperial Army along the coast. The battery then supported the reconquest of that village by naval infantry. Later, a concentration of fire by armed barges halted another Austro-Hungarian effort at Grisolera. After this double repulse the Imperial Army decided to concentrate its efforts against the Cortellazzo beachhead, considered the only suitable road to Venice.

Given this supreme endeavour the k.u.k. Kriegsmarine had to be fully committed. Its mission was to silence the troublesome battery and to provide flank cover for the army's advance over the Piave and the final push to Venice.

The Imperial Navy conducted an initial reconnaissance on 15 November, resulting in a brief duel between the battery and seven torpedo boats, which were sweeping the area for mines to permit the deployment of the coastal battleships from Trieste. Italian fire caused the sweepers to retreat before they had finished clearing all the channels. On 16 November, after such a fair warning, the Austro-Hungarians began their final decisive assault against Cortellazzo. Five old torpedo boats swept ahead of *Budapest* and *Wien* and their escort of eight torpedo boats. The coastal battleships engaged the battery at 1045 from about 9,000m. This extreme point of the Italian line was under the Regia Marina's protection. The 240mm battleship guns fired one round every three or four minutes; Austro-Hungarian, Italian and French aircraft sparred overhead, while on land the Italian guns replied briskly. They hit *Wien* seven times but the 6in rounds bounced off the ship's armoured hull. *Budapest* took one shell beneath her waterline. Then, at 1230, the k.u.k. Kriegsmarine squadron sighted seven Italian destroyers approaching and turned to meet this foe, giving the battery a welcome respite. The Italian flotilla attempted to lure the Austro-Hungarians toward Cortellazzo, where three MAS boats waited in ambush. Suspecting a trap, the Austro-Hungarian squadron let the Italians go and returned to its primary mission. *MAS 13*

Table 1: Naval Surface Actions, Bombardments and Raids: 1917-1918

Year	Date	Nation		Forces		Result
1917	22 Feb	IT	N	MAS 1	L	Parenzo. Aborted.
1917	12 Mar	AH	S	DD Csepel, Orjen, Tátra, Balaton	SE-N, Rd	Otranto barrage. Failed attack on FR MV Gorgone.
1917	21 Apr	AH	S	DD Uskoke, Streiter, Warasdiner, Réka; TB 84F, 92F, 94F, 100M	SE-N, Rd	Otranto barrage. MV Japigia+.
1917	11 May	AH/IT	N	AH DD Csikós; TB Tb 96 F, 78 T, 93 F. IT DD Animoso, Ardito, Ardente, Abba, Audace II.	SE-N	Chase. AH retire to cover of minefield.
1917	15 May	AH/IT	S	AH DD Csepel, Balato. IT DD Borea; 3xMV	SE-N	Borea+, MV Carroccio+ Verità, Bersagliere.
1917	15 May	AH	S	CLS Novara, Helgoland, Saïda	SE	Otranto barrage. BR trawlers: Admirable+, Avondale+, Coral Haven+, Craigoon+, Felicitas+, Girl Gracie+, Girl Rose+, Heleonora+, Quarry Knowe+, Selby+, Serene+, Taits+, Transit+, Young Linnet+. Gowan Lee, Floandi, Union AH Novara (1).
1917	15 May	AH/IT/FR/BR	S	AH CLS Novara, Helgoland, Saïda; DD Csepel, Balaton. AH support group AC Sankt Georg; DD Warasdiner, TB Tb 84, 88, 99, 100. Entente: IT DDL Mirabello, FR DD Cdt. Rivière, Bisson, Cimeterre; BR CL Dartmouth, Bristol; IT DDL Aquila; IT DD Acerbi, Schiaffino, Pilo, Mosto; CLS Marsala; DDL Racchia; DD Insidioso, Impavido, Indomito	SE-D	Chase follows raid on the Otranto Straits. Aquila (1), Dartmouth (3), Bristol (splinters). Novara (10), Helgoland (2), Saïda(2), Csepel (splinters). Dartmouth later torpedoed by UC 25 and FR DD Boutefeu+ mined.
1917	24 May	BR/IT	N	BR BM Earl of Peterborough, Sir Thomas Picton; IT cover force	SB-D	vs AH front lines.
1917	25 May	IT/AH	N	IT MAS 18. AH tug	SE-N	Off Trieste. Brief chase to Sbat.
1917	3 Jun	IT/AH	N	IT MAS 14, 15, 19. AH DD Wildfang, Csikós, TB Tb 93 F, 96 F	SE-N	Off Tagliamento R; brief encounter.
1917	18 Aug	BR/IT	N	BR BM Earl of Peterborough, Sir Thomas Picton; IT Pbat Cappellini, Faà di Bruno, TB 40, 41, 42 PN, 46, 48 OS, 58 OL. IT covering force: DDx13, TBx13, 9xMAS	SB-D	vs AH front lines.
1917	22 Aug	IT/BR/FR	S	DDL Riboty, Aquila, Sparviero. Also minelayers DDL Mirabello; DD Schiaffino. Cover BR CL Weymouth, IT CLS Bixio, mixed IT and FR DD flotilla	SB-D	Ragusa. SB to cover minelaying mission.
1917	26 Aug	IT	S	MAS 6, 91	HF	Durazzo. No results.
1917	10 Sep	IT/BR	S	DDL Riboty, Racchia; TBs 55, 56, 57 AS, 61 OL. Strategic cover force BR CL Bristol and IT DDx2	SB-D	vs. Pojani, Stula, Fieri and Punta Semeni in Albania.
1917	18 Sep	IT	N	IT TB 13, 15, 16 OS	LRd	Failed landing near Parenzo.
1917	22 Sep	IT	N	MAS 1 escorted by TB13, 14, 15, 16 and 18 OS	L	Parenzo. Aborted.
1917	23 Sep	IT/AH	N	IT MAS 16. AH TB 77 T, 78 T	SE-N	Off Grado. AH boats on mine mission. Brief engagement.
1917	24 Sep	IT/AH	N	IT TB 9, 10, 11, 12 PN. AH TB 95 F+3xTB	SE-N	Leme Channel; brief encounter.
1917	29 Sep	IT/AH	N	AH DD Turul, Velebit, Huszar, Streiter; TB 90F, 94F, 98 M. IT DDL Sparviero; DD Abba, Orsini, Acerbi, Stocco, Audace II, Ardente, Ardito	SE-N	Chase between Punta Maestra and Parenzo. Sparviero (5), Orsini (1), Velebit (1), Huszár (1), Turul (splinters), Tb 94 F (splinters).
1917	27 Oct	AH	N	TB T4, 6, 9	SB-N	vs. Punta Sdobba.
1917	14 Nov	IT/AH	N	AH TB Tb 84 F, 92 F, 94 F, 99 M and 100 M. IT DD Animoso, Ardente, Audace, Abba	SB-D, SE-D	Mouth of Piave. AH TBs retreat behind minefield.
1917	16 Nov	AH	N	BBC Budapest, Wien; TB Tb 6, 9, 84 F, 92 F, 94 F, 98 M, 99 M, 100 M; minesweeping force Tb 23, 27, 30, 61, 65	SB-D	vs Cortellazzo 6in Sbat. Wien (7), Budapest (1.)
1917	16 Nov	IT/AH	N	AH as above. IT DD Orsini, Acerbi, Animoso, Stocco, Ardente, Abba, Audace II, MAS 9, 13, 15.	SE-D	MAS 15. AH force suspends bombardment and withdraws.

Continued overleaf:

1917	18 Nov	BR/IT	N	BR BM *Earl of Peterborough* and IT cover force	SB-D	vs AH front lines. Also 20, 30 Nov.
1917	18 Nov	IT	N	DD *Abba, Ardente, Audace, Animoso*	SB-D	vs Revedoli.
1917	19 Nov	IT/AH	N	IT DD *Stocco, Orsini, Ardito, Sirtori*; covering force BBO *Saint Bon, Emanuele Filiberto*; DD *Animoso, Audace II, Ardente, Abba*; MAS *9, 13, 15*.	SB-D	vs troops between Revedoli, Santa Croce and Caorle. AH DD *Triglav II, Reka*; TB *Tb 78 T, 82 F, 86 F, 87 F, 91F* turn away.
1917	21 Nov	IT	N	DD *Animoso, Audace, Ardente, Abba*	SB-N	vs Grisolera. Repeated 23 Nov SB-D.
1917	25 Nov	IT	N	Aux MS *Capitano Sauro, Folgore*	SB-D	vs AH front lines. *Folgore* (stranded), later salved by IT.
1917	25 Nov	IT	N	DD *Animoso, Audace, Ardente, Abba, Stocco, Orsini, Ardito, Sirtori*	SB-D	vs Grisolera.
1917	28 Nov	AH	N	DD *Triglav II, Réka, Dinara*; TBs *79 T, 86 F, 90 F*	SB-D	vs Senigallia.
1917	28 Nov	AH	N	DD *Triglav II, Dukla, Streiter, Huszár, Reka, Dinara*; Tb *78 T, 79 T, 82 F, 86 F, 87 F, 89 F, 90 F, 95 F*	SB-D	vs Porto Corsini, Marotta, and Cesenatico.
1917	9 Dec	IT	N	MAS *9, 13*	HF	Trieste. BBC *Wien*+
1917	13 Dec	AH/?	S	AH DD *Tátra, Balaton, Csepel*. Four Allied DDs	SE-N, Rd	Otranto barrage. TT action against 4 unidentified DD.
1917	14 Dec	IT	N	MAS *16, 19*	SB-D	Piave R. MAS *16*+
1917	19 Dec	AH	N	AH BBC *Budapest*, CLS *Admiral Spaun*; BBO *Arpád*; DD *Triglav II, Lika II, Dukla, Scharfschütze, Streiter, Turul*; TB *Tb 3, 5, 6, 7, 84 F, 92 F, 94 F, 98 M, 99 M, 100 M*; MS force: *Tb 20, 23, 27, 30, 32 34, 37, 61, 65, 66*.	SB-D	vs 6in SBat at Cortellazzo.
1917	19 Dec	IT	N	DDL *Sparviero, Aquila*; covering force: DD *Animoso, Abba, Audace II, Ardente*; TB *9, 10, 11, 12 PN, 13, 16 OS*; MASx5	SB-D	vs Grisolera.
1917	10 Dec	BR/IT	N	BR BM *Sir Thomas Picton*; IT cover force	SB-D	vs AH front lines. Also 9 Feb, 21 Mar 1918.
1918	10 Feb	IT	N	TB *12 PN, 13 OS, 18 PN*; MAS *94, 95, 96*	HF	Buccari, Gulf of Quarnero. No results.
1918	5 Apr	AH	N	DD *Uskoke*; TB *Tb 26*	L	Ancona. Landing party captured.
1918	22 Apr	AH/BR/FR	S	AH DDs *Csepel, Uzsok, Dukla, Lika (II), Triglav (II)*. RN DDs *Jackal, Hornet*, AU DD *Torrens*; FR DD *Cimeterre*.	SW, SE-N	Otranto barrage. *Hornet* (major) and *Jackal* (slight).
1918	8 May	AH	N	DDs *Turul, Huszár II, Réka, Pandúr*	L	at Silvi. Aborted.
1918	13 May	IT/AH	S	IT MAS *99 100*. AH DD *Dukla*; TB *Tb 84, 98*.	HF, SE-N	Durazzo. AH MV *Bregenz*+.
1918	14 May	IT	N	Attack craft *Grillo*	HF	Pola. The attack craft was scuttled. Wreck recovered by AH.
1918	15 May	IT	S	MAS *99, 100*	HF	Antivari, unsuccessful.
1918	16 May	IT	N	MAS *94, 95*	HF	Trieste, unsuccessful.
1918	10 Jun	AH/IT	N	IT MAS *15*, MAS *21*. AH BB *Tegetthoff, Szent István*; DD *Velebit*; TB *Tb 76 T, 77 T, 78 T, 79 T, 81 T, 87 F*	SE-N	BB *Szent István*+.

The self-propelled pontoon *Valente* armed with a 305mm/46 gun. (Enrico Cernuschi collection)

and *15* then charged (the other boat suffered an engine defect). Habsburg lookouts sighted them immediately, and the elusive little boats became the target of the whole Austro-Hungarian force. The MAS boats launched their four torpedoes at 900m but the two battleships dodged these weapons. Meanwhile the Italian destroyers returned to support the small motorboats.

By this time the confused action was at its height. The smoke of gunfire dominated the stage; dogfighting aircraft criss-crossed the sky; the Italian battery was straddling the enemy battleships; then Austro-Hungarian flying boats reported that the Italian battleships *Saint Bon* and *Emanuele Filiberto* had emerged from Venice and were steaming at flank speed towards Cortellazzo.

1918	21 Jun	IT	N	TB *3, 9, 10, 12 PN, 40, 41, 42, 64, 65 OS, 46, 48, 58 OL*	SB-N	vs Punta Tagliamento and Caorle.
1918	1 Jul	AH/IT	N	AH DDs *Balaton* and *Csikós*, TBs *83 F* and *88 F*. IT DD *Missori, La Masa, Audace, Orsini, Acerbi, Sirtori* and *Stocco*.	SE-N	Chase off Caorle. **Balaton**(1), **Csikós** (1), TBs **83 F** (3) and **88 F** (1); **Stocco** (3) and **Orsini** (splinters).
1918	2 Jul	IT	N	TB *3, 40, 64, 65, 66 PN, 15, 18, 48 OS*. Minesweeping force TB *Climene, Procione*; Aux MS *Grado, Capitano Sauro*	SB-N	vs mouths of the Piave R.
1918	7 Jul	BR/IT	S	BR BM *Earl of Peterbourgh, Sir Thomas Picton*; IT cover force	SB-D	vs AH lines at Malacastras.
1918	7 Jul	IT	N	TB *3, 40, 65, 66 PN, 48 OS, 58 OL*	SB-N, L-Rd	vs targets of opportunity between Caorle and Cortellazzo and raid at Punta Semana.
1918	19 Jul	IT	S	*MAS 97, 101*	HF	Antivari, unsuccessful.
1918	5 Sep	IT/AH	S	*MAS 100, 218*	HF	Durazzo, unsuccessful. Also 9 Sep.
1918	5 Sep	AH/IT	S	IT DDL *Aquila, Nibbio, Sparviero*; TB *6 PN, 12 PN*. AH TB *Tb 86, 19, 38*	SE-D	Gulf of Drin AH break off. **Tb 86**.
1918	15 Sep	IT	N	TB *3, 4, 65, 66 PN*; MAS *94, 95*	L	Istria coast. Insertion of agents. Also 16 Sep. Removal 17 Sep.
1918	1 Oct	IT/AH	S	*MAS 100, 218*	HF	Durazzo, unsuccessful.
1918	2 Oct	IT/BR/ US/AU/ AH	S	IT AC *San Giorgio, San Marco Pisa*; BR CL *Lowestoft, Dartmouth, Weymouth*; DD *Nereide, Ruby, Nymphe, Cameleon, Acheron, Goshawk, Jackal, Tigress*; IT TB *8, 10, 11, 35, 36, 37, 38, 42, 67 PN*; MAS *92, 97, 98, 102, 202; 210*. Covering force IT BB *Dante*, BR CL *Glasgow, Gloucester*; IT CLS *Marsala*; DDL *Aquila, Nibbio, Sparviero, Racchia, Riboty, Rossarol, Pepe, Poerio*; DD *Nievo, Schiaffino*, AU DD *Swan, Warrego*; BR DD *Acorn, Lapwing, Tribune, Shark, Badger, Fury*; US SCx11	SB-D, HF	Durazzo. AH DD *Scharfschütze* and *Dinara*, TB *Tb 87* MV *Stambul+* (3,187 GRT) MV *Graz, Hercegovina*. MAS *98*. **Weymouth** (TT) by *U 31*.
1918	21 Oct	IT	S	TB *12, 42 PN, 57 AS, 58 OL* covered by DDL *Aquila, Nibbio, Sparviero, Mirabello, Racchia, Riboty*	SB-D	vs Medua.
1918	31 Oct	IT	N	TB *41, 64 PN, 56 AS*	SB-D	vs targets of opportunity between Caorle and Revedoli.
1918	1 Nov	IT	N	*S2* semi-submersible attack craft	HF	vs Pola. BB *Viribus Unitis+*.

Note: A full list of the table abbreviations was published with Part 1 of this article, *Warship* 2015 p.171.
Ships in **bold italics**: damaged (number of hits in parentheses).
Ships by +: sunk.

The Austro-Hungarian commander thus decided to return to Trieste. If Cortellazzo was the focal point of the Italian war and, in perspective, of the European military balance at the time, this decision to retreat was fatal and the commander's superiors agreed, relieving him from duty a few days later. Indeed, the Habsburg Navy had acted at the decisive point but doomed its effort by giving the enemy forewarning and by employing its forces in a half-hearted manner. It sent only two small and slow coastal battleships, while six better battleships, more than enough to overwhelm the Italian Venice squadron, remained at Pola. Given the Navy's ability in the upper Adriatic to assemble a dozen destroyers and torpedo boats, it was not even a lack of escort forces which tipped the balance. 'Too little, too late' was becoming the by-word for the k.u.k. Kriegmarine's strategy, and the Regia Marina's sea denial campaign was paying dividends by reinforcing the conservative mind-set of the k.u.k's high command.

The Skirmishes Continue

Italian destroyers bombarded Austro-Hungarian positions near the front on 18, 19, and 21 November. On the 21st two k.u.k. Kriegmarine destroyers and five torpedo boats, returning from a minelaying mission, sighted an Italian flotilla of four destroyers but did not engage. Such episodes further troubled the relations between the

Imperial Army and the Navy. On 28 November, in response to these blows, the Habsburgs attacked at dawn the coastal towns of Porto Corsini, Cesenatico, Marotta and Senigallia with six destroyers and eight torpedo boats grouped into two flotillas.

The mission began badly when *Dukla* and *Tb 89 F* collided, holing *Dukla* and forcing her to abort. The attack continued nonetheless and damaged the railway line. However, the coastal batteries replied and the operation had to be abandoned. During the return to Pola, Italian aircraft attacked the Habsburg ships with no results. Next the Austro-Hungarians encountered an Italian flotilla of two leaders and seven destroyers. The light cruiser *Admiral Spaun* sortied to support the returning group. When the Austro-Hungarian warships reached the cover of their own coastal batteries and minefields the Italians turned away. The Austro-Hungarians planned a similar attack against Ancona with seven destroyers and four torpedo boats in two groups for 8 December. When Italian aircraft sighted the warships off Pesaro the bombardment groups aborted their mission. Air attacks on the retreating ships were unsuccessful.

On the night of 9/10 December the Regia Marina finally executed its long-planned action against Trieste. Two torpedo boats towed *MAS 9* and *13*, under the command of Lt. Cdr. Rizzo, to their target. The Italian crews cut through the seven protective cables with special cutters fitted to their bows, a process which took two hours. After penetrating the harbour *MAS 13* fired two torpedoes at *Budapest* which exploded harmlessly, but *MAS 9* put two torpedoes into *Wien*, which capsized and sank within five minutes with the loss of 46 men. The boats then roared out of the harbour unscathed.

The loss of *Wien* sparked a crisis in the k.u.k. Kriegsmarine staff. This was the greatest Habsburg loss in the war to date, but the naval high command suspected treachery rather than enemy initiative. Moreover, things were going no better in the lower Adriatic, where the declining mechanical efficiency of the light forces was forcing a reduction in activities. After the May Battle of the Otranto Strait, surface forces conducted only four sorties before the end of the year. In fact, hope for victory lay with the Army, and the major focus of the Austro-Hungarians was a final push against the Piave before the departure for France of the seven German divisions temporarily transferred to the Italian front that autumn. The Habsburg army would again strike towards Cortellazzo while the German XIV Army would assault Mount Grappa. The Navy's job was to support the flank, but bad weather forced a postponement of the launch of the mission, scheduled for 12 December, until the 19th. The k.u.k Kriegsmarine squadron was formed by the old coastal battleships *Budapest* and *Arpád*, the cruiser *Admiral Spaun*, six destroyers and ten modern torpedo boats, while ten more elderly torpedo boats swept for mines. The importance the Austro-Hungarians placed on this attack was so great that for the first time their aircraft used gas bombs.

By chance, that morning *Sparviero* and *Aquila* were also out shelling enemy lines off Grisolera, near Cortellazzo. The two vessels were covered by four destroyers, six torpedo boats, and five MAS boats. The Austro-Hungarian flotilla sighted the flashes from the Italian bombardment, so the commanding officer altered the planned course to avoid a confrontation. This upset the complex` schedule required to coordinate with the aircraft which were to spot the fire of his ships. Nor did the weather cooperate. When the bombardment against the battery of Cortellazzo began at 0900 there was little impact as a dense fog covered the area, making accurate fire impossible.

The Italian flotilla in turn spotted the blasts from the Austro-Hungarian guns and headed towards the enemy, while the battleships *Saint Bon* and *Emanuele Filiberto* sortied from Venice. However, the squadrons missed each other in that foggy day of 'blind man's buff', and the action ended with everything falling through, as the Austro-Hungarian Army offensive against Cortellazzo had been crushed by 1400 under the fire of Italian Navy pontoons which had been assembled using a series of artificial channels furiously dug in the Laguna Veneta during the previous weeks. Neither the Italians nor the Austro-Hungarians could know that the few ineffective shells fired by the Imperial Navy that day would be the last from a k.u.k. surface vessel to hit the Italian coast.

Austria-Hungary Feels the Pressure

On 1 February 1918 a mutiny erupted at Cattaro and some k.u.k. warships raised the red flag. Although order was quickly restored, this rebellion confirmed the poor morale and worn material condition of the Austro-Hungarian forces. On 28 February the k.u.k. Kriegsmarine's commander Admiral Njegovan resigned; he was replaced ten days later by the young (49 years) and ambitious Rear-Admiral Nikolas Horthy de Nagybánya. Horthy intended to revolutionise the navy's strategy. His first act was to pay off the five elderly battleships of the *Monarch* and *Habsburg* classes, the armoured cruisers *Kaiser Karl VI* and *Sankt Georg*, whose speed was 18 knots at best, and the protected cruisers *Aspern* and *Szigetvár*. He sent the three 'Erzherzog' coastal battleships to Cattaro. By the stroke of a pen the Imperial Navy thus surrendered any future hope of collaborating with the Army as a flank force along the mouths of the Piave River. The battle force was reduced to the four dreadnoughts, the three pre-dreadnoughts of the *Radetzky* class, and the scouts of the Cattaro squadron.

Meanwhile, the declining logistical efficiency of the Entente fleet required similar action on the other side of the hill. On April 1918 the Regia Marina placed three of the four pre-dreadnoughts of the *Vittorio Emanuele* class in reserve. Only the *Napoli* remained at Brindisi as a floating battery. The old armoured cruisers *Francesco Ferruccio* and *Varese* with their 18/19-knot top speed were considered no longer fit to operate with the battle fleet after April 1916. The French situation was even

The Austro-Hungarian protected cruiser *Aspern*. (Courtesy of Zvonimir Freivogel)

Table 2: The Balance of Naval Power in the Adriatic, April 1918

	Dreadnought Battleships	Pre-dreadnought Battleships	Armoured Cruisers
Italy	5	–	3
France	7	–	–
Britain	–-	–	–
Total	12	0	3
Austria-Hungary	4	3	0

Notes:
1. Six of the French dreadnoughts were normally based at Corfu.
2. The table does not include the two old Italian pre-dreadnoughts *Emanuele Filiberto* and *Saint Bon* at Venice, nor the three Austro-Hungarian coastal battleships of the 'Erzherzog' class at Cattaro.

worse. In 1917 the battle fleet based at Corfu had been reduced from its average of six dreadnoughts, eight pre-dreadnoughts, and eight armoured cruisers to six dreadnoughts and five semi-dreadnoughts of the *Danton* class (four after the sinking of *Danton* on 19 March 1917, by a German submarine off Sardinia) and eight armoured cruisers. From the spring of 1918 the poor condition of the five pre-dreadnoughts of the *Patrie* class caused the surviving *Danton*s and the armoured cruisers to be transferred to the Aegean. The French lacked a scouting force and even the necessary destroyers, almost all of which were being employed as convoy escorts in the Mediterranean and Aegean, including those of the Adriatic flotilla, by late spring 1918. The value of the Marine Nationale's dreadnoughts as an effective battle force was thus compromised, and their potential role in the event of an Austro-Hungarian sortie from the Adriatic significantly reduced.

The British had already retired their four pre-dreadnoughts based at Taranto in January 1917, and so the strain of war had finally reduced all the navies involved in the Adriatic theatre to their real rather than their 'paper' strength (see Table 2).

Meanwhile, the skirmishes between the flotilla craft continued apace. In the upper Adriatic the Italians sent a force deep into the Gulf of Quarnero to attack shipping in the small harbour of Buccari on the night of 10/11 February. The attack failed, but the little warships returned safely despite a vigorous Austro-Hungarian reaction. On 12 March the Regia Marina's armed pontoons conducted a heavy bombardment along the Piave Nuovo River, destroying four bridges just built by the Austro-Hungarian army and causing the Habsburg generals to lament once more the absence of their navy and its inability to do anything similar. Then, on the night of 4/5 April, a platoon of Austro-Hungarian sailors landed near Ancona in an attempt to seize by *coup de main* an MAS and a torpedo boat. However, customs guards captured the whole force within a few hours. Mirroring the intense activity in the northern Adriatic, the war in the south was far from over. On the night of 22/23 April there was a night engagement in the Otranto Strait between five Austro-Hungarian destroyers and an Allied force of four destroyers: two British, one Australian, and one French. The Habsburg warships damaged both Royal Navy vessels and escaped unharmed.

Target Pola

Thaon di Revel had pursued from the first the concept of attacking Pola to force the Austro-Hungarian fleet to retreat to less well-defended and -equipped ports, thereby compromising its efficiency. Thirty-one Italian air raids between 1915 and 1918 did not do the job, which led to consideration of more imaginative ways to hit the base. In 1916 Lieutenant Angelo Belloni proposed a type of cruise missile whose technology was ahead of its time. Ansaldo advanced the idea of a heavy battery similar to the ones used by the Germans in 1918 to hit Paris. This option was considered unrealistic given Italy's limited industrial capabilities. A flying bomb named 'Crocco-Ricaldoni', launched by aircraft beyond enemy defences, was tested in 1918. A project for a submarine armed with a 6in gun that could fire submerged, except for the gun's muzzle and a short-base periscopic rangefinder, was dropped after 1918. More practical was another Belloni plan based on two modified submarines which would force the booms protecting the harbour at Pola with the help of swimmers equipped with respirators and wet suits he had just invented. The project was awaiting final trials when the war ended.

The first of the new ideas to be implemented was a June 1917 proposal of an 8-tonne torpedo-armed motor boat equipped with caterpillar chains. This materialised as the attack craft *Grillo*, *Cavalletta*, *Locusta*, and *Pulce*. The first 'next generation' attempt to penetrate Pola occurred on the night of 8/9 April using a *Grillo*-type attack craft. The first endeavour failed, as did another made on the night of 13/14 April. One of the two commanders of the tiny boat was the then-Lieutenant Alberto da Zara, one of Italy's more successful admirals during the Second World War. The Regia Marina persisted with further attempts against Pola by the *Grillo* craft on the nights of 6/7 and 8/9 May. A few hours after the last Italian effort, the Austro-Hungarians tried to land a platoon just north of Pescara to mine the railway, but the force was sighted and a squad of customs guards repulsed the raid.

On 10/11 May another attempt to penetrate the Pola defensive system failed. A sixth attempt on 13/14 May almost succeeded. Despite being sighted by the enemy, *Grillo* passed four of the five booms before being stopped and scuttled. The Austro-Hungarians recovered the wreck and tried to copy the boat, but their prototype was incomplete when the war ended. The teething troubles affected other innovative Austro-Hungarian attack craft, including a 7.6-tonne *Versuchsgleitboot* hovercraft and the more conventional 33 knot, 6.5-tonne *Motorboot Szombathy* which was still experiencing problems with her four aircraft engines when the war ended.

On 13 May an unusual event occurred off Durazzo when *MAS 99* and *100* attacked the steamer *Bregenz* (3,905 GRT), escorted by the destroyer *Dukla* and two torpedo boats. It was to have been a typical harbour forcing mission, but the boats took the chance opportunity presented and scored the first sinking of a ship underway by a motor torpedo boat. The MAS withdrew without damage despite the escort's counterattack. Unsuccessful attempts by MAS against Antivari and Trieste followed on 15 and 16 May respectively.

It was clear that, after two years, the tactics used to force harbours were no longer effective. After the repeated failures with the *Grillo* craft, Regia Marina staff promoted two, quite different plans. The first, conceived in June 1918 after the British action against Zeebrugge, was, like that harbour-blocking attempt, based on brute force. The ancient battleship *Re Umberto*, used as a floating battery since the beginning of the war, was to be modified to serve as a ram to breach Pola's defences and open up a passage for a dozen MAS. It was an almost desperate endeavour, but the war concluded just as the naval dockyard of Venice was completing the ship's adaptation for that mission. The other, much more discrete and economical plan, was proposed in April 1918. A self-propelled mine named 'Mignatta', manned by two divers, was to penetrate the enemy base. This promising idea sparked a round of tests, training and modifications throughout the spring and summer of 1918.

The Final Bill

Since the start of his tenure, Admiral Horthy had considered a grand replay of the 15 May 1917 attack against the Otranto barrage and finally settled on a plan using nearly the entire fleet: four dreadnoughts, three pre-dreadnoughts, four light cruisers, 13 destroyers, and 32 torpedo boats. He was not seeking a decisive battle, just the chance to cut off and destroy the Italian armoured cruisers and the Entente light units which would be pursuing, as usual, the raiding force. A success would have boosted the Imperial Army's morale before the last, decisive offensive along the Piave River.

Horthy divided the force into nine units according to a complex plan that required each group to follow a strict schedule. Before dawn on 10 June, while steaming south near the small island of Premuda, *MAS 15* under Commander Rizzo sighted the battleships *Tegetthoff* and *Szent István* escorted by a destroyer and six torpedo boats. Rizzo, section leader with *MAS 21*, decided to attack even though doctrine dictated that the motor torpedo boat should conduct only furtive night attacks. Luck helped, of course, as the Austro-Hungarian force was running behind schedule, but the hard truth was that the episode capitalised on the sum of years of Italian initiative in the northern Adriatic. *MAS 21* missed with two torpedoes fired at *Tegetthoff* but both of *MAS 15*'s weapons struck *Szent István* and caused flooding which, coupled with ineffective damage control, eventually proved fatal.

The k.u.k. Kriegsmarine battle force had been harbour-bound for too long. Prior to that fateful morning, *Szent István* been 937 days in commission, but had remained at her moorings 94.4 per cent of that time. She had spent a total of 54 day underway, with just one real sortie to Pago Island; the other excursions had lasted only 45 minutes and had involved steaming to and from the protected Fasana Channel for firing exercises. Despite an intense training program instituted by Horthy, her crew was inexperienced and lacked not only the ability to

The crews of *MAS 15* and *21* after the sinking of the battleship *Szent Istvan*. (Aldo Fraccaroli collection)

The Italian motor torpedo-boat *MAS 15*. (Courtesy of Erminio Bagnasco)

The Italian destroyer *Granatiere*. (Enrico Cernuschi collection)

The Italian *esploratore leggero* ('light scout') *Sparviero*. (Enrico Cernuschi collection)

effectively control damage but, like the escort, to sight those two small grey boats heading towards her.

After years of sudden attacks inside and outside their bases, the Austro-Hungarian tendency was to credit the Italians with an espionage system far more capable than it really was. Horthy thus cancelled the whole operation, believing that he was steaming into a trap. He could not know that those two MAS boats were there as part of a regular patrol and lacked radio sets due to a lack of space on board. Thus, years of hard and mostly routine naval warfare paid their dividends. The Italian Army collected the morale boost from that sortie and the last Imperial offensive failed on its first day, 15 June, under a deluge of pre-emptive counter-battery fire which included, of course, the Regia Marina's pontoons.

The Last Fires

The dramatic loss of the *Szent István* did not affect the routine of the naval war. Italian torpedo boats shelled positions between Punta Tagliamento and the Piave on the nights of 20/21 June, 1/2 July and 6/7 July. On the night of the second mission, two Austro-Hungarian destroyers and two torpedo boats were at sea to support an air raid against Venice. At 0310 a flotilla of seven Italian destroyers sighted the enemy force south of Caorle and attacked. An inconclusive thirty-minute action followed before night intervened and the k.u.k. Kriegsmarine warships retreated to Pola. Both sides suffered damage. On the Albanian front, also on July 7, the British monitors *Earl of Peterborough* and *Sir Thomas Picton* shelled the Habsburg lines in support of an Italian offensive, the first on that front since 1916. The attack succeeded and the Italian troops advanced more than sixty kilometres in a week.

MAS *97* and *101* tried another forcing action against Antivari on the night of 19/20 July, but the defenders detected the approach of the boats and forced them to abort their mission. On 5, 9 September, and 1 October MAS *100* and *218* tested the defences of Durazzo, but their attempts to penetrate the port failed.

If the nocturnal waters of the lower Adriatic were open to all who could practise stealth, by day the Entente navies and air forces exercised total control. On 5 September, while *Sparviero* and the torpedo boats *6* and *12 PN* were recovering a flying boat downed by an engine defect, the leaders *Aquila* and *Nibbio* sighted three Austro-Hungarian torpedo boats on a minesweeping mission in the Drin Gulf. The fire of the Italian warships damaged *Tb 86* before the fleeing k.u.k. Kriegsmarine warships reached the cover of Medua's shore batteries. On 14 September an Entente offensive from Salonika started to crack the Bulgarian, German, Austro-Hungarian, and Turkish line. On 30 September Bulgaria signed an armistice. Meanwhile, allied forces advanced into Serbia and Turkey. On 29 September the Italians renewed their Albanian offensive. The next day the Austro-Hungarian lines broke and a 250km advance began which lasted more than 32 days.

On 2 October a strong Entente force which even included a US Navy contingent bombarded Durazzo. This effort sank one Austro-Hungarian steamer and damaged two destroyers, a torpedo boat and two freighters. The shore batteries hit only MAS *98*, wounding one sailor. The underwater menace confirmed, however, its enduring validity when the Austro-Hungarian submarine *U 31* torpedoed and damaged the light cruiser *Weymouth*. On 4 October an Italian column, coming from Macedonia, entered Elbasan in Albania. On 14 October Italian cavalry supported by a Regia Marina torpedo boat flotilla from Valona occupied Durazzo; on the 15th the Italian army entered Tirana, and on the 27th Medua. A landing party from the torpedo boat *57 AS* captured Dulcigno in Montenegro, while Antivari and Vir Bazar fell the next day. The torpedo boats *37* and *38 PN* occupied Pelagosa, a 'no man's land' since September 1915, on 3 November.

During the final battle of Vittorio Veneto, fought between 24 and 31 October, the Italian Navy's gun-armed pontoons played an important role along the lower Piave, covering the crossing of the river at Cavetta by two naval battalions on the 30th. The k.u.k. Kriegsmarine fleet in Pola had been handed over on 31 October to the newly-formed State of Slovenes, Croats, and Serbs; the Cattaro squadron followed this same agreement on 1 November. A few hours later, during the night of 31 October/1 November, the 'Mignatta' semi-submersible attack craft *S2* penetrated Pola, having been ferried there by the torpedo boat *65 PN* and MAS *95*. *S2*'s two-man crew mined the dreadnought *Viribus Unitis*

The Italian battleship *Regina Elena*. (Enrico Cernuschi collection)

which, unknown to the Italians, was no longer under Austro-Hungarian control. The ship capsized with the loss of over 300 men.

The action was helped by fog and the chaos attending the dissolution of the Austro-Hungarian Empire. Between 27 and 1 November the Germans, observing that everything was crumbling, scuttled ten U-Boats and two torpedo boats (sent by railway and assembled in the Adriatic) at Pola, Fiume, Trieste, and Cattaro. Fourteen other submarines were ordered to sail for Germany. On 30 October the Ottoman government signed an armistice with the Allies. An armistice with Austria-Hungary followed on 4 November, and on the 11th Germany finally bowed to the inevitable. That same day Emperor Karl, the last Habsburg sovereign, relinquished power.

Conclusions

The conduct of the surface naval war fought in the Adriatic suggests several conclusions. The Austro-Hungarian k.u.k. Kriegmarine was greatly outnumbered even before Italy entered the war. It was fighting a true 'war of poverty'. Warships larger than a destroyer were completely irreplaceable. Even destroyers were nearly beyond its capacity to construct. Coal was a constant problem. The entire fleet would consume coal at a rate of 1,000 tonnes an hour in a general sortie. There were 400,000 tonnes on hand at the start of the war but only 95,000 tonnes available by the beginning of 1918. Given these basic facts, there was little the surface fleet could accomplish to affect the outcome of the war with one exception: it could support the submarine campaign. Thus, in the south, much of the effort of the scouts, destroyers, and torpedo boats was focused on attacks against the Otranto antisubmarine barrage. Other than that, the Navy's missions were defensive. However, military choices are never absolutely black and white, and the leadership of the Habsburg navy fell into the trap of acting as if they were. Thus, for example, when the time arrived for the fleet to support the Army in its final push to capture Venice, the Navy withheld coastal battleships that might have turned the tide. In the ultimate irony, these very same ships were retired as superfluous just a few months later.

Italy's naval strategy in the Adriatic was conservative because Rome enjoyed the superior position from the beginning of the war, with little to gain and much to lose. This was true politically as well as militarily. As long as the Austro-Hungarian battle fleet remained in Pola, Italian superiority was a given. The Italian naval leadership recognised what Haus seemingly missed: that for a day or two, before the Entente battle fleet could deploy north, the Habsburgs could assert battleship superiority in the Northern Adriatic. Venice was only 70 miles from Pola and the Italians could easily imagine a damaging sortie by the Imperial fleet. Such a scenario defined Italy's naval strategy in northern waters. The Regia Marina subjected the Austro-Hungarians to a death of a thousand cuts in the form of constant patrolling, mining, and bombardment by small craft, and persistent efforts to penetrate Austro-Hungarian harbours and attack k.u.k. warships in their sanctuaries. The fact that the largest Habsburg warship on a special and unique mission was sunk by the smallest Italian warship on a routine patrol symbolises the success of the Italian strategy.

Naval intervention in the land war was another important feature of the Adriatic surface campaign, especially in the north. Austria-Hungary applied sea power by conducting hit-and-run bombardment raids on poorly-

The Austro-Hungarian destroyer *Velebit*. (Courtesy of Zvonimir Freivogel)

The Italian armoured cruiser Pisa at Constantinople in 1919. (Enrico Cernuschi collection)

defended towns of little military value. Even the surprise bombardments of 24 May 1915 conducted by the entire fleet (including the dreadnoughts) ignored Italy's important bases, particularly Venice. It was as if Japan opened its war against the United States by attacking Dutch Harbor in Alaska instead of Pearl Harbor in Hawaii. Unable to win the war through action at sea, the Empire needed to defeat Italy on land. The endeavour – a surprise offensive from the Tyrol designed to capture Venice and knock Italy from the war, as well as punish the kingdom for its betrayal of the Triple Alliance – was attempted in May 1916. Before the grand offensive the Austro-Hungarian Chief of Staff asked the Navy to undertake 'a major action aimed at inflicting serious damage on the enemy.' Admiral Haus refused on the basis there was nothing the Navy could do that would seriously harm the enemy. If there were, 'the fleet would not have remained inactive for nine months.'

The increasingly serious crisis along the Isonzo River finally led the Austro-Hungarian Army to ask for German assistance in September 1917. The following offensive at Caporetto again failed to knock Italy out of the war. This time the k.u.k. Kriegsmarine tried to support army efforts, but the two irresolute and under-strength attempts made at Cortellazzo fell far short of what was required.

With the Entente naval blockade strangling the Empire, final defeat was only a matter of time barring a decisive military victory on the field. Submarine warfare, in such a scenario, was the only choice, and the German campaign in the Mediterranean was the only reasonable naval option. However, the Austro-Hungarian Navy, its industries and dockyards, were never able to support more than three dozen German submarines at any time between 1916 and 1918. Even the most important strategic result achieved by the k.u.k. Kriegsmarine during the war – the tying down in the Adriatic Sea of some thirty modern Italian and French destroyers – was cancelled out, from the spring of 1917, by the arrival in the Mediterranean of a further thirty destroyers from Japan, Australia, and the United States.

The renewed efforts made on June 1918 during the starved Empire's last offensive marked a further disaster for the Austro-Hungarian battle fleet, which lost its final rendezvous with history, even though, by this time, it would have been little more than a propaganda gesture. The battleship squadron, by now in a state of near mutiny, was also absent during the final battle of Vittorio Veneto in October 1918, when the Empire's last hope of survival, at least in a truncated form, was crushed by Italian and Allied divisions which destroyed the Imperial Army, the last pillar of Habsburg power. By that time in the southern Adriatic, k.u.k. Kriegsmarine's guerrilla campaign had died out for want of an adequate logistical infrastructure.

Sea power is a matter of using the sea when and where it best meets a nation's requirements. Italy and its Entente partners did this in the Mediterranean. They supported their armies; they confined the enemy fleet to its harbour; they blocked its passage into the open seas and, in general, made the narrow waters of the Adriatic a dangerous environment for their foe. Austria-Hungary, meanwhile, survived with its fleet largely intact until it was eroded by mutiny and finally handed over to a secessionist movement during the final days of the war. In terms of Austria-Hungary's strategic position this survival may be regarded as a victory of sorts, but not of the sort that wins wars.

The Austro-Hungarian battleships Tegetthoff and Erzherzog Franz Ferdinand (closest to the camera) at Venice on 24 March 1919. (Rivista Marittima)

POST-WAR AIO AND COMMAND SYSTEMS IN THE ROYAL NAVY

In the third in a series of articles on technical developments in the Royal Navy during the post-war era, **Peter Marland** describes the sequence of AIO and Command Systems.

This article builds on the previous articles on fire control and weapons systems, across the same post-war period. More detailed explanation about the processes can be found in the Annex, but the principal definitions are:

– **Action Information Organisation (AIO)**: a team of officers and ratings based in the Operations Room and outstations, who collect and collate information, and display it in a format which enables decisions to be taken (BR1982).
– **Command and Control (C2)**: the processes through which a commander organises, directs and co-ordinates the activities of the forces allocated to him (BR1806).

The RN Operations Room is the equivalent of the US Navy's Combat Information Center (CIC), and the whole suite of displays and their operators are now usually referred to as the Command System, which links sensors and communications through the centralised system, and outputs data to other consorts, aircraft, and to the associated weapons systems. Whilst the Command System displays radar, sonar, and electronic warfare (EW) information about the measured position of contacts in 'real time', combining both video and synthetic data, the computer also handles 'non-real time' information that might be 'stale' or include a degree of uncertainty.

The key officers' duties are laid down in Queen's Regulations for the Royal Navy. The Officer of the Watch (OOW) remains responsible for ship handling, even when under 'advisory' control by the ops room. The Principal Warfare Officer (PWO) is responsible for tactical employment of weapons (ordering 'take track…', 'engage' or 'shoot'), whilst the Weapon Engineer Officer (WEO) is responsible for system technical performance, including maintenance & repair. Overall effectiveness therefore rests on a partnership between the Warfare and Engineering personnel.

Personnel: The formation of the Navigation and Direction (ND) branch of the RN has complex roots which include the introduction of the Radar Plot (RP) branch of ratings from 1940, and the later creation of the new ND officer structure (see Schofield *op. cit.*). This endured through to 1975 for officers, and to 1993 for the ratings, when the new Warfare branch was formed.

Equipment: Individual RN equipments are described by a range of 'Nomenclatures' that can often confuse. This is a somewhat idiosyncratic scheme (see Annex), but generally antennae and electronics have a three-letter Outfit title (with variants in parentheses) whilst radio, radar and sonar sets have a Type number.

This article follows the rough chronology of early (analogue, CDS and ADA), middle-period (ADAWS 2-10, CAAIS and CACS), then modern systems with a cut-off date at 1994.

Background

Pre-War: Ships were reliant on visual sightings, surveillance by organic aircraft such as the Walrus, and by asdic. Pre-war, the plot was kept on an ARL table by the Navigator or a staff officer. Ships were fought from the bridge, using the 'captain's sight' to point the main armament at the desired target. During defence watches, the ship's Gunnery and Torpedo officers alternated as the Principal Control Officer.

Wartime: War saw the widespread introduction of an anti-submarine (A/S) plot, with the conning officer able to look down onto the picture using a 'viewplot'. The plot was kept by an instructor or any other 'spare' officer. This evolved into a below-decks space with radar displays and communications which was further developed for Fighter Direction.

Plan Position Indicators (PPI) only appeared from mid-1943, when centimetric radars offered continuous, all-round rotation. There are several sources for early wartime development which describe the Ops, Air Defence and Gun Direction rooms.[1] The impression is that for in-service ships this was an *ad hoc* process that evolved rapidly 1942–44, adding radar displays, internal voice pipes and radio circuits, but that bespoke provision

POST-WAR AIO AND COMMAND SYSTEMS IN THE ROYAL NAVY

was made after 1944 for the late-war cruisers (*Swiftsure* onwards), the new light fleet carriers and the battleship *Vanguard*.

Post-war: The RN was now using second-generation radars Types, 277, 293 and 960, with a wide range of dedicated cathode ray tube displays in PPI, azicator, or height position indicator formats. These used external cursors, overlays or light strobes to extract information; they did not have integrated or interlaced electronic markers, which only came in the mid-1950s.

Given the increasing scope of the 'plot', which developed into the ops room, changes in the size and location of the compartments mirrored changes in the way Captains 'fought' their ship. This was a progressive move from a simple asdic 'hut' to a larger plot close to the compass platform, to separate Radar Display Room (RDR), Aircraft Direction Room (ADR), and Target Indication Room (TIR) or Gun Direction Room (GDR) in larger ships.

This development continued to an enclosed bridge with the ops room a deck below, onto *Leander* class frigates with another deck's separation. For the missile destroyers (DLG) and the new carriers there was a captain's lift from close by the bridge down to the deeper ops room complex. This sequence is shown in Fig.1.

There were two drivers for this sequence: firstly the need for situational awareness in challenging anti-submarine actions with intermittent contacts that required 'furthest-on' circles, integration of DF and radar information plus manoeuvring consorts, all of which went beyond what one officer on the bridge could reasonably keep in his head; secondly the move towards a deeper ops room for 'closing down' under nuclear conditions where the deck-levels above the ops room provided additional screening, particularly against radioactive fall-out.

Operations Room Development: Across the period, there was significant growth in AIO compartment sizes, which started with RDRs of 12ft x 10ft, but grew to combined spaces measuring up to 30ft x 20ft. Whilst a late-war cruiser required 600 sq ft of ops room, there was a steady increase from the post-war Type 12 onwards; the ops rooms of later frigates (Type 21 onwards) had a surface area of 1100 sq ft (c.100m^2), and also incorporated EW and sonar sets, thereby subsuming separate EW Office and Sonar Control Room compartments.

Collingwood notes show that a battleship or large cruiser was fitted with up to 26 displays, while a destroyer or sloop had 7 displays. These were supported by a number of perspex plots and tote boards, often marked up from the rear.

JE was the prototype of all subsequent PPIs, but there were special purpose displays such as:

– Outfit JH: IFF displays associated with another radar.

Fig.1: Progressive Development of RN Ops Room Location/Facilities.

Pre-war destroyers and sloops: OOW standing at centre-line pelorus on open compass platform; asdic operator seated in hut or shelter, with verbal dialogue between them. Captain controls tactical battle from the bridge.

Some wartime destroyers: OOW standing on open compass platform, but able to see compiled AS plot via '*Viewplot*'; asdic operators elsewhere. (In post-war *Daring* and 'Ca', this plot space expanded into full ops room.)

Wartime cruisers: OOW standing on open compass platform; ADO on GDP for visual TI, and RDR+ADR+GDR below decks. Ship fought from compass platform by Captain or PCO.

Post-war frigates (T12/41/61/81): OOW standing in closed 'T-shaped' bridge; ops room one deck away; TASO exercising 'below control', with steering information relayed to OOW by AI broadcast.

1960s onwards (Leander/DLG): OOW in closed 'turret' bridge; ops room more than one deck away; 'below control' exercised by PWO, with OOW on Command Open Line. One-man captain's lift in DLGs and larger ships, down to deeper ops room complex, 3–6 decks below.

WARSHIP 2016

Fig.2
HMS *Ocean* Operations Room 1953.

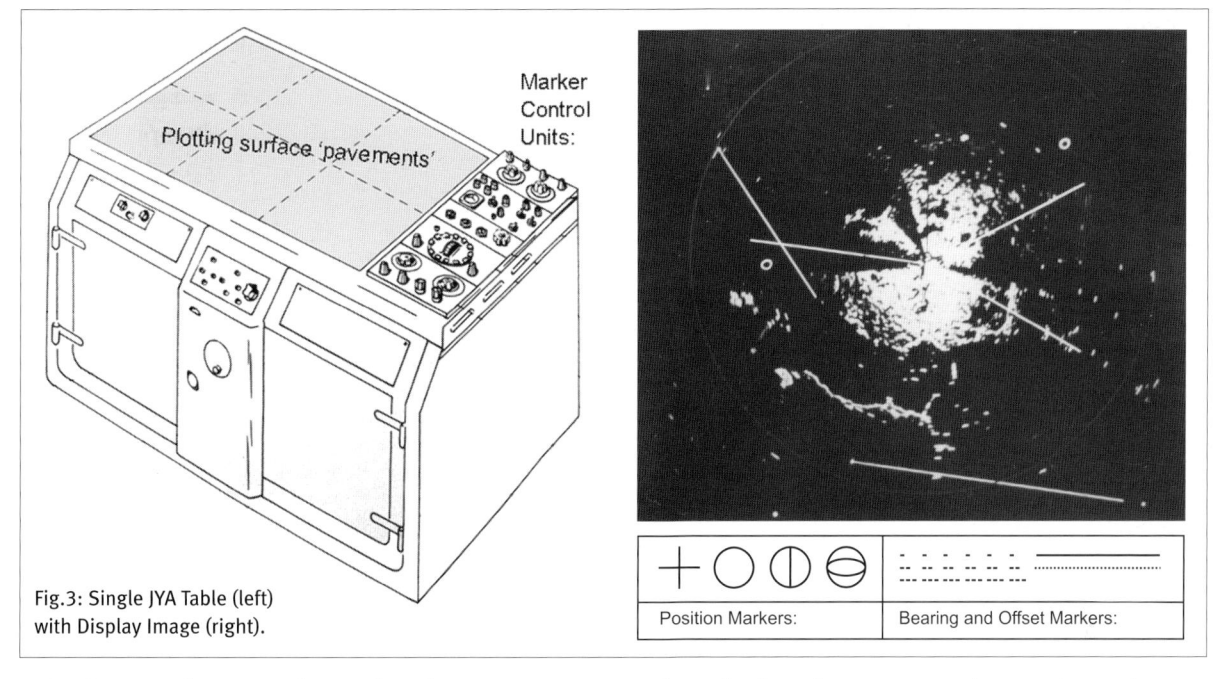

Fig.3: Single JYA Table (left) with Display Image (right).

- Outfit JK (Skiatron): featured a long-persistence picture from slower rotating early warning radars such as Types 279 and 960.
- Outfit JR: a prototype auto-radar plot for surface plotting.

There was transition from a wartime large navy supported by conscription (and from 1948 by National Service), down to a reduced-size 1950s navy with a large reserve fleet of unmodernised ships. With the advent of new-construction ships with increasingly sophisticated equipment, the retention of skilled technical personnel became a serious issue, reducing the effectiveness compared to wartime expectations.[2]

Naval AIO doctrine allowed for two separate procedures: Surface (SU) plotting (including underwater contacts), and air plotting for Air Defence (AD) purposes. There were two parallel development strands: increasing analogue display integration, and the first proper command system (CDS).

Analogue Displays and Manual AIO

There were new JC series displays for general purposes, a JW display (replacing the *Skiatron*), plus the JDA fixed deflection coil and JUA rotating coil displays that became the main workhorses in ships' operations rooms. These were supplemented by 'Project Cambria' (known initially as the Automatic Surface Plot, and later as Outfit JYA), which had its roots in a 1951 proposal for SU and ASW plotting systems.[3] The prototype was trialled in 1953–54 as a single ship system. The radar display carriage was moved to offset ship's motion, giving a ground-stabilised display projected upwards onto the plotting surface. Contacts were plotted by hand, using a chinagraph pencil under UV light, with periodic tick marks, in order to measure the target's speed.

The plotting surface comprised six 15in square perspex 'pavements'. Because the action could move beyond the edge of the plot, JYA allowed the radar carriage to be 'indexed', with the operator shuffling the pavements to reassemble the track history. The electronic markers allowed the sonar to 'cut through' contacts, consort information could be entered via the offset markers, or the radar operator could use the 'tracker' to point out an intermittent contact (like a periscope 'riser') to the plotting team. Production systems were manufactured by EMI and entered service from 1959 through to the last *Leander* in 1973. Most first-rate A/S frigates had two JYA: one for the surface plot, the other for the A/S plot. JYA was relatively expensive and was supplemented by the simpler JYB modification of older ARL tables that omitted the electronic markers.

Concurrently with the RN developing 'Cambria', Canada was developing DATAR, which was demon-

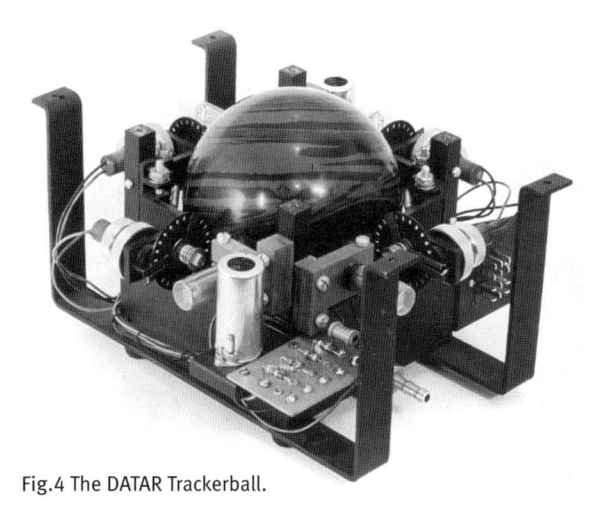

Fig.4 The DATAR Trackerball.

strated to UK and US observers during August-September 1953. DATAR used 3800 valves in the first seagoing digital computer, had the first operational tracker ball (made from a five-pin bowling ball!), and an inter-ship data link. The RN view by DCNS (R-Ad. G Barnard) in November 1953 was:

> Canadian DATAR has some very interesting techniques which may have wide application ... but in its present form – weight 20 tons and 5000 electronic valves for digital computation – is quite unacceptable to us for fitting in A/S frigates. The Canadian Naval Board are however confident that the weight can come down ... and some of the valves be replaced by magnetic drums ...[4]

The manual MATCH (Medium range Anti-Submarine Torpedo-Carrying Helicopter) system relied on markers to link cuts from sonar Type 177 through to the helicopter controller's display, where he directed the attack by a Wasp helicopter armed with a Mk 44 torpedo. Other components included the RRA transponder to enhance the radar echo and code the helo's response (the initial simple Luneborg lens reflector was inadequate). The first application of MATCH was in the frigates of the 'Tribal' class from 1961, then *Leanders*, with *Rothesay* conversions from 1968.

JW was eventually replaced by JHB, a display-only version of the CAAIS display, whilst the venerable JUA (for Type 978 radar) has successively been replaced by JUD (for Type 1006) and JUF (for Type 1007). The General Operations Plot (GOP) was usually kept on a traditional ARL plotting table, with a light-spot driven for own ship position, and using intelligence reports to form the wide-area picture, as a time-based historical record.

Comprehensive Display System (CDS)

CDS (codename 'Peevish') was the first automated Command & Control (C2) system, and had its roots in ASE's work on interlaced markers, patented in 1947. The development contract went to Elliott Brothers' research laboratory for a shore-based X-model to demonstrate the concept. This was later re-engineered into a production CDS, manufactured by Pye.

X-model CDS:[5] The project ran from 1 December 1946 until at least 16 October 1951. Elliott's built the development prototype, and the work was led by CA Laws and MV Needham; having been given the problem, the company developed the concept, hardware and architecture. The initial X-model[6] was demonstrated at Borehamwood in June 1950, and delivered to ASE Witley as X1. The subsequent X2 model went to the USA, being installed at the Naval Research Laboratory's Chesapeake Bay annex and trialled in 1952.

Displays: The consoles had combinations of PPI and analyser displays plus a larger conference display, tracking joysticks, code-setting switches and auxiliary displays. The displays had extensive zoom and offset facilities (see Fig.5).

Information Storage: The CDS X-model handled:

– Position information: stored as a voltage on a potentiometer, variable between ±150V, with separate 'x' and 'y' channels for each contact. Each position was updated by the tracking operator's joystick, and a motor drive was engaged by magnetic clutch to move the potentiometer wiper up or down to align the marker with the radar paint.
– Code information: discrete numbers were set by hand switches, and held as a fixed voltage (15–150V in 15V steps) on a uniselector. These covered track number (tens and units), height, identity/ hostility, and number of contacts in the raid.

The code information was shown on an auxiliary display once a contact had been 'hooked', but was mirrored in part on the main PPIs. The conference display information was selectable by 'tabs' to clean up the presentation by eliminating superfluous detail not relevant to decision

Fig.5: CDS X1 Display Equipment.

making. The initial aim was to include asdic tracks; however, this was not carried into production and all CDS variants only handled air tracks.

Track Symbology: Three separate monoscopes each selected a character for track number and height, or a symbol for ID. These were combined into a single composite video feed, which was displayed on separate 'code-reading displays' (a small CRT), or onto the main tactical display. Elliott's used different shapes and fills to represent ID/hostility, height and raid size:

– **Colour:** Up to the mid-point of system development (December 1948), colour was being considered as a way to highlight a few specially-selected tracks, using a colour filter disk spinning in front of the PPI screen and synchronised with the 'bright-up pulse' for the nominated target.
– **Symbology:** The shapes in the ID code plate (see Fig.6) represented low, medium or high contacts (each with single, few or many in the raid). The ten columns used different symbols that represented each type of aircraft contact.

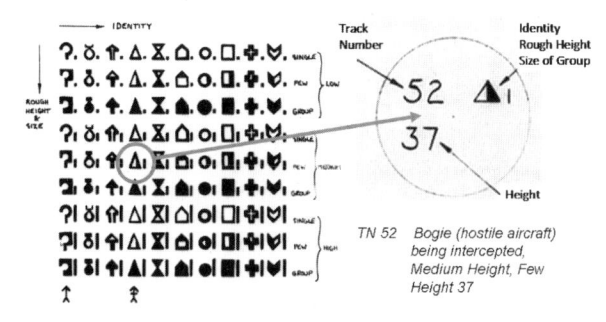

Fig.6: CDS X-model ID Code Plate (left); appearance of Code Reading Display (right).

This was demonstrated in 1950, but was not taken into the Production CDS or ADAWS, although it was implemented by the US Navy for Naval Tactical Data System (NTDS) as a structured set, initially with three-tier hostility (see Fig.7) plus some additional symbols for tracks of special interest.[7]

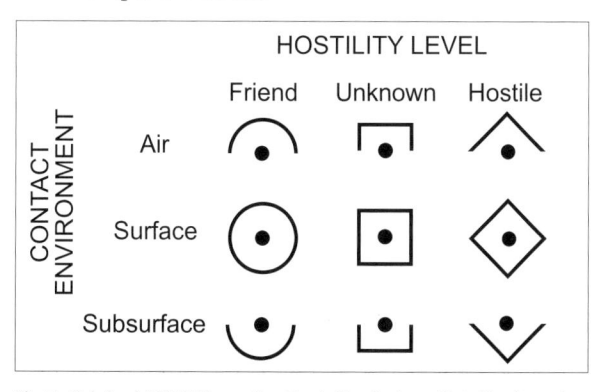

Fig.7: Original NTDS Three-tier Track Symbology Set. The 'spot' marks the track centre. (Drawing by John Jordan, based on information supplied by the author)

The Production CDS simplified the portrayal of track information. A Seaslug Mk 1 training film shows a carrier with Type 984/CDS passing tracks out to the accompanying 'County' class destroyer using DPT (Fig 8). The separate track and store numbers allowed for CDS with different capacities to be linked via DPT, while maintaining a single Force track number across all the receiving units.

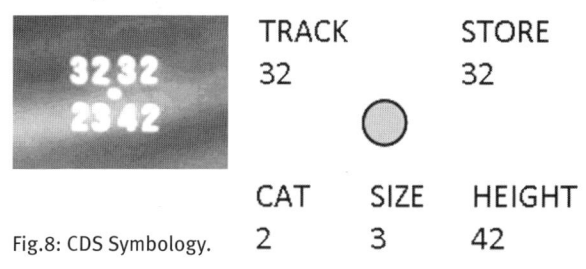

Fig.8: CDS Symbology.

Development: The architectures of the X-model and Production systems were broadly similar; the design could support 96-tracks, but the X-model included only one tracking group (detector, analyser, supervisor, and three trackers) able to handle 24 radar tracks plus another 8 from other sources, making 32 in total, and was therefore a one-half scale representation of the system fitted in RN carriers. The production system replaced the electromechanical position memory by capacitor storage, but kept the same uniselector method for code data. The X-model did not offer rate aiding for the joystick/marker tracking of contacts. This came later, where integration of the positional voltage gave the rate of change (velocity) to be added back into the tracking loop, requiring the operator simply to fine-tune the predicted position.

The facilities lost on transition to the production CDS were symbology and the large conference display, while the semi-automatic tote board, air intercept computers and DPT were added.

US Views: The US views on CDS were based on the X2 model. This was not the same as the later in-service CDS, however it 'did much to sell the concept' to the USN, who were 'wowed' by the combination of PPI, joystick markers and symbology during their visit to Borehamwood in 1950. After acquiring the X2 model, the US Navy then initiated their own EDS (Electronic Data System) and NTDS programmes. The US view was that these allowed a tracker to manage eight contacts, compared to only two in the previous manual system; it therefore had about four times more capability.[8] Despite criticism of the X-model, Production CDS was much closer to EDS than NRL's report suggested, since both used capacitor storage and early tactical data links. EDS had a slightly more modern user interface, using a pen to pick off positions rather than a joystick.

Production CDS

Benjamin (*op. cit.*) describes the development of the Comprehensive Display System and the associated

Type 984 three-dimensional radar, operational at sea from early 1958 in *Victorious*, then *Hermes*. CDS was also fitted in the first four 'County' class DLGs with an early data link, the Digital Plot Transmission (DPT) system.

Type 984 was the first complex multi-function radar, and used stacked beams with a microwave lens to provide volume search, heightfinding, long range warning and fighter control. Codenamed 'Postal', a complete development model was installed ashore in 1955 prior to the first ship fit in *Victorious*. Its subsequent success against the USN in Exercise 'Riptide' (15–20 June 1959) was cited by Benjamin as evidence of its operational worth. Type 984 was tightly coupled to CDS; it was far more successful for 3D volumetric cover than the contemporary AN/SPS-2, and led the equivalent US AIO technology by about five years. The basic CDS architecture (including representative types of display) is shown in Fig.9.

CDS had all the features of a modern AIO system, including numeric information written onto the CRT tube in addition to the radar video, except that the tracking was manual with rate aiding to ease the operator's job. The analogue information (and its retrieval by joystick 'hooking') was tied together by a unified track number scheme, and information was also presented on a large semi-automatic tote visible to both levels in the ops room.

Operations Room Layout: The operations complex in the 1950s carrier reflected WW2 lessons and RAF practice ashore, with a two-level ops room which allowed force staff involved in battle management to view from the upper gallery. The main tote could be viewed through the open well down to the lower level, which was occupied (in the RDR) by the picture compilation team, and in the ops room lower-level by the fighter controllers managing air intercepts (see Fig.10).

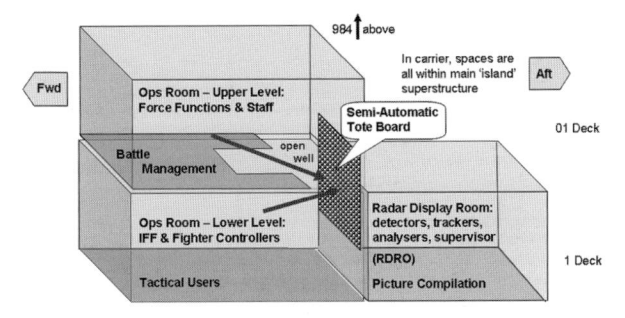

Fig.10: Two-Level Operations Room Layout.

The CDS handbook gives full details of architecture and displays in 48-track and a reduced 32-track version in *Hermes*.[9] ADM 1/29140 shows that the Batch 1 DLGs had a full 24-track CDS, albeit with only a receive DPT; this matches the report on proposed variants (ADM 220/647). The full scope of the Royal Navy's CDS programme was therefore limited to:

- 48-track system (JZA) in *Victorious* (Type 984); may have been in *Ark Royal* (Type 965).
- 32-track system (JZB) in *Hermes* (Type 984): intended to have a smaller 'footprint'.
- 24-track systems (JZC?) in DLG 01-04 (Types 278, 965 and 992).

Maintenance: CDS/984 was a large system; Table 2 gives the overall component count.

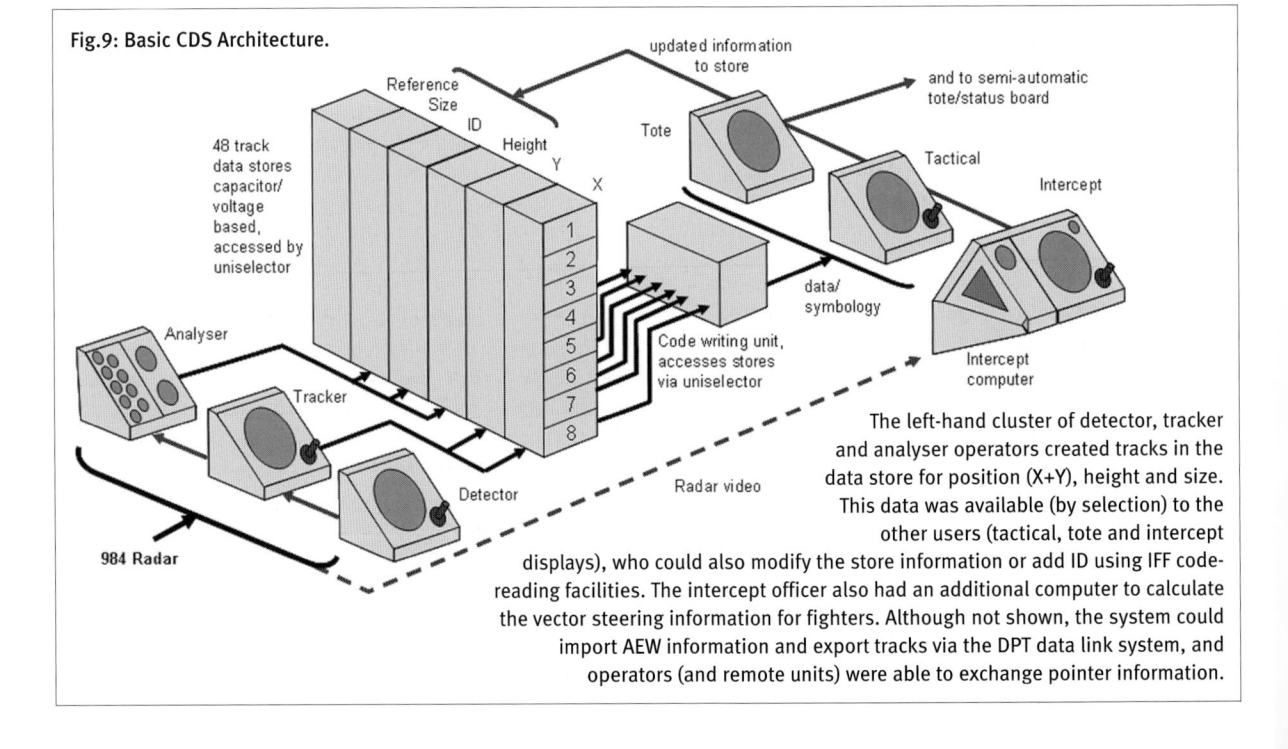

Fig.9: Basic CDS Architecture.

The left-hand cluster of detector, tracker and analyser operators created tracks in the data store for position (X+Y), height and size. This data was available (by selection) to the other users (tactical, tote and intercept displays), who could also modify the store information or add ID using IFF code-reading facilities. The intercept officer also had an additional computer to calculate the vector steering information for fighters. Although not shown, the system could import AEW information and export tracks via the DPT data link system, and operators (and remote units) were able to exchange pointer information.

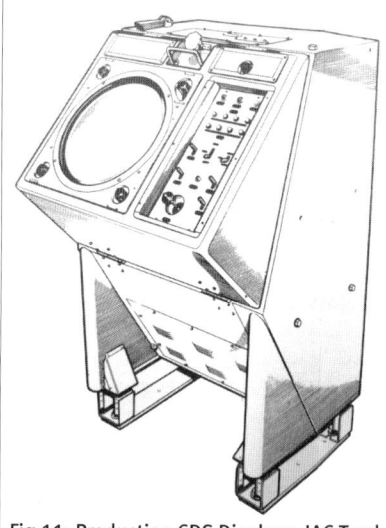

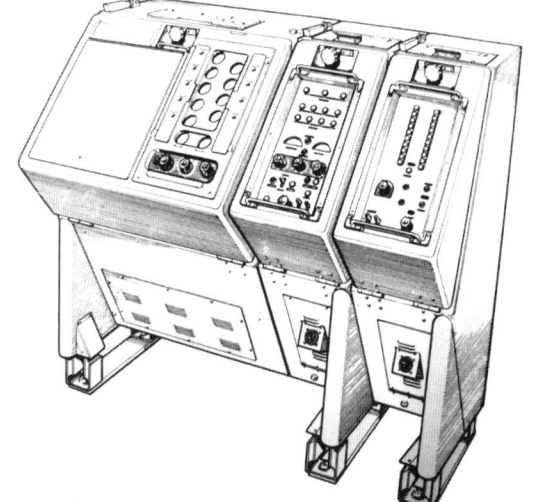

POST-WAR AIO AND COMMAND SYSTEMS IN THE ROYAL NAVY

Fig.11: Production CDS Displays: JAC Tracker; JAD Analyser; JAH Intercept Officer; JAK Tote Reporter.
The illustrations show the modular nature of the production hardware, with largely common PPI and analysis (heightfinder) displays, plus code input panels, and the aircraft intercept computer (the triangular panel to the left of the PPI).

Table 1: Principal Components of CDS variants

Description:		JZA	(JZC)
Displays:	Warning Beam Operator JAA	1	
	Detector JAB	1	1
	Tracker JAC	6	3
	Analyser JAD	4	
	Track Correlator JAE	1	1
	RDR Officer JAF	1	
	Direction Officer, Staff DO JAG and JAJ	6-7	
	Intercept Officer JAH	4-6	2
	FDO, Air/Sea Plot, ID Officer JAJ		2
	Tote Reporter JAK	2	
	Maintenance Display JAL	1	
	Control equipment radar JAM		1

Note: Later DLG displays have suffix (2) or (4). Group Units (electronic and uniselector) came in 24-track size, giving the DLG system 24 tracks, in contrast to JZA's 48.

Ship's staff from *Victorious* ('CDS - A Maintainer's View', NER Vol.13 No.4 dated Apr 1960, pp.124–40) made it clear that Type 984 with CDS was a challenge, but could be managed. Support documentation included 'divide-by-two' fault finding logic, and the article stressed lessons learned about the ageing of valves, monitoring video performance, and block changes of components such as relays prior to wearout.

Reports from the mid-1950s show that the RN was facing very serious problems in retaining skilled technicians, and this may account for the aversion to large-scale valve technology in escorts. Despite this, by the end of the 1950s the RN were looking after CDS with 8000–9000 valves. However, it should be noted that this was in a carrier, and CDS was designed for reliability and had a lot of redundancy in the track stores and the displays. This would not necessarily have been true for a valve-based digital computer (such as the Canadian DATAR), where a single failure could well lead to a total stop.

Other Systems: Air Defence units (carriers, cruisers, 'Battles' and the Type 61 frigates) also had the US 'Bellhop' system (SRR-4A) that downlinked video from the AN/APS-20 radar in an AEW Skyraider (later the Gannet). The ship's SPA-8A displays were able to offset the picture to give the force the relative position of the inbound air raid, and this was supplemented by FU1 UHF/DF and TACAN for positive control of defensive fighters, replacing the YE beacon used by wartime 'Pylon' units to marshal returning friendlies.

Table 2: 984/CDS Component Count

Valves	Resistors	Capacitors	Relays	Switches
8912	47400	11750	2115	1372

Note: In comparison, the early digital computers (152 and DATAR) used 3456 to 3800 valves.

ADA and ADAWS

The need for a computer-based tactical data handling system became apparent in the 1950s and led to a series of studies. The requirement for Action Data Automation (ADA) in all major warships (endorsed 2 June 1959) prompted the development of the Ferranti Poseidon computer,[10] whose software was produced by ASWE. The prototype Poseidon was completed in 1961, but the memory had to be redesigned several times. The first system was installed in HMS *Eagle*, which re-commissioned in May 1964 with Outfit DAA; this was a purely tactical picture compilation system for Air Defence using fighters.

Before ADA entered service, work had begun on a derivative with full weapon control, as the Action Data Automation Weapon System 1 (ADAWS 1). This was developed for the second batch of 'County' class ships (DLG 05-08) and entered service when HMS *Fife* commissioned in June 1966. In *Eagle*, Type 984 fed through its own detection computer (ADAC-D) to three general-purpose Poseidon, with a total programme capacity of 16,384 words; while in the DLGs the radars fed through SPADE (Simple Processing And Detection Equipment) to two Poseidon. ADAWS 1 integrated multiple radars (Types 278, 965 and 992), ESM and sonar, with weapon direction outputs to Seaslug, the gun and Seacat systems.

Displays: Each 12in CRT display could be switched to: raw radar, markers, raw radar and synthetic markers interlaced, remote AEW data, or a tabular ('tote') display. The central unit allowed either operator to input up to eleven alpha-numeric characters (selected from the horizontal rows of ten pushes, followed by an inject push to enter the data, and also featured a joystick. The console is shown in Fig.12. *Eagle* had fifteen JGA workstations, while the DLGs had eight to ten.

JYC was the surface plotting system in DLGs, and used a wet film mechanism to photograph the radar picture. The 35mm film was developed, washed and fixed using a chemical and water spray, dried, then projected up onto the horizontal screen. The process was automated, set to repeat every 2, 4, 8 or 16 seconds. This was ultimately replaced by a JYD horizontal display, a version of the later JZ tactical plot.

Project Management: The initial emphasis when ADA was approved was on saving personnel through automation, compared to the previous semi-automatic system, CDS. The vision was credited to Benjamin (by then the Head of Division). Hardware development was relatively uneventful, but software was a different story, and progressive slippage meant that *Eagle* and the first DLGs sailed with civilian programmers onboard and did not have full capability for at least two years after acceptance.

The problems were compounded by the catastrophic failure of the ADAC:D autodetector, which required reinstatement of manual picture compilation. Project management was exacerbated by: the 'back-to-back' development for *Eagle* and the DLGs; knock-on from slippage; and resource limitations, especially the shortage of junior RNSS scientific personnel able to carry out analyst/programme duties, due to a civil service manpower freeze.

Benjamin's publications give a somewhat 'rose-tinted' view of this period, and ADA only delivered a pale shadow of the concept sold in the 1959 staff requirement for a largely automatic AIO with 'gold braid modulation' of recommendations offered by the TEWA. It also failed to deliver the promised manpower savings.

ADA left a somewhat raw aftertaste amongst the RN personnel involved, but a 1969 symposium showed the lessons had been learned, and that ADAWS 2 for Bristol should be significantly better (ADM 220/2155). With hindsight ADA was a technical success, and was in advance of the equivalent RAF system ('Linesman') by RRE Malvern.

The episode points out how much more difficult large-scale digital system integration into a single software operational programme was to become. Whilst ADA/ADAWS 1 does not rank as a scandal, it was a very close call, but contributed to the relatively smooth running of the subsequent ADAWS 2-10 development. However, it marked the decisive shift towards Ferranti dominance of naval C2 business that was to endure for another 23 years, through to 1987.

Transition from Poseidon to Later Systems: After Poseidon, Ferranti first moved to the F1600 using discrete transistors and drum memories; launched in 1964, it was first applied to West Drayton ATC and the RN Stage 1 Combined Tactical Trainer. Ferranti then 'pulled' F1600 and substituted FM1600, which used Micronor II integrated circuits (1965), multi-layer printed circuit boards and solid-state memory to give a computer that was twice as fast at a quarter of the size and at reduced cost. In September 1966 five FM1600 systems were ordered for ASWE as the Seadart and ADAWS 2-3 development testbeds for the Type 82 destroyer and CVA-01 (at a cost of over £1m), plus the associated display suites from Plessey.[11]

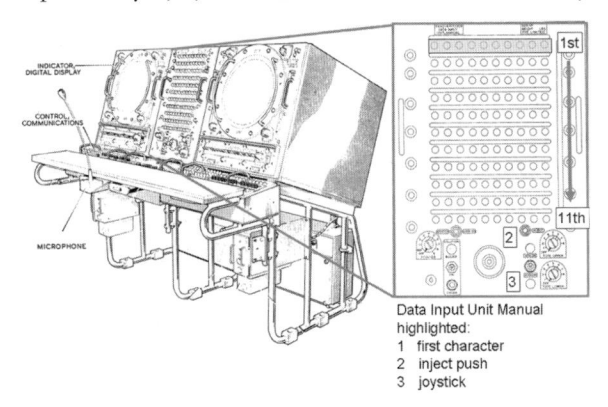

Fig.12: JGA Console.

Data Input Unit Manual highlighted:
1 first character
2 inject push
3 joystick

Table 3: ADAWS Variants (circa mid-1980s)

ADAWS	fitted to:	Data Processing/Display System Outfits + Notes:
2	*Bristol*	DAC/JZP, with 2 plots, 8 totes and 15 LPD.
3	CVA-01: cancelled by Healy review in Feb 1966.	
4	Type 42 B1	DAD(1)/JZQ(1) with 2 plot, 4 tote and 13 LPD. Updated to DAD(2)/JZQ(2) at refit,
	Sheffield	adding 2 extra totes and full de-slaving.
5	Ikara *Leander*	DAE/JZR(1) or (2), one FM1600 with 2 plot, 2 tote and 4 LPD.
6	*Invincible, Illustrious*	DAF(1)/JZS(1) with 4 plot, 7 tote and 18 LPD in *Invincible*, updated to JZS(2) in *Illustrious*, by adding 3 mini-tote.
7	Type 42 B2	DAG(1)/JZT(1) with 2 plot, 4 tote and 13 LPD. Updated to JZT(2) in *Manchester* (first stretched
	Exeter	B3 hull) with 2 plot, 5 tote and 14 LPD, with partial de-slaving.
8	Type 42 B3	DAG(2)/(JZT(3) with 2 plot, 8 tote, 2 mini-tote and 16 LPD.
	Gloucester	
9	Reserved for T42 batch 1 on major refit as DAH(1)/JZU(1) - see ADAWS 4 refit above.	
10	*Ark Royal*	DAF(2)/JZS(3), plus 8 mini-totes - display was to be JZV(1).
11/12	Earmarked for upgrade of earlier ADAWS, but overtaken by ADIMP.	
ADIMP	Type 42 B2/B3 and CVS	ADAWS Improvement Programme.

Note: *Bristol* and CVA-01 were both to have been fitted with Type 988, the Anglo-Dutch Project 'Broomstick' 3D radar, which eventually became the Dutch MTTR radar in HMNLS *Tromp*. CVA-01's ADA would have included 2 plots, 18 totes and 24 LPD, supported by three F1600 series computers and four 'drums', at a UPC of £1.34M (1964).

ADAWS 2 Series: ADAWS 2 was based on a 1964 staff requirement,[12] and entered service in HMS *Bristol* in March 1973. This was followed by a scalable family of variants (up to ADAWS 10), all with Outfit DA and JZ series nomenclature.

ADAWS was essentially a 'star wired' system, with all components linked to the central data processing node as a single 'hub' (see Annex). In this configuration, the computer was the single link between all the input devices (keyboards and rollerballs) and output devices (totes and labelled plan displays), though with a reversionary capability to display radar without the symbology. ADAWS was based around the FM1600 computer and Plessey series 8 displays:

- ADAWS used two FM1600 computers in all ships except the Ikara *Leanders*. The original core store (1000 series) had 64k words. This was increased to 128k words in the new Plessey 858 series core stores in the mid-1980s, further increased to 192k with the 858 Mod. Capacity growth was therefore from 64k to 384k.
- The contemporary CAAIS (see below) ran on the broadly similar FM1600B with 16k memory blocks, and the largest variant used 64k in total. Software for both ADAWS and CAAIS was written in the FIXPAC low-level language, also known as FIDS ABC (the instruction set format). Later ADAWS versions were coded in a higher-level language (Coral 66), in a super-modular from.
- The ops room equipment was based around three main items: labelled plan display (LPD) and Tote display, both set in vertical consoles, and the larger tactical plot with a horizontal LPD display and three keyboards, sharing a single tote at the rear (see Fig.13). The user interface was an ABCD layout keyboard and a roller ball, plus special purpose keys. During later life there were a number of supplementary mini- or micro-totes fitted above the LPD, thereby increasing the overall number of user positions in the Operations Room.

Computer Hardware: In the early 1970s memory was the single most expensive item in the computer, and affected the scalability of the design. The desire to 'save digits' through economical use of this scarce resource was exemplified by the use of a flat earth grid system and $\pm \Delta$ delta coordinates with reference to a fixed reporting point, rather than absolute (Lat/Long) coordinates that would require more digits to achieve the same precision. Memory size therefore affected the space available to the Main Track Table, which held the tracks and other information such as reference points and bearing lines.

None of the early military computers used hard drives (though these were available in the civil sector) because of concern about their shock resistance. All systems ran with the operational programme (and tactical data) resident in the main memory, usually a ferrite core store. Removable media was available, with programmes loaded (or data recorded) via reel-to-reel magnetic tape, punched paper or mylar tape. Both FM1600 and 1600B

Table 4: ADAWS MIG Qualifiers

The full list included 84 qualifiers (not applicable to every system); samples are:

BD	Bearing and Distance	LT	Link 11 transmission
BR	Bearing and Range	OC	Off Centre (display)
CF	Console Function	PC	Picture Common (compilation)
GM	Greenwich Mean Time	RD	Recording
IF	IFF	TN	Track Number
LL	Latitude and Longitude	XT	Link 10 transmission

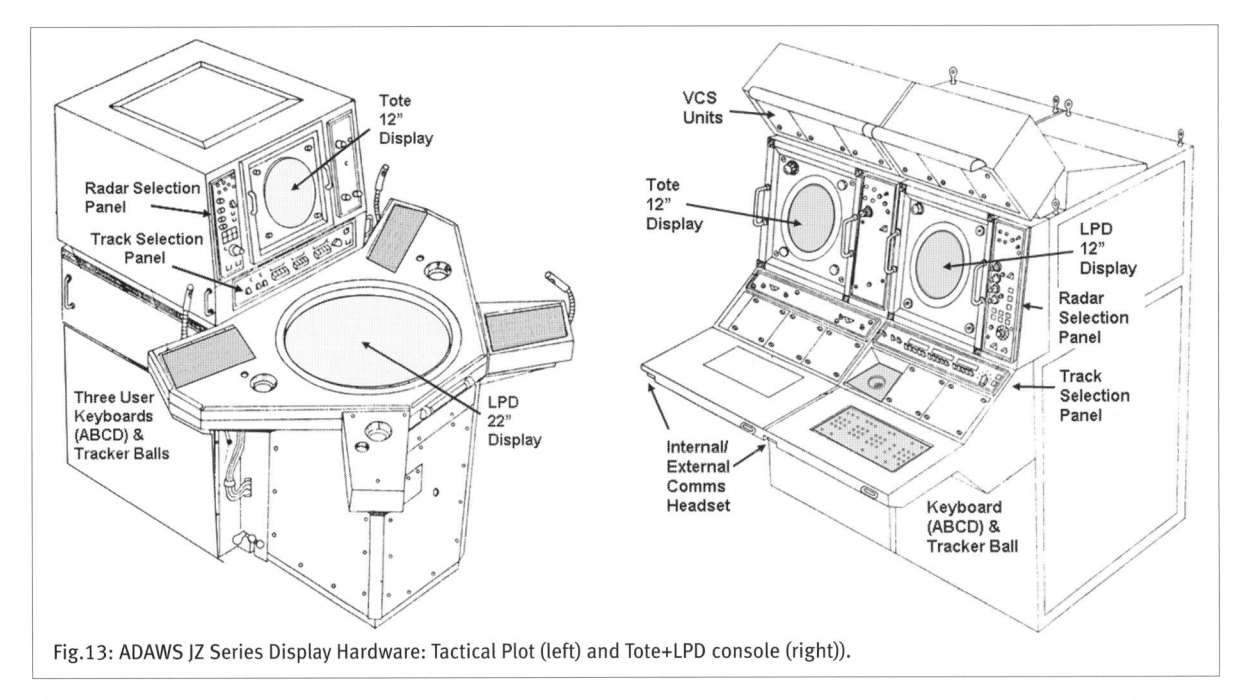

Fig.13: ADAWS JZ Series Display Hardware: Tactical Plot (left) and Tote+LPD console (right)).

also used a BIAX store (the predecessor of the modern BIOS) for the 'permanent' micro-programme instructions.

Software: Programme interrupts occurred every 1/64 sec, and the overall cycle time to carry out the full Op Programme once (including servicing all the inputs) was approx 1/8 sec. With the original ADAWS 2-6 systems, good practice required the system to be 'dropped' nightly during a quiet period to clear out any track fragments; this avoided creeping corruption which might lead to a system freeze or crash at an inopportune moment. By the late 1980s, the larger 858 core stores and new software assemblies based on Coral 66 led to much-increased reliability, and reduced this housekeeping chore to a weekly routine.

Fig.14: ADAWS Consoles in the Operations Room of HMS *Bristol*.

Fig.15: ADAWS LPD and Tote Symbology.

The schematic shows the ADAWS LPD, with user injections via the associated keyboard. User injections were constructed around a qualifier, + (is), - (cancel or delete) and ? (query), track numbers, plus special purpose keys defined by the user's console function or the tracker ball, followed by inject.

For example, PC ? 2020 In asks: What is the position and course of track number 2020? The output from the manual injection was given on the code reading line of the tote display. The response to the above question might be:

2020	HB	C325	S600	Z4	LOW	EW1266	} for an air track – similar format
TN	Identity	Course	Speed	Size	Height	Cross-referred Tracks	} for surface or sub-surface
J3	GW13440430		I2344	GARB	NONE		} tracks, adds name, confidence
Source	CCG reference		IFF Mode 1	Mode 2	Mode 3		} level and depth fields.

ADIMP: ADAWS was eventually succeeded by the ADAWS Improvement Programme (ADIMP) under SR(S)7900. In contrast to the single 'hub' ADAWS, ADIMP added the Combat System Highway (CSH) to distribute or 'federate' the combat system components, and allowed sub-systems to communicate directly with each other, all linked via the CSH – see Annex. ADIMP introduced a federated architecture, and also introduced distributed auto-extractors (LFA for Type 996, LFB for Type 1022, LFC for Type 1007, and Outfit LFD as the radar track combiner), all replacing the central LAX and LFX.

The display hardware in the Ops Room was updated as Outfit JFP, a colour Labelled Plan Display. The system used a QWERTY keyboard for manual injections. The main flat panel display surrounding the LPD incorporated the tote and readout areas. ADIMP then replaced the large horizontal tactical plot with a cluster of three flat panels. ADAWS Mod 1 has subsumed all the other installations, and became a single family covering the remaining CVS, LPH and LPD hulls with a common ADAWS 20 Edition 3 software package. System nomenclature included computer outfits DAH and DAK.

CAAIS

The Computer Assisted Action Information System (CAAIS) evolved in the early 1960s during development of ADAWS. At one point in 1963 the naval staff looked for a simple data processing system in all *Leander*s at a unit procurement cost (UPC) of £300k.[13] This was superseded by NSR7934 for a smaller, more versatile Action Information Organisation for all ships of frigate size and above, drawn up in 1966 by David Armytage. The original ADA goal was a UPC of £300k, but CAAIS had a target unit cost of £100k, of which 20% was for the tactical data link.

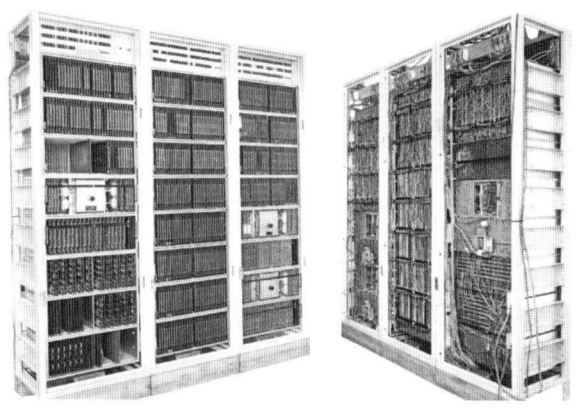

Fig.16: Ferranti FM1600 Cabinets: Front and Rear, Doors Removed. The image shows the principal cabinet layouts for ADAWS to give the relative scale (12 bays of full-height cabinets in dual-computer ships such as Type 42/82 and the *Invincible*-class CVS, but only 6 bays in Ikara *Leanders*. Note the separate control desk, and compare to the much smaller 2-bay CAAIS DBA-1 at Fig.20. (John M Peters, *Echoing Down the Years – ASRE Remembered*, Elgar Press (2002), p.51.)

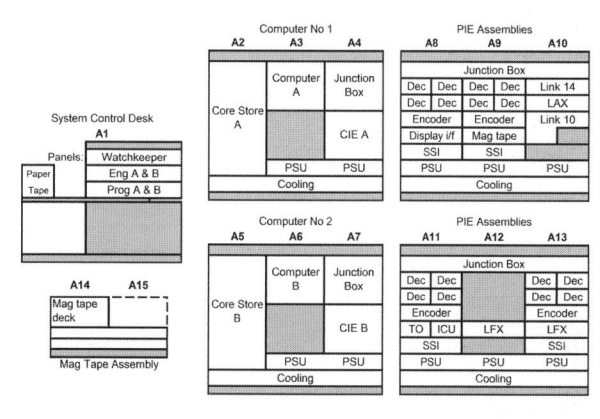

Fig.17: ADAWS Cabinet Footprints.

Table 5: CAAIS System Variants

Ship	Data Process Outfit/size		Displays/Outfit	Date:
Leander B2	DBA(1)	16k	six JHA(1)	Nov 75
Type 21	DBA(2)	64k	six JHA(1)	May 74
Hermes	DBA(3)	32k	nine JHA(1)	Jan 77
'Hunt' class MCMV	DBA(4)		two JHA(3)	Mar 80, replaced by Nautis 3
Leander B3A Seawolf	DBA(5)	64k	six JHA(1)	Nov 80
Type 22 B1 *Broadsword*	DBA(5)	64k	six JHA(1)	May 79

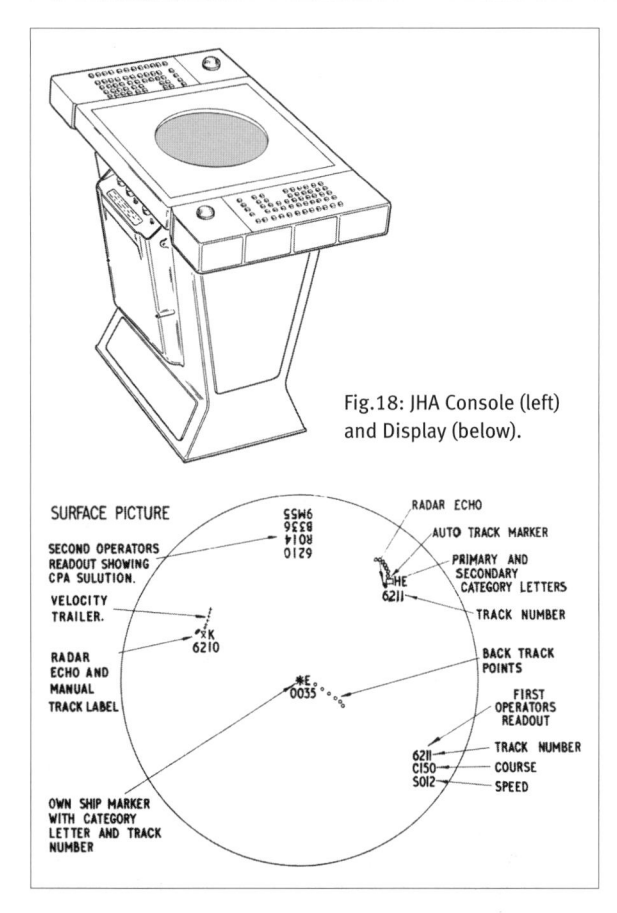

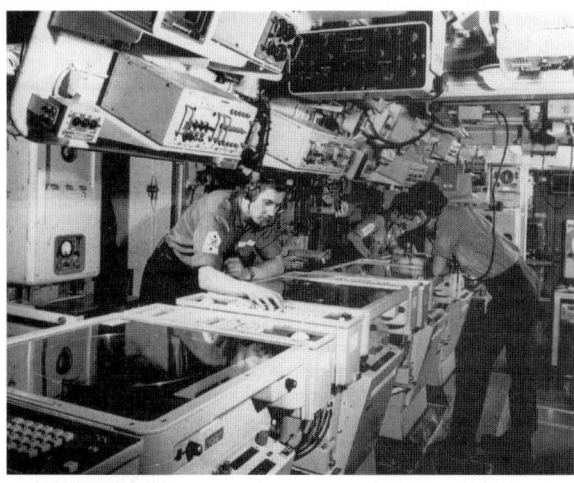

Fig.18: JHA Console (left) and Display (below).

Fig.19: CAAIS in HMS *Torquay* (Ferranti, p.387).

Ferranti, Plessey and Elliott's were competing for the CAAIS contract, but tension within Government caused by the Bloodhound enquiry led to a preference for an Elliott 920-based solution. In 1967 Ferranti pulled out of the consortium, but by September 1968 won with a separate tender based on the new FM1600B, sharing the overall system contract with Decca (for the CA1600 displays, derived from their commercial Deccascan). The FM1600B was one fifth of the size of FM1600 but offered half the computing capacity. Shipfitting started in 1970: in the frigates *Torquay* and *Londonderry*, the carrier *Hermes*, the *Leander* B2 and 3A (13), Type 21 (8) and Type 22B1 (4) frigates, and the 'Hunt' class MCMV (13).

CAAIS provided automatic tracking of air and surface targets, the processing of sonar and EW inputs, a digital datalink, target designation, and vectored attack for ASW helicopters. In frigates the six JHA consoles were dedicated to specific functions. The JHA display had a 16in PPI, shared between two seated or standing operators (referred to as North and South). Each had a keypad and rollerball, with a broadly similar (but reduced scope) set of manual injections compared to ADAWS. The display had no separate tote, and used a small readout area on the main display to show system responses to each operator query, whilst the top formed a reflection plotter (see Figs.18 & 19).

The CAAIS architecture included a single FM1600B, JHA displays, and either LAX or LFX plot extractors, plus Link 10. The footprint of DBA(1) in a two-bay cabinet is shown in Fig.20. The system limitations (compared to ADAWS) were 32k (rising to 64k) core store sizes (about a quarter of the equivalent ADAWS

Fig.20: Cabinet Footprints: CAAIS Outfit DBA 1, WSA1 Outfit DBD(1) in Type 21. The shaded area is common to all FM1600B computer suites.

Table 6: CS Track Number Limits

System	CDS	ADAWS 1	CAAIS	ADAWS 2	4	5	6/10	7/8
Track Capacity	48	256	60/120	500+	300	180	500+	500+
Memory	n/a	44k	32k › 64k	64k › 128k › 256k 192k › 384k				
Display Range	200	256	144	512				
Earth Model	Flat							
DPT	Yes	*Eagle*	No					
L10	No	DLG	Yes					
L14 TX	No		?	Yes				
L11	No			Added	No	No	Added	Yes
L16	No						DLPS	

System	CACS	ADIMP	DNA1
Track Capacity	500+	500+	1000
Display Range	2000		
Earth Model	Round - WGS84		
L10	No		
L14 TX	DLPP	DLPS	
L11	DLPP	DLPS	
L16	No	DLPS	Stand alone

system), and therefore a reduced maximum main track table size (60 or 120 compared to 500+ in ADAWS). This gave a pro-rata reduction in physical footprint (down to two cabinets).

In ADAWS the 4.5in gun (GSA1), Seadart (GWS30) and Ikara (GWS40) ran as part of the single operational programme on the central FM1600; they did not have stand-alone or reversionary modes.[14] In contrast, CAAIS only provided a tactical picture, giving target indication to the autonomous weapons: the Mk 8 gun, Seacat, Seawolf and Exocet. Therefore the Type 21 required another three-bay cabinet as Outfit DBD(1) to handle both gun and Seacat. This situation only changed with ADIMP, in which the radars were independent of the C2 system, and with Outfit DNA(1) where radars, vertical-launch Seawolf (GWS26) and GSA8 were full highway members largely independent of the central C2 suite.

Track Numbers: During the development of RN command systems there was usually a physical limit on the maximum number of tracks available due to finite memory capacity, and the increasing computational demands within the operational programme. This is illustrated (in roughly chronological sequence) in Table 6. Note that from CACS onwards the core store technology shifted over to distributed solid-state memory, and CPU size was less relevant.

Other Systems: Computer Aided Navigation Equipment (CANE) from Racal was fitted in 'Island' and 'Castle' class patrol vessels and in RFA *Argus* from 1977. This system was based on a KS500 Decca Kongsberg computer, and known as Outfit DEA(1), or DEB(1) in *Argus*.

The Naval Autonomous Information System (NAUTIS) was developed by Plessey for the *Sandown* class Single Role Mine Hunter from 1989, and was closely integrated with the Type 2093 variable depth sonar. Nautis-F through -P were proprietary versions, also fitted to HMS *Fearless*, or exported as AN/SYQ-15. Subsequently, an upgraded Nautis 3 has been retrofitted to all surviving MCMV. These systems used a Mil-Std 1553 highway and autonomous consoles, each with a shared copy of the main track table.

CACS

The Computer Assisted Command System (CACS) was introduced for Type 22 Batch 2 frigates onwards, based on NSR7933 dated November 1978. The driver was the perception that the combination of two new sensors (RN Outboard and sonar Type 2031) would generate a quantum increase in the number of contacts and tracks across a much larger tactical area of interest than CAAIS was capable of handling.

Table 7: CACS Variants

	CACS fitted to:	Data Processing/ Display System Outfit:
1	Type 22 Batch 2	DFA(1)/JJA(1)
2	Type 42 on refit	Projected, but not taken forward
3	Type 43 or 44	T44 cancelled by Nott review
4	Type 23 (and AOR)	DFB/JJB – terminated and superseded by SSCS
5	Type 22 Batch 3	DFA(7)/JJA(5)

Note: An AOR fit was to mirror Type 23, with CACS 4 and GWS26 VL Seawolf, to allow it to conduct unescorted 'delivery boy' operations, servicing widely dispersed Towed Array escorts, but was abandoned as the cold war threat receded in the early 1990s. This AOR fit was known as DNA(2), nomenclature subsequently reused for the upgraded DNA(1).

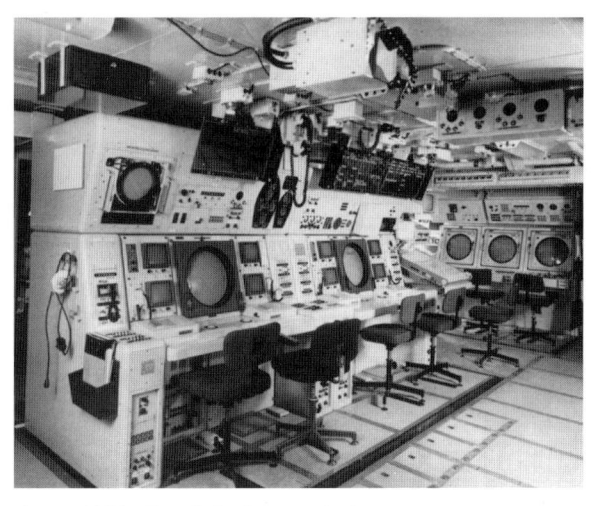

Fig.21: CACS 1 Type B Conference Displays.

CACS was initially based on FM1600E as the central computer node and Argus M700/20 in the displays, all linked by an ASWE Serial Highway (ASH). Displays were Plessey series 9 with mini-totes and light pens, using a slightly clumsy menu-driven user interface. CACS introduced the Data Link Pre-Processor (DLPP) as a federated system handling Links 11 and 14 (but not Link 10), and used Outfits DFA and JJA, with both Type A single-user and larger Type B conference displays (see Table 7 for variants).

The first Type 22 Batch 2 with CACS 1 (HMS *Boxer*) was commissioned on 14 January 1984, with the remaining ships being progressively added through to May 1990 (last Batch 3 with CACS 5). CACS had a troubled introduction to service and required multiple upgrades: the underpowered FM1600E was replaced by a compatible (but faster) F2420, extra Argus processors were added to the displays, and some of the software scheduling was optimised by the MASCOT kernel.

CACS 1 was to have been followed by CACS 4 in the Type 23: a significant re-design of the architecture added a triple Combat System Highway (CSH) in order to distribute the combat system. This is shown in Fig.23, and contrasts with the 'star' functionality of CACS 1.

There were multiple reasons why CACS 4 'hit the buffers' in 1987:

- The CACS requirement was only effectively a 'half generational' improvement on its predecessor, yet it consumed a full quota of project effort, finance and timescale. Its technical approach predated the exponential growth in COTS performance.
- The display clarity lagged well behind acceptable quality, and the overall system hardware was essentially at the end of a technical cul-de-sac.
- The projects deliveries repeatedly failed time, cost and quality goals, and the system was reluctantly accepted against a reduced requirement.

CS Federation: Both CACS and ADIMP, which were developed in a similar timeframe, began the process of 'federation' by moving specialist functions out into sub-systems that allowed the parent CS to be isolated from downstream changes behind a standard interface across the Combat System Highway (Outfit RJL). These included:

- Navigation: managed by the Ship's Navigation and Plotting System (SNAPS) which interfaced to all the radio navigation aids and chart plotting tables, giving a single Best or Computed Position to the command system for picture stabilisation. SNAPS was Outfit JNA-Z, depending on the number of consoles and plotting tables fitted.
- Tactical Data Links: managed by the Data Link Pre-Processor (DLPP) or Processing System (DLPS) – this development was largely due to bitter experience with *Bristol*'s Link 11 integration into a monobloc ADAWS 2 programme. DLPP was Outfit RJH, while the later DLPS was RJJ.
- Captain's Combat Aid (CCA) and EW Control Processor (EWCP), both implemented in ADIMP ships to improve command advice about weapon arcs, hard- and soft-kill co-ordination, and EW picture management. These were Outfits JZZ and UCB respectively.

Post CACS Period: The pronounced problems led

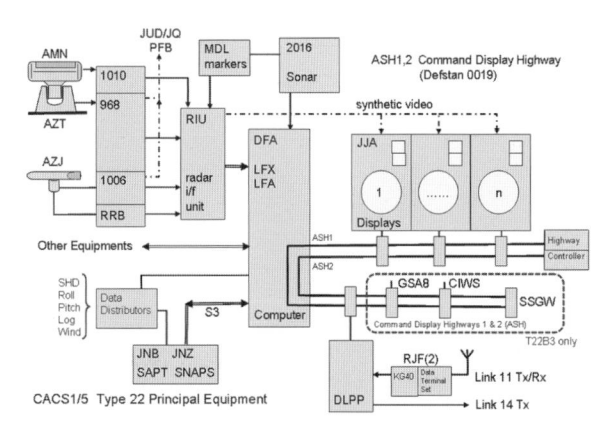

Fig.22: Schematics for CACS 1.

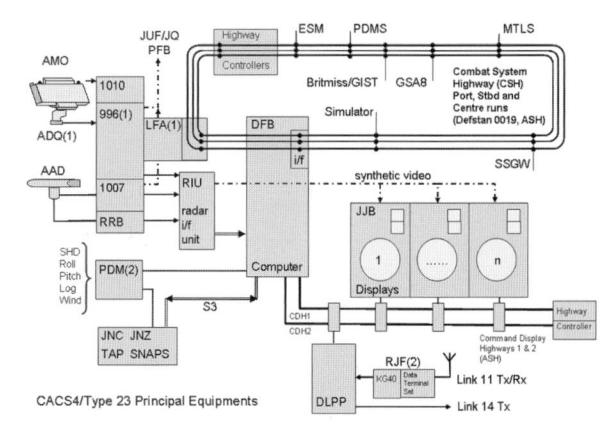

Fig.23: Schematic for Cancelled CACS 4.

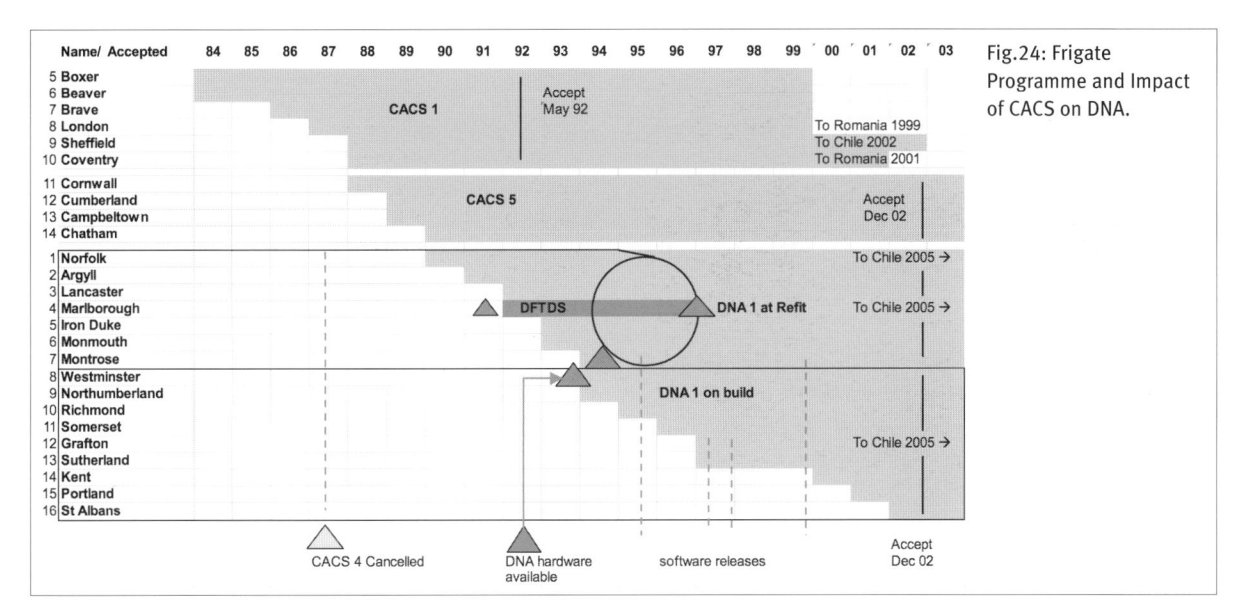

Fig.24: Frigate Programme and Impact of CACS on DNA.

Controller of the Navy to cancel CACS 4 in July 1987 and to re-compete the development. He accepted that the first seven (out of sixteen) Type 23s would be delivered without any command system, although the Combat System Highway (CSH) did allow a measure of interaction between the sensors and other combat system components, and Seawolf acceptance firings were carried out using an additional stand-alone console to control the trackers. Ships were given two additional Colour Tactical Displays in the ops room, but were not deployable to front-line settings like the Gulf.

The get-well programme became DNA(1), with a revised requirement approved in July 1987 as SR(S)7874 and a new procurement competition, leading to a contract awarded to Dowty-SEMA in August 1989, with first hardware fits from HMS *Westminster* (1992) onwards, plus a staged software development. This delivered operational phased releases from mid-1995 onwards, with formal acceptance in December 2002.

DFTDS: In the interim period before DNA(1) was available, DRA's Maritime C2 division brought forward previous research as the Data Fusion Technology Demonstrator System (DFTDS). This was initiated in 1987 and fitted in HMS *Marlborough* between March 1991 and November 1996. DFTDS did much to introduce modern commercial off-the-shelf (COTS) components and large colour displays. In addition, it also incorporated a rule-based inference fusion engine. There was no direct 'pull through' from DFTDS into Outfit DNA, given that they were largely parallel development activities, and exploitation was also thwarted by intellectual property rights (IPR). The most charitable interpretation is that DFTDS significantly 'raised the bar' of the expectations held by the RN community by showing what was possible using modern technology.

Surface Ship Command System (SSCS): DNA(1) was based on the previous Submarine Command System; the Type 23 frigate was fitted with a total of twelve multi-function consoles running on a fibre optic local area network, and linked to the wider triple-redundant combat system highway though input/output nodes. The system used a large number of COTS processor cards, including transputers in some of the consoles to ease the processing load caused by analysis of passive towed array

Table 8: Growth in Processor Hardware

CACS 1/5	DNA(1)	DNA(2)
Original system in 1984 had dual FM1600E as CPU, plus dual M700/20 as: ASH display highway controllers, in each of seven consoles, and in DLPP.	Original 1994 system had two Input Output Notes, with 11 processors, and twelve Multi-Function Consoles, each with 5 processors, reinforced by another 27 processors in MFC's used for TMA work. Intel 386/486, 68020 and T800 processors.	Each MFC now based in single commercial PC. Operating system 'locked down' versions of Windows.
Get-well upgrade in 1992 to 2*F2420 in each CPU node (dualed), plus 3 M700/20 in each display console.	Initially Intel 286 and Motorola 6500, but moved to Pentium 100 and Motorola 68040 by 1996.	Hardware in batch staged releases.
Overall upgrade from 20 to 29 processors, at 4X CPU power, and +50% display processor capacity.	Total 226 processors. COTS at the chip level, on bespoke boards.	Total of 41 networked commercial Pentium dual-core processors. Overall, 10% of DNA(1) hardware cost.

contacts. The rapid shift in computing hardware requirements can be gauged from Table 8, which compares CACS 1 as designed with the 'get well' programme and its successors.

Since DNA(1) in Type 23 completed development, the system has been enhanced to form the core of CMS 1 in the Type 45 destroyer. DFTDS was followed by a range of research demonstrators which have not been translated into actual service use.

Tactical Data Links

The current UK position on Tactical Data Links (TDL) is traceable back to USN and RN experience at the end of the Second World War in the Pacific theatre. Fleet Air Defence against large air raids of increasingly fast *kamikaze* aircraft or manned *Baka* bombs used voice telling of information from radar picket destroyers with a standing Combat Air Patrol (CAP). In the early post-war period, this led the RN to adopt the US Navy's Airborne Early Warning (AEW) aircraft with the 'Bellhop' downlink, and to develop CDS with Digital Plot Transmission (DPT), the forerunner of all present data links.[15]

Apart from the Canadian DATAR demonstration in 1953 (which included a UHF pulse code modulation link handling 128 tracks), the first TDL to enter service was the RN's DPT, which was the father of Link 1, used by NATO's Air Defence Ground Environment.[16] CANUKUS work on Tactical International Data Exchange (TIDE Link Roman II, latterly Link 11) started in 1956. The UK participated in the early programme but ADA and ADAWS 1 systems lacked the required computer power and were unable to integrate the terminal.

The Allies were concerned about the expense of the initial Link 11 terminal equipment, which used complex DQPSK modulation, and the Dutch proposed a cheaper Link 13, leading to very successful France-Germany-Netherlands-UK trials. However, the UK unilaterally terminated the programme in 1965, only to reintroduce the very similar national Link X (later Link 10) with a simpler serial coding scheme for CAAIS in 1968 (tested in 1971). Link 10 went into all UK ADAWS and CAAIS ships, plus some Dutch and Belgian units. It had extensive operational use during the Falklands, but was withdrawn during the mid-1990s as ADAWS ships were converted and CAAIS-fitted frigates withdrawn. Link 14 is a separate computer to non-computer unit broadcast of tracks using a 75-baud teleprinter circuit and manual plotting (updated as the Link 14 autoplot, most recently as Outfit PDT).

The third-generation Link 11 terminal (AN/USQ-59) finally entered service in HMS *Bristol* as Outfit RJF, but the parent ADAWS operational software structure caused numerous integration difficulties. Trials started in 1977, were put in abeyance by the ship's fire, re-started in 1980-81, only to be interrupted by the Falklands Conflict in 1982; ADAWS Ed 2 only achieved formal acceptance in 1984, as additional Type 42 Batch 2 joined the Fleet.[17] Criticism focused on the 20-year gestation period, and the success of other countries such as France and the Netherlands, with relatively small 'in-house' software development teams, in achieving Link 11 operational capability with much less effort than the UK.[18]

Table 9: NATO Tactical Data Links

NATO Link:	Function and Status:
1	Ground-based Link SOC-SOC. Derived from DPT as Link Roman I; in service.
2	Link for radar and associated data between land radar stations cancelled (included in Link 1).
3	Slow speed warning link connecting evaluation centres and Operations Centres.
4	Interceptor aircraft control link – in service USN/USAF and some other NATO services. Abandoned by RN/RAF. TADIL C.
5	Proposed as Ship-Shore-Ship link with Link 11 characteristics; overtaken by events.
6	Link connecting Main Control Centres, weapon systems or sites, primarily for missile control.
7	Civil and Military Air Traffic Control data. In use by France only.
8	Proposed as Ship-Shore-Ship link with Link 13 characteristics. Abandoned.
9	Air Defence Control Centre/Air Base link, primarily for scrambling interceptors. Abandoned.
10	Originally Link X. Ship to ship link, as simpler alternative to Link 11. Was in service with RN, RNIN and BeN, but withdrawn from mid 1990's.
11	Originally Link Roman II. Primary maritime link for use on HF or UHF. Widely fitted in NATO forces. Evolved as CLEW and SLEW waveforms, due out of service shortly.
12	Proposal for 9600bps UHF TDMA link based on early USN link in 1960s. Abandoned 1965.
13	Alternative HF link to Link 11. Abandoned after successful sea trials in 1965, but resurrected as Link 10.
14	75 baud teletype link for broadcast from computer fitted ships to non-computer fitted ships.
15	75 baud teletype link for non-computer ships to computer ships. Abandoned in 1965 by RN. Not adopted by other NATO services.
16	High speed ECM resistant multi-function TDMA data link in widespread use by NATO, running on JTIDS radio network.
22	NATO Improved Link 11 Equipment (NILE) on HF and UHF to give increased data rate and some ECM resistance, F message set gives commonality of Data Elements with Link 16.

Notes: Grey – abandoned TDL. TADIL A, B, C and J are alternative US designations for Link 11, 11B, 4 and 16
Source: Godel (1984)

The experience led the UK to resort to a Data Link Pre-Processor (DLPP) as a federated sub-system, able to insulate the command system from link changes. Outfit RJH was first fitted in Type 22 with CACS and was refined as the Data Link Processing System (Outfit RJJ) in Type 23 and ADIMP ships. The US contemporary is the Command & Control Processor (C2P).

Operational Usage: From a UK perspective, the 1982 Falklands campaign was fought exclusively with Link 10, whilst the later Gulf Wars were largely fought with Link 11 (1991), supplemented by JTIDS/Link 16 (2003).

Layout of Operational Spaces

Ships' operations rooms up to and including Type 23 were laid out by Captain Naval Operational Command Systems. Layouts are relatively dense, with operators seated side-by-side in rows, with vertical displays each surrounded by the clutter of dedicated overhead units or stateboards, and with supervisors 'roving' on long-lead headsets. The layout is by functional area, with operators or compilers near their respective controllers, supervisors and directors.

All the current systems that refitted flat panels into existing consoles have significant 'dead space' behind the display due to the depth of the former CRT tube. In theory, wider reconstruction of legacy operations rooms could free-up space, but this would require new consoles and significant work-in-wake.

Wartime and early post-war ships used sound-powered telephones as the main method of internal communications. This was replaced in the 1960s new construction by Composite Communications Units and then, from HMS *Bristol* onwards, by RICE (rationalised internal communications equipment). This included interphone (point-to-point), intercom, broadcast, and external comms (radio circuits), all accessed via a common headset. RICE also allows users to dial up numbered extensions via the telephone exchange, or to talk to magnetic loops inside machinery spaces or on the flight deck.

Fig.25: Contemporary Operations Room.

The Versatile Console System (VCS) was developed by Aish in a similar timeframe, and included consoles in the ops room, bridge, and as freestanding units throughout the ship, designed around a common 6in modular cube unit (the 1x1), able to include internal communications, dial indicators, and other ancillaries. Larger units are multiples of the basic format, while increased component density has permitted smaller ½x1 units.

Conclusions

During the early post-war period, apart from a Canadian demonstrator that fizzled out, the RN initially had a five year lead, based on full operational capability of the analogue CDS in *Victorious* from 1958, whilst the USN only reached an equivalent NTDS digital system in 1963. Thereafter progress alternated:

- NTDS was exported to Australia, France (triggering SENIT), Germany and Japan. Initially it supported AAW in carrier, cruiser and destroyers, but did not include surface or underwater sensors, nor was it extended down to ASW frigates until much later.
- The UK developed a scalable family of ADAWS systems and led with the small CAAIS computer systems for frigates, well before the USN. The UK then took a relatively pedestrian step to CACS, whilst the US moved ahead with both Aegis and New Threat Upgrade (NTU).
- In parallel, the Dutch remained more innovative with their in-house Government TNO organisation, developing a series of DAISY command systems.
- RN progress of TDL waxed & waned, with early success on DPT and Link 10 but prolonged delays in making Link 11 work.

When asked why ASE was more innovative and productive, Benjamin's view was that it had a 'younger culture' compared to its predecessors. When contrasting the RN with other Navies, the development of AIO and C2 generally shows the RN in very positive light, but:

- ASWE led much of the early digital innovation. Since award of the single hardware/ software contract to Ferranti in February 1965, there was a progressive estrangement between science and acquisition. Where research projects have delivered demonstrators there has been little or no 'traction' or *exploitation* into later acquisitions.
- This was compounded by not having an 'in-house' software development capability, and unlike the USN, the Netherlands and France, the RN were less able to be 'intelligent customers'. The USN was able to prototype JOTS-CSS, TLAM mission planning and CEC; and because they owned the IPR, they were able to re-compete later phases.
- Long relationships with a prime contractor may lead to a degree of 'lock-in' that could be characterised as unduly comfortable, leading to less interest in IPR.

MoD was hands-off until forced to act by CACS, but then lacked leverage; there is a risk that this sequence could be repeated in future.
- Requirements cannot be reality free; there is a need to forecast future technology trends and to identify opportunities. After being a market leader in real-time systems with mainframes in the 1970s, the RN became a market follower by the 1980s, once single-board computers appeared. CACS was a disaster because it was only a half-generational step during a period of rapid improvement in computer technology.
- The latter part of the post-war period has been marked by an accelerating pace of change in computer hardware, but increasing dependence on commercial operating system software represents a potential cost escalator. The USN experience with Aegis *vs* CEC integration reinforces the need for modular programmes, regression testing, and a drive towards open standards.
- The evidence is that running sequential projects has real benefit as 'lessons learned' are carried forward, while the CDS X-model and DFTDS are examples of the value of technology demonstrators. Although not taken forward, they significantly 'raised the bar' of user expectations.

Overall, the RN had a roller-coaster ride with its Post-War AIO and C2, with pronounced highs and lows, but this was a period of real optimism, rooted in a belief the Navy was at the leading edge of what was practicable.

Annex

This annex contains definitions, principles and architectural information for AIO and C2 systems.

Nomenclature

Individual RN equipments are described by a range of titles or different 'Nomenclature' schemes. Generally Type numbers cover radio, radar and sonar transmitters, while three-letter outfits are aerials, receivers and all else. Generally, older equipment was identified by two letters and a number, but this evolved into the current three-letter outfit title, with variants in parentheses, eg Outfit ABA(2), or in the case of radio, radar and sonar sets, by a type number, eg radar Type 996. Types M-P-Q-R refer to sequential modifications, while Types X-Y-Z were development models; numbers in parentheses, eg (2) identify variants in configuration. This is described in CB03329 (ADM 239/627), but has no easily decodeable structure – unlike the US JAN system, where for AN/SPY-1 the SPY explains the system functionality. The most recent introductions have reversed the sequence (eg Outfit 4KMA).

Systems may also be known by their project open name or colour codeword (eg 'Orange Crop'); in some cases missiles or weapons had Ministry of Supply trigrams, for example CF299 (Seadart). Finally, there are a few cases where the UK numbering scheme mirrors the equivalent US structure (UK/URC-613 was the Code Division Multiple Access modem used by RN satcoms). Airborne systems tend to use both a codeword name and an ARI (airborne radio installation) number. Finally, some of the UK outfits were an acronym for the original name, eg Wide Band Amplifier became Outfit WBA(1).

All these different systems co-exist side-by-side in documentation; there are a few unexplained gaps where a nomenclature was issued for planning purposes but the project was subsequently cancelled before delivery.

C2 Principles

Table A1 lists the categories of objects held within a command system.

Display Types: The most predominant type of display is the Plan Position Indicator (PPI). This is the classic radar picture where, as the aerial rotates, the scan 'paints' the contact on a CRT tube with a persistent phosphor, so that targets remain visible until refreshed by the next rotation. This display can either be with respect to ship's head, or stabilised North-up. In addition, the picture can originate at the centre, or be 'true motion' so that the scan origin is moved at scale course and speed, and all

Table A1: Command System Information

Tracks	An item with a position/course/speed and ID/hostility, based on real-time sensor reporting of a contact. Track may also be dead-reckoned onward, when sensor looses contact, or expressed as an uncertainty ellipse if there is a time-late element to the reporting chain. Also possible to set up a synthetic 'virtual' track.
Lines of Bearing	Bearing information without range, for example visual input, ESM or passive sonar. Drawn on display as a line from the point of origin, with associated reference TN and supporting information.
Reference Points	Established reference, can be real object (eg Eddystone light) or entirely virtual, and can be geographically fixed, or floating with the tide (eg man overboard, for a search datum).
Tactical Areas or Constructs	Lines, zones or boundaries, for example air lanes or corridors, which represent guidance about restrictions.
Map or Charting	Backdrop information, often available via Additional Military Layers that provide a synthetic coastline or charted depth contours, beyond the coverage of own sensors. Could also include wreck information, if not entered as a reference point.

Note: The 'environment' of a track has to be declared as surface, air or sub-surface. Adjacent tracks can be associated (linked together) when they are believed to be the same real world object, or can be correlated (joined or fused).

POST-WAR AIO AND COMMAND SYSTEMS IN THE ROYAL NAVY

Fig.A1: PPI Display.

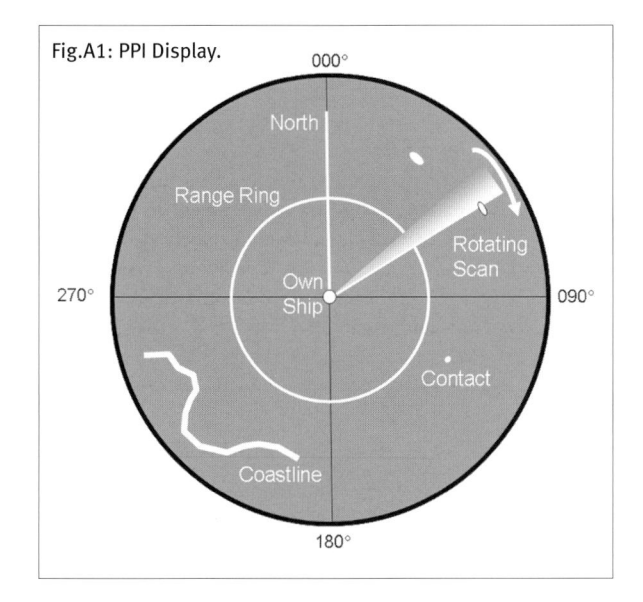

three consecutive scans through the track extractor, which may delay a first appearance on the command system. Tactical Data Links then add up to 10–12 seconds net cycle time to the reporting process. In contrast, near-real time is represented by minutes latency, whilst non-real time reporting like the Recognised Maritime Picture might well include items last updated 4 to 24 hours ago.

Command System Architectures: There are three main families of combat system architecture. The most basic analogue system was used from WW2 up to the 1960s, and linked sensors directly to displays, requiring some additional equipment to give a 'fan out' of one to many (see Fig.A2).

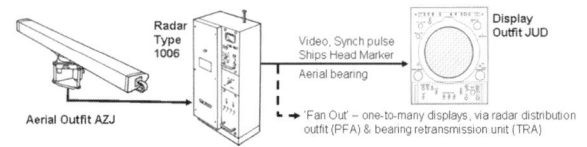

Fig.A2: Basic Sensor to Display Architecture.

contacts are shown at true (rather than relative) course and speed. A portion of the display can be selected for a zoomed view by off-centring.

In analogue systems, the 'flyback' time when not painting the raw video is sometimes used to draw additional marker symbology. However more sophisticated pictures with a lot of added text information, such as the Labelled Plan Display (LPD), tend to use cursive or raster strokes. The LPD adds synthetic symbology on top of the radar picture, correctly registered so that the contact details (eg track numbers) are aligned with the contact that they refer to.

A tote display is an alpha-numeric list of the requested information; the format is equivalent to an airport flight departures board. Modern systems add tote window areas around the LPD on a single flat panel display.

Track Numbers: There are several schemes available: Force Track Numbers, Common Track Numbers, and NATO track numbers use up to 5 characters, with a mix of octal numbers and alpha-numeric characters, ranging from 00001–77777 up to OA000–ZZ777.

Identity and Hostility: A tracks ID covers the type of platform it represents, eg 'bomber' or 'destroyer', while the target's hostility has evolved progressively via three and five levels to the current seven-tier structure (see Table A2).

The second reflects the early to middle period digital command systems, which introduced auto extractors and a computer & display suite in a 'star-wired' configuration (see Fig.A3). The computer system was the single link between all the input devices (keyboards, rollerballs), and output devices (totes and LPD), though with some reversionary capability for radar only (no symbology). Additional sensors (ESM and sonar), weapon systems, tactical data links may be integrated, and peripherals like a magnetic tape deck hold the operational programme or record events.

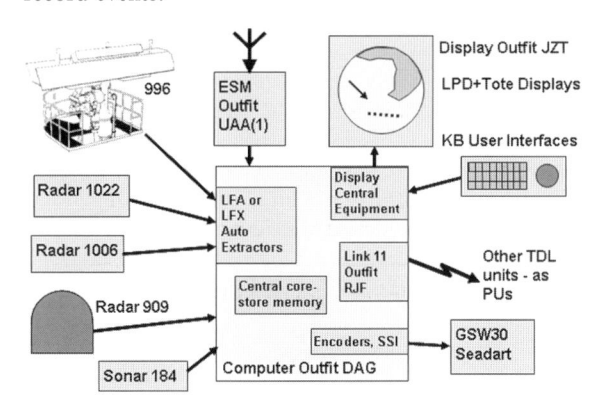

Fig.A3: Star-wired CS Architecture (eg ADAWS 7).

Table A2: Seven-tier Hostility Categories

Unknown	Pending	Assumed Friend	Neutral	Suspect
		Friend		Hostile

Latency: The definition of real time, near or non-real time information depends on the sensor used. Radar tracks usually have a latency of 2–6 seconds, based on the rotation rate of the radar that updates the position. Establishing a track in the first instance often takes up to

The third version (see Fig.A4), in general use for modern command systems, distributes or 'federates' the components, which are linked via a combat systems highway (CSH) or, more recently, a Data Transfer System (DTS). The key difference is that all the common information is on a (usually dual- or triple-redundant) highway, and performance is possible even in the event of command system failure, because the highway has combat system track numbers which are independent of Outfit DNA.

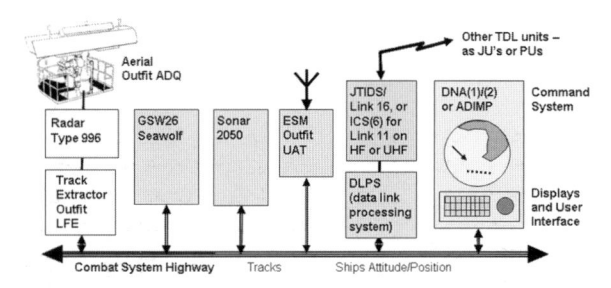

Fig.A4: Distributed or Federated CS Architecture (eg ADIMP or DNA).

Tactical Data Link

As an extension to the command system, a tactical data link (TDL) allows the same information about a track (track number, ID/hostility, position, course and speed) to be transferred across to other unit(s) via a networked communications bearer. This enables the remote unit's command system to display the same information as the first. The sensor information can provide supplementary coverage, and the information (eg bearings-only ESM data) can be correlated between platforms. TDL 'bridges' the demarcation between command systems and external communications, being tightly integrated with the CS.

Examples of classical TDL include NATO Links 11, 16 and 22. In Link 11, units reported track positions (as Delta offsets) with respect to an imaginary Data Link Reference Point (DLRP) which is defined in absolute (Lat/Long) terms. This 'flat earth' representation reduced the number of digits to be transmitted for a given granularity, and was typical of military computing in the 1960s. The later links use more sophisticated reporting (absolute WGS84 coordinates), more robust network designs, and protection against jamming or spoofing. Bearers include UHF (for line of sight) or HF (for extended ground wave cover), and in some cases satcom (potentially global coverage):

In contrast, the Recognised Maritime Picture (RMP) process used formatted (machine readable) messages that were originally sent by teleprinter. The Form Gold message set includes the same track number, ID/hostility, position, course and speed information, but also allows for sensor uncertainty ellipses and the time of the fix, so that receiving units can draw furthest-on circles to offset

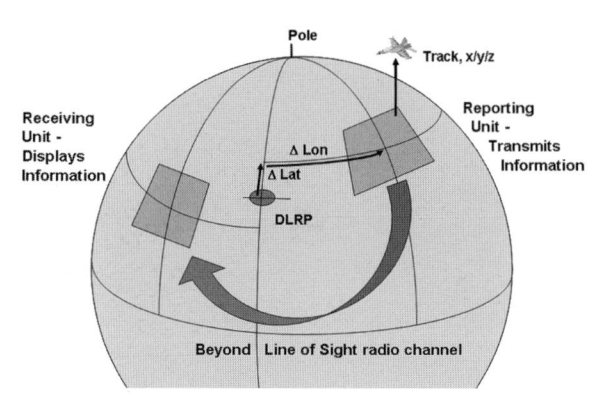

Fig.A5: Link 11 Transfer via DLRP.

latency that may range from many minutes through to hours (or days). Classically, the RMP has added-value from intelligence assessment, and forms the 'backdrop' to the real time command system picture, so that a unit has all the information to help categorise an Unknown track that enters its sensor coverage for the first time.

Naval Personnel

There has been progressive evolution of RN branch manpower structures and titles across the period (see Table A3). The Weapons & Electrical branch (1966) became Weapon Engineering (WE) from 1979.

Personnel training followed the sequence of new-entry and then career training, undertaken at either HMS *Collingwood* (engineers) or HMS *Dryad* (operators),[19] both now merged as a single Maritime Warfare School. Following career courses, personnel undertook Pre-Joining Training as preparation for specific ships, and then consolidation through On-Job Training onboard. Ships' Command teams could rehearse in the *Dryad* 'models', and ships were worked up by Flag Officer Sea Training as Basic or Continuation Operational Sea Training. Finally, whole-ship performance was enhanced through Area Capability Training during Joint Maritime Courses.

Organisations

The main post-war research establishment was ASWE at Portsdown; the smaller independent outstations (Haslemere, Slough, Teddington, and West Drayton)

Table A3: RN Manpower changes

Epoch	Pre 1975	1975-90	1990 onwards
Operator	Separate Gunnery, TAS, Communications and Navigation & Direction branches.	PWO introduced in 1972 and Warfare officers changed 1975: to (A),(U),(C) or (N), plus AWO. Rating titles changed, but structure not significantly altered.	Warfare branch rating structure re-shaped in 1993 to Operator Maintainer (OM); simplified structure pulling in most of the WEM(O) and (R). Leaves rump of ET technicians to WEO, plus separate IT branch created from Instructors.
Maintainer	Electrical 'L' branch from 1946. Ordnance and Electrical branches merged as WE branch in 1966, with O, C and R mechanic and artificer streams. WE Officer.	EBD transferred high power to MEO as ME(L) ratings in 1979. Remaining WE categorised as WEM(O) or (R), and technicians as WD, OC, AD or CEW. WEO inherited explosives from 1981.	

were gradually closed or merged together. There were trials ranges at Aberporth and Funtingdon. Gunnery was initially led from AGE Teddington, but moved via Portland to Portsdown as part of ASWE.

In the immediate post-war period, Admiralty (later MoD) control was exercised via London-based Naval Staff divisions, with strong linkage to the main establishments (*Dryad, Excellent, Mercury* and *Vernon*), whose experimental trials Commanders (X, T, P) led system acceptance. Overall ship design was led by the Ship Department at Foxhill, whilst the Weapons Department (DGSW(N)) at Portsdown (and the similar DGUW(N) at Portland), managed projects and production.

The ADA Rule Writing Group eventually morphed into Captain Naval Operational Command Systems, who also led on cross-system issues like total ops room layouts, which were outside the brief of individual equipment projects. In parallel, the User Requirements and Trials Section at Leydene carried out similar work on layouts for communications spaces.

During the bulk of the post-war period, technical development was Government led, via the combination of research scientists (ASWE), procurement & development engineers (DGSW(N)), and serving RN staff embedded in the projects as Naval Applicators. This was a creative partnership, responsible for most of the major improvements. In this era, ASWE-sponsored research led to an X, Y or Z model for sea trials, which was then 'production engineered' by a commercial firm before the main shipfitting programme could begin.

Acknowledgements:

Lt Cdr Clive Kidd RN Rtd, HMS *Collingwood* museum, plus Professor Simon Lavington for access to the Elliott's microfiche records (to be archived at the Bodleian library). Discussion with Maurice Needham (died 11 Feb 2015). Interview with Dr James Berry 15 May 2015. Taped interview by Captain David Armytage (died 2 Feb 2015, but held by IWM).

Abbreviations:

ABCD	Older RN AIO keyboard layout
ADA	Action Data Automation
ADAWS	Action Data Automation and Weapons System
ADR	Air Defence Room
AIC	Action Information Centre (early name; became the ops room)
AIO	Action Information Organisation
ARL	Admiralty Research Laboratory (early plotting table)
ASE/ASRE/ASWE	Admiralty Signals/Signals & Radio/Surface Weapon Establishment
ASH	ASWE Serial Highway
BR or CB	(RN) Book of Reference, or Charge Book (classified BR)
C2	Command and Control
CAAIS	Computer Assisted Action Information System
CACS	Computer Assisted Command System
CDS	Comprehensive Display System
COTS	Commercial Off-The Shelf equipment or software
CS	Command System (or Combat System)
CSH	Combat System Highway
DATAR	(Canadian) Digital Automatic Training and Remoting
DFTDS	Data Fusion Technology Demonstrator System
DGSW/DGUW(N)	Director General Surface (or Underwater) Weapons (Navy)
DLG	Destroyer Light Guided Missile
DLPP/DLPS	Data Link Pre-Processor/Processing System
DLRP	Data Link Reference Point
DPT	Digital Plot Transmission (the first RN tactical data link)
DRA	Defence Research Agency (successor to ASWE; led to Dstl and QinetiQ)
EW	Electronic Warfare
FD/FDO	Fighter Direction/Fighter Direction Officer
GDR	Gun Direction Room
ID	Identity
JTIDS	Joint Tactical Information Distribution (radio bearer for Link 16)
IPR	Intellectual Property Rights
LAX/LFX	Limited Area Auto-extractor/Limited Full Auto-extractor for tracks
LPD	Labelled Plan Display
MATCH	Medium Range Anti-Submarine Torpedo Carrying Helicopter (Wasp, then Lynx)
MCMV	Mine Counter Measures Vessel
MIG	Manual Injection Guide (for AIO computer system user)
MWC	Maritime Warfare Centre (tactical school at HMS *Dryad*)
ND	Navigation & Direction (RN branch)
OOW	Officer of the Watch (on the bridge)
OPV	Offshore Patrol Vessel
PDM	(Outfit) solid-state data transmission system for ship's attitude
PFA/PFB	(Outfit) video distribution systems
PPI	Plan Position Indicator (radar display format)
PWO	Principal Warfare Officer (in the ops room)
QWERTY	Modern commercial computer keyboard format
RDR	Radar Display Room
RICE	Rationalised Internal Communications System, including interphone, intercom & broadcast, with access to telephone and radio circuits
RP	Radar Plot (rating)
RRA/RRB	(Outfit) ships receiver, for I band transponder fitted to helicopters
SSCS/SMCS	Surface Ship Command System/Submarine Command System
TASCO	TAS Control Officer (marker used by JYA)
TASO	Torpedo and Anti-Submarine Officer
TDL	Tactical Data Link
TIU	Target Indication Unit (part of Gun Direction system)
TN	Track Number
TRA – TRN	(Outfit) bearing or data distribution systems
VCS	Versatile Console System (built from 6" cubes for dials, and comms)
WEO	Weapon Engineer Officer

Principal Sources:

F Kingsley (Ed), *The Development of Radar Equipments for the Royal Navy*, Naval Radar Trust, Palgrave Macmillan 1995.

H Derek Howse, *Radar at Sea: The Royal Navy in World War 2*, Macmillan 1993.

BB Schofield, *The story of HMS Dryad: Navigation & Direction*, Kenneth Mason 1978.

R Benjamin, 'The Post-War Generation of Tactical Control Systems', *Journal of Naval Science*, Nov 1989, Vol.15 No.4, pp.262–276. Largely reproduced by E Grove as Chapter 9 in *Cold War, Hot Science*, Ed R Bud and P Gummett, Harwood 1999. See also Benjamin's autobiography *Five Lives in One*, Parapress 1996.

DL Boslaugh, *When Computers Went To Sea: The Digitisation of the USN*, IEEE Computer Society Press, Piscataway NJ 1999. [For the equivalent USN developments, and US view of British efforts.]

EN Swenson, JS Stoutenburgh and EB Mahinske. 'NTDS – A Page in Naval History', *ASNE Journal*, Vol.100 No.3, May 1988, pp.53–61.

J Vardalas, 'From DATAR to the FP-6000 Computer', *IEEE Annals of the History of Computing*, Vol.16 No.2, 1994. [Early 1953 Canadian work and interaction with both US and UK.]

Footnotes:

[1] The AIO committee formed in Jun 43 led to a pamphlet on the AIO (CB 04357) issued Sep 1944, later replaced by the Radar Manual 1945 (CB 4182), ADM 239/307. Other sources include Fanning (1995), Monograph 3, 'The Action Information Organisation', pp.147–72, in FA Kingsley, *op. cit*. Fanning describes destroyers completing in 1942 without any AIO but also includes later (and larger) layouts in new-construction ships.

[2] See JNSS for comments about Exercise 'Mariner' in 1954.

[3] ADM 220/416 Electric Ops Plots. This emphasised systems smaller than CDS, using interlaced markers with offsets, projected up onto the underside of the plotting surface. The author had a clear understanding of CDS, and wished to avoid monoscope symbols, but would use joystick and potentiometer storage for a small number of contacts (ie he cherry-picked the X-model CDS, but thought DPT would be needed to exchange contacts with other ships in company.)

[4] ADM 1/24605, p.19, para 30

[5] This summary is based on the surviving Elliott's microfiche project records, plus the author's telephone conversation on 3 Nov 2014 with Maurice Needham, then aged 94 and the senior surviving member of the CDS development team 1947–51.

[6] X model = a laboratory version; Y model = initial sea trials; Z model = semi-/pre-production standard.

[7] This was eventually expanded to the current seven-tier structure, giving the basic 7x3 matrix represented by Mil-Std 2525: Friend, Assumed Friend, Neutral, Unknown, Pending, Suspect, Hostile.

[8] D Boslaugh, *op. cit.*, citing Louis A Gebhard, 'Growth of Naval Radio & Electronics', NRL report 7600, June 1976, pp.379–84.

[9] ADM 234/608 BR 4245 Part 1: 'Preliminary HB for JZA/JZB'.

[10] Poseidon built on Ferranti's work on the Pegasus computer from 1953–56. This spans the period when medium-sized computer architecture and instructions sets gelled, and technology moved from packaged valves and mercury or nickel delay line memories to transistors and drum memory (see Simon Lavington, *The Pegasus Story: A History of a Vintage British Computer*, Science Museum 2000).

[11] John F Wilson, *Ferranti: A History*, Carnegie Publishing Ltd. 1999, p.486.

[12] NSR7898 for ADA in Type 82, endorsed 9 Dec 1964. Under cover of an older NSR7868 for ADA in all major warships (endorsed 2 June 1959), this was to be followed by individual submissions for each platform: NSR7903 for ADA in all *Leanders* dated 15 Jul 1963, and NSR7909 for ADA in CVA-01 class, approved 15 Nov 1965, see ADM1/28513.

[13] See ADM1/28514: NSR 7903 ADA in *Leanders*, dated 15 Jul 1963. Framed round a single F1600 computer with a tactical plot and 5 consoles and the hope that the system might also supersede the below-decks elements of the MRS3 fire control system (other papers mention the 30 bit Univac CP667 as an alternative), the proposal was aimed at *Leander* long refits due to start in 1969, but the project was not fully funded in 1964/65, and by Feb 1966 DGW opinion was that Confessor (Seawolf) would probably need a different (not ADAWS) computer.

[14] Although GSA1 retained an analogue emergency fire control box for direct surface fire.

[15] R Benjamin, *op. cit.*

[16] Link 1 for exchange of air surveillance data between CRC and CAOC/SOC.

[17] Early ADAWS software had a monobloc structure, leading to the need to fully re-test the entire assembly after every change (regression testing). From 1984 assemblies onwards, the new Coral 66-compiled programmes were based on super modules.

[18] NA Godel in XCC 80005, 'An Introduction to RN Tactical Data Links', Oct 1980, and XCO 85014, 'Tactical Data Links in the Royal Navy up to 1984 – the First 35 years', Nov 1985.

[19] Operator training was originally based at branch schools (HMS *Excellent*, *Dryad*, *Mercury* and *Vernon*), but gradually coalesced to *Dryad* alone, before being merged as MWS on the *Collingwood* site.

THE SOVIET *FUGAS* CLASS MINESWEEPERS

In the latest in a series of articles on the interwar Soviet Navy, **Vladimir Yakubov** and **Richard Worth** give an account of the conception and construction of the minesweepers of the *Fugas* class, together with a service history of the ships completed.

Given the Russian Navy's success in the First World War in using mine warfare for both offensive and defensive purposes, it is little wonder that one of the first new construction projects of the reborn Soviet Navy was a minesweeper. As the resumption of warship construction approached in 1925, defence planners formulated the tactical-technical requirements for a 500-tonne successor to the 190-tonne *Klyuz* of 1916: equipment for fast and slow sweeps, a speed of 16 knots, and weaponry including two 47mm guns, one 7.62mm Maxim quad mount, thirty mines, and twenty depth charges. The six-year construction plan of 1926 included no minesweepers, but the concept continued to evolve. In 1927 a new minesweeping manual was published; it called for a minesweeping force built around big vessels possessing speed, seaworthiness, and a non-distinctive silhouette. The ships would undertake a variety of sweeping missions, including intelligence trawling outside the coastal strip, boundary definition minefields in remote areas, and the destruction of identified barriers. High-seas sweeping would take place only in special cases. Along with depth charges and minelaying gear, weaponry would include guns for self-defence against aircraft, submarines, and light surface forces. With oil fuel and two shafts, they would range to 1500nm, operating at sea for five to six days with a maximum speed of 18–20 knots to match fleet movements. Lastly, the ships would need equipment for towing damaged ships, radios powerful enough to communicate with the fleet, and minesweeping launches to perform sweeps in shallow waters.

On 13 June 1930, army Chief of Staff B M Shaposhnikov issued a report for adjusting the second half of the First Five-Year Plan (1928–1932). His proposals included construction of five minesweepers for each of the Baltic and Black Sea fleets, with the first six scheduled for completion in 1933. Cheapness and simplicity of the design would allow for mass production in wartime.

Characteristics: staff requirements (1930)

Displacement:	330 tonnes
Draft:	1.9m–2.4m
Machinery:	two-shaft diesels, 17–19 knots (with sweeps 12–15 knots)
Range:	400–450nm
Armament:	one 102mm gun

Project 3

The Scientific-Technical Committee (NTKM) under Yu. A Shimanskii completed initial studies for this new fast minesweeper (BTShch) in October 1930, and the prelim-

A Finnish reconnaissance photo of the Baltic Fleet's T.206 *Verp'* taken before the Winter War in 1939. Paravanes and marker buoys are seen amidships. (Boris Lemachko Collection)

inary design received approval from the Revolutionary Military Council on 25 April 1931. Work then transferred to the newly-formed Central Construction Bureau for Special Shipbuilding (TsKBS, renamed TsKBS-1 in 1932), where L V Dikovich undertook final design work – his first major assignment – with help from F P Muragin and G I Veraksa. The project was monitored by an experienced naval engineer, N M Alexeev. Titled Project 3, the design reached completion by the end of the year, having grown to 400 tonnes with a simplification of the hull to ease construction. The Revolutionary Military Council gave its approval on 27 March 1932.

Characteristics: Project 3 (as designed, 1932)

Displacement:	383 tonnes standard, 433 tonnes full load
Dimensions:	62m wl x 7.2m x 2.0m normal
Machinery:	two 42 BMRN-6 turbocharged diesels; 2800shp = 18.8 knots (with sweeps 15 knots)
Range:	1360nm at 18.8 knots
Sweep gear:	two K-1 paravanes, one MTShch sweep, one MZT sweep, one electro-magnetic sweep, Schultz sweep
Armament:	one 100mm B-14 gun
	one 37mm gun
	31 1926 mines
	20 B-1 and 20 M-1 depth charges
Complement:	42

Although the Russians had commissioned the world's first diesel warships in 1910 (*Kars* class), Project 3 represented the first Soviet effort in that field. The head of machinery at TsKBS, A V Speransky, assigned the motor design work to A K Greiner. A I Maslov at TsKBS-1 had responsibility for the working drawings, which reached completion in April 1934.

Official orders for the first batch of ships, known unofficially as Series I, were placed from November 1931 to July 1932. They included only eight ships: four from the Severniy Yard in Leningrad for the Baltic Fleet, plus three at Marti Yard and one at Sevastopol for the Black Sea Fleet. Conflicts with other programs resulted in a shuffling of yards, and a lack of materials caused further delays. The first keel-laying, scheduled for December 1932, actually took place almost a year later. The hulls progressed to launch in a reasonable time, but the supply of diesels became a major problem. An import deal for turbochargers from Germany fell through, forcing a switch to a domestic supplier, and changes in the diesels themselves brought further complication.

Characteristics: Project 3
(*Fugas* as completed, 1936)

Displacement:	428 tonnes normal, 476 tonnes full load
Dimensions:	62m wl x 7.2m x 2.03m normal
Machinery:	two 42 BMRN-6 turbocharged diesels; 2856shp = 18.5 knots (with sweeps 14.9 knots)
Bunkerage:	96 tonnes diesel
Range:	1900nm at 18.5 knots, 2900nm at 14 knots
Sweep gear:	two K-1 paravanes, one MTShch, one MZT
Armament:	one 100mm/56 B-24 gun (206 rounds)
	one 45mm/46 21-K gun
	(200 rounds normal load/
	800 rounds overload)
	two single 12.7mm DK MG
	(10,000 rounds per gun)
	31 1926 mines
	20 B-1 and 20 M-1 depth charges
Complement:	42

The 100mm gun and its crew, 1943. (Boris Lemachko Collection)

THE SOVIET *FUGAS* CLASS MINESWEEPERS

Zaryad ran factory trials between 9 and 13 September 1936, observed by a committee headed by K L Nikandrov. The acceptance report on her day of commissioning noted some problems. The ship had was 9 tonnes overweight, a modest figure in comparison with other Soviet projects at the time. The report attributed 4.7 excess tonnes to structural modifications and the rest to underestimates of the weight of the electrical and propulsion systems. The ship was 0.5 knots slower than designed due to the disparity between the design of the screws and the diesel power levels. The speed of 14.75 knots with sweeps deployed fell short of the Navy's hoped-for 16 knots. The operation of the Schultz sweep (a type of the sweep that was hung between the two ships and allowed them to sweep large areas of the sea) proved difficult; proper operation required a speed of 6–7 knots by both ships, but *Zaryad* could not maintain such a low speed using both diesels, and using only one diesel caused an overload that spewed large quantities of soot and sparks from the funnel. The paravane winch performed

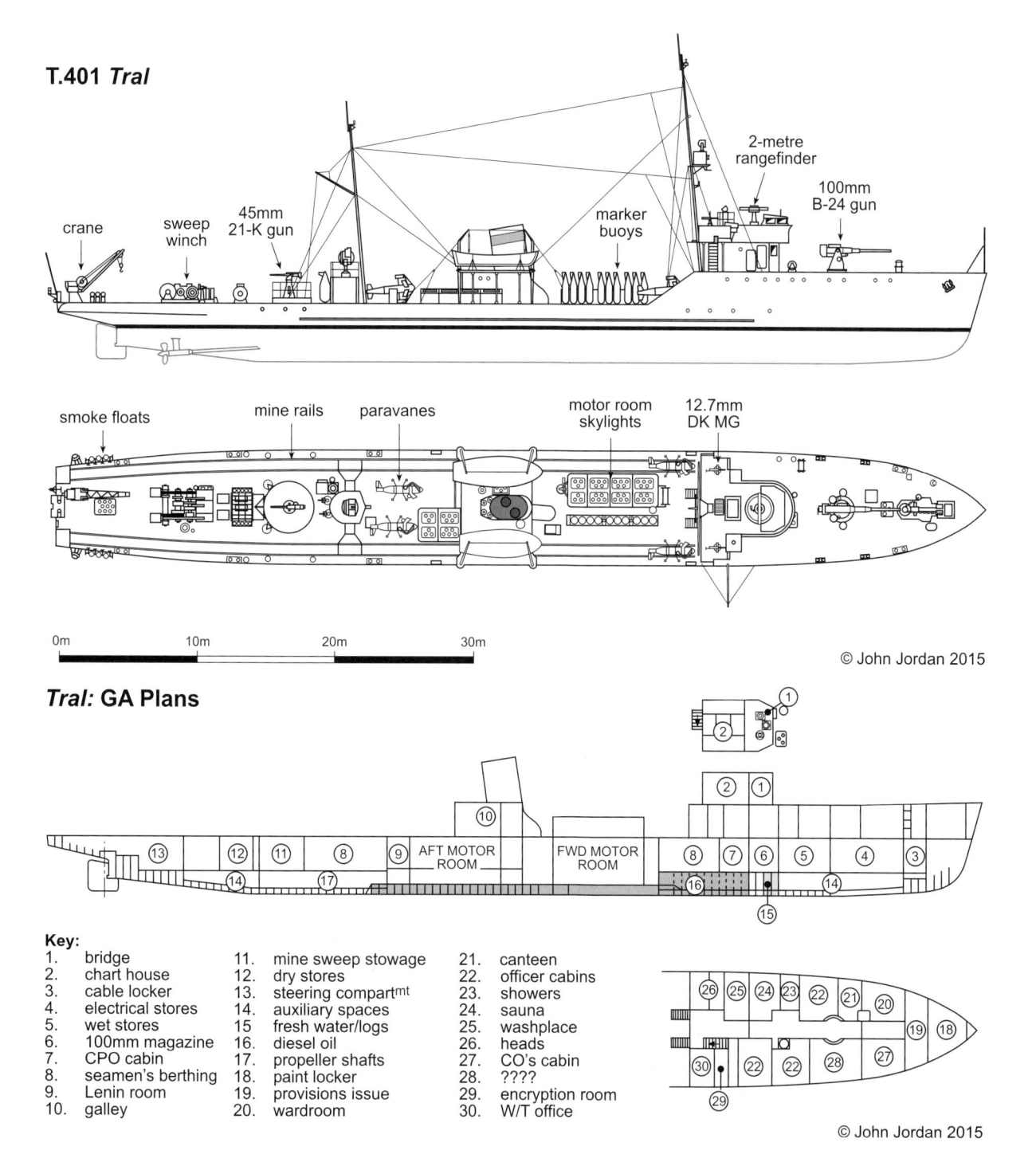

T.401 *Tral*

Tral: GA Plans

Key:
1. bridge
2. chart house
3. cable locker
4. electrical stores
5. wet stores
6. 100mm magazine
7. CPO cabin
8. seamen's berthing
9. Lenin room
10. galley
11. mine sweep stowage
12. dry stores
13. steering compart^{mt}
14. auxiliary spaces
15. fresh water/logs
16. diesel oil
17. propeller shafts
18. paint locker
19. provisions issue
20. wardroom
21. canteen
22. officer cabins
23. showers
24. sauna
25. washplace
26. heads
27. CO's cabin
28. ????
29. encryption room
30. W/T office

© John Jordan 2015

101

T.217 *Kontr-Admiral Yurkovskiy* is seen approaching the camouflaged T.205 *Gafel'* at the forward base of Lavensaari in the Gulf of Finland, c.1943. (Boris Lemachko Collection)

poorly, and a new design was prepared by K M Kashin. However, the overall minesweeping system was highly rated, as was the ship's accommodation. The weaponry prompted mild debate from those who felt the main gun should have an aft mount, since it was most likely to operate during withdrawal; but in this matter no change was made.

The most significant problem discovered during trials focused on the metacentric height, which was 0.72m as designed but in reality was a mere 0.42m at 411 tonnes. With mines embarked the figure declined to 0.23m, which was considered unsafe. The ships were not allowed out to open seas until the issue was resolved. The remedy on completed ships involved reducing the size of water tanks in the superstructure, adding a bulkhead to the fuel tanks to reduce free-surface effects, and shipping 15 tonnes of ballast. The ships that were still under construction underwent more significant modifications. Equipment for the electro-magnetic sweep was relocated from the main deck down to the bottom of the ship. The small superstructure on the stern that housed equipment and supported a 45mm gun was eliminated, with the gun repositioned on on a special upper-deck platform. The process of restoring metacentric height was protracted and eventually involved the NKVD – never a good idea in the Stalin-era USSR – whose Project 43 patrol ships derived from the Project 3 design. The NKVD refused to accept the ships until the factory fixed them from its own budget. Eventually the issues were resolved, and the ships went on to serve without any critical problems in that area.

Project 53

On 11 June 1933, before the first Series I ship had begun building, the Council of Labour and Defence, along with the Council of People's Commissars, adopted the 'Programme of Naval Construction 1933–1938' which called for the construction of twenty-seven fast minesweepers. However, new units would introduce some variations, as on 29 October 1934 the Naval Command instructed TSKBS-1 to refine Project 3 to cure various issues that had emerged during construction. Changes included a lighter stem and rudder, improved access to hull voids, a raised bridge, lowered engine skylights, and a modified deck hatch construction. This variant became Project 53, also known as Series II. Construction began in spring 1935, before any Series I ships reached completion. Plans to build ten units at the inland Izhorsky shipyard did not work out, because transit under the various river bridges on the way to the sea would have necessitated removing the entire superstructure from the hull, inflating construction costs; instead the ten units were reassigned to Ust'-Izhorsky.

With problems becoming apparent in completed Series I boats, designers produced the Series III (Project 53U), which included all existing modifications as well as a 0.2m widening of the hull. Most Project 53U ships were built at the Ust'-Izhorsky yard which, unlike Izhorsky, was downstream of the offending bridges so that removal of the superstructure was not necessary.

In addition to the usual difficulties, these ships encountered the biggest hurdle of all: the onset of Operation 'Barbarossa'. T.220 and T.221 required more than five years of building time, entering service in 1946. The outbreak of fighting caused the cancellation of two units, T.222 and T.223 (yard no.134 and yard no.35 from Petrozavod), before construction began.

Project 58

With Project 53U construction ongoing in the Baltic, the Black Sea shipyards decided to build new ships with the

Characteristics: Project 53 (T.406 as of 1943)

Displacement: 418 tonnes normal, 494 tonnes full load
Dimensions: 62m wl x 7.2m x 2.21m normal
Machinery: two 42 BMRN-6 turbocharged diesels; 2800shp = 18.45 knots
(with sweeps 13 knots)
Bunkerage: normal 59 tonnes, maximum 94 tonnes diesel
Range: 3300nm at 16 knots
Sweep gear: two K-1 paravanes, one MTShch, one MZT
Armament: one 100mm/56 B-24 gun (150 rounds)
one 45mm/46 21-K gun
(1500 rounds normal load + 700 rounds overload)
three single 37mm/70 70-K
(1500 rounds normal load + 1500 rounds overload)
one 12.7mm DShK machine gun
two single 12.7mm Browning MG
31 1926 mines
20 M-1 depth charges
Complement: 66

Characteristics: Project 53U (T.217 as of 1943)

Displacement: 446 tonnes normal, 490 tonnes full load
Dimensions: 62m wl x 7.62m x 2.37m normal
Machinery: two 42 BMRN-6 turbocharged diesels; 2800shp = 18.5 knots
(with sweeps 14.5 knots)
Bunkerage: normal 59 tonnes, maximum 94 tonnes diesel
Range: 1630nm at 16 knots, 2360nm at 18 knots
Sweeps: two K-1 paravanes, one MTShch, one MZT
Armament: one 100mm/56 B-24 gun
(206 rounds + 94 rounds overload)
one 45mm/46 21-K gun
(500 rounds normal load + 700 rounds overload)
two single 20mm Oerlikon
(7000 rounds normal load + 3000 rounds overload)
four single 12.7mm DShK MG
28 1926 mines
30 B-1 depth charges
Complement: 60

original Project 3 fixes within the original beam (ie without the 0.2m widening). This became known as Project 58. Construction in Sevastopol proceeded with relative alacrity with ships reaching completion in less than two years. Wartime pressures plagued the final batch of four ships, sometimes referred to as Project 58-*bis*, and aborted two of them.

Like every other fleet, the Soviet Navy found its prewar anti-aircraft standards grossly inadequate, so the *Fugas* series received various weapons upgrades. During the war, most units shipped three automatic 37mm 70-K cannon: one replacing the stern semi-automatic 45-K and the other two in various locations on the upper deck or forecastle. Baltic ships typically mounted two 20mm Oerlikons, while the Pacific fleet ships received four 12.7mm DShK or two twin Brownings. These measures were not entirely adequate, but issues with stability and speed prevented more elaborate rearmament.

The minesweeping gear was diverse but only with regard to mechanical sweeps. Electro-magnetic sweeps never entered production, although T.204 operated a solitary test outfit in early 1941. Through most of the war the ships had no ability to fight EM mines, and not until early 1944 did T.401 and T.407 receive British LL sweeps via Lend-Lease. Other surviving ships received EM gear after the war.

In January 1943 two Black Sea units, *Garpun* and *Iskatel'*, performed trials with torpedo armament to boost their value as shipping raiders. Limits on deck space and topweight prevented installation of a standard torpedo launcher, so the ships received two single side-mount launchers from MTBs. The experiment managed to fling torpedoes in almost every unintended direction, and the torpedo gear quickly found its way ashore.

T.401 *Tral* leaving Sevastopol in 1942. Note that she still has the inadequate AA armament of one 45mm 21-K gun on the stern and a pair of DK machine guns behind the superstructure. (Boris Lemachko Collection)

M-26 Mine

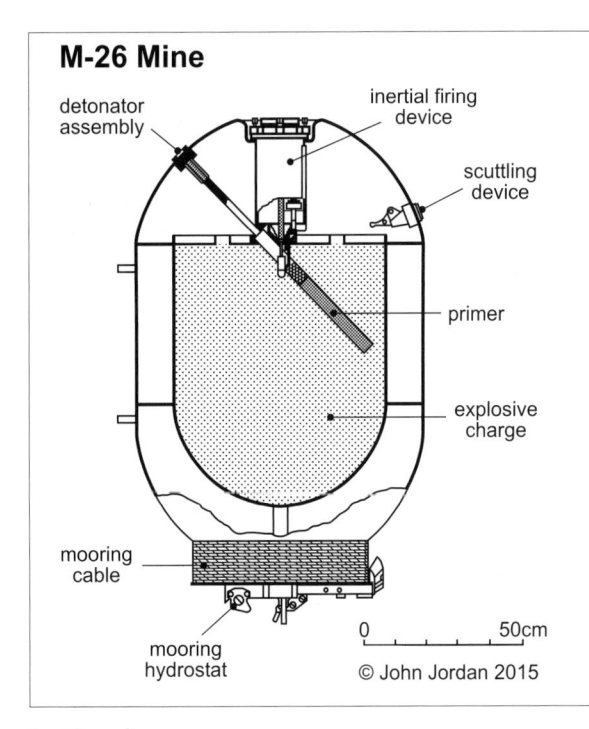

- detonator assembly
- inertial firing device
- scuttling device
- primer
- explosive charge
- mooring cable
- mooring hydrostat

0   50cm

© John Jordan 2015

The 1926 mine carried more than twice the explosive of Great War models. In order to lower its centre of gravity and reduce the chance of shipboard accidents it was stowed on its side.

Max. weight:	960 kg
Dimensions:	1840mm x 900mm x 1000mm
Explosive weight:	242-254kg
Anchor line:	130m
Max. deployment depth:	130m
Min. depth:	18m
Max. depth below the water:	6.1m
Min. depth below the water:	1.2m
Max. deployment speed:	24 knots
Max. height for deployment:	4.6m
Depth accuracy:	0.6m
Arming time:	15–25 minutes

In Service

The *Fugas* class, in all its variants, had an active war. Operating under a Soviet high command that was reluctant to risk its large ships, the minesweepers were saddled with tasks that went well beyond mine work, such as raiding enemy shipping, landing marines and conducting fire support. Such useful warships filled a key niche, being strong enough to accomplish many tasks but small enough to lose without endangering a fleet commander's head. Seventeen of the forty-two ships fell prey to combat: one to air attack, two to submarines, one to land-based artillery, one to torpedo boats, and twelve to mines – not a surprising loss rate for a class which saw such intensive service. Strong hulls enabled the ships to survive close explosions of mines, bombs and artillery shells. Overall they were successful in their extensive array of assigned tasks.

Their postwar careers were mundane, with the class participating in the cleanup of the seas. The installation of the new minesweeping gear involved additional topweight, which led to the removal of most of the weaponry, leaving just one or two 37mm anti-aircraft guns. Most units were retired from the Soviet Navy

Soviet sailor in waterproof gear climbing into the crow's nest of a minesweeper. (Boris Lemachko Collection)

The 100mm B-24 gun was not a DP weapon, elevating only to 45 degrees. Note the canvas cover on the bridge to protect the crew from bad weather. (Boris Lemachko Collection)

between 1955 and 1961. Three ships were transferred to the North Korean Navy, becoming the largest and most powerful ships in that fleet and enjoying lengthy careers. One of them lasted into this century and may still be extant, which would make it one of the oldest active warships in the world.

Project 3

T.201 *Zaryad*: Based in Tallinn, 22 Jun 41; hit a mine near Ristna Lighthouse and sank, 30 Jul 41.

T.202 *Bui*: Based in Tallinn, 22 Jun 41; damaged by three close mine explosions during minesweeping operations, 6 Jul 41; hit a mine and sank near Cape Yumindanina during evacuation of Tallinn, 15 Aug 41.

T.203 *Patron*: Based in Tallinn, 22 Jun 41; damaged by a near-miss bomb near Hanko, 30 Jul 41; further damaged by artillery fire, 1 Aug 41; participated in laying defensive minefields in the Gulf of Finland; damaged in Tallinn by artillery near misses, 25 Aug 41; survived the evacuation of Tallinn but sunk by a mine near Keri Island during Hanko evacuation, 25 Oct 41.

T.204 *Fugas*: Under repair at Libava, 22 Jun 41; with one working engine, laid six minefields totalling 206 mines, 23 Jun 41; towed damaged destroyer *Storozhevoi* to Kuresare, 27 Jun 41; laid several more minefields, Jul 41; participated in defence and evacuation of Tallinn, Aug 41; laid more minefields around Gogland, Sep 41; involved in three collisions in the space of two weeks: bow damaged in collision with another minesweeper at Kronshtadt on 28 Oct 41, side holed in collision with subchaser *MO.303* on 1 Nov 41, and forecastle damaged in collision with minesweeper T.218 during evacuation of Hanko on 9 Nov 41; hit a mine while covering MTBs leaving for a minelaying operation and sank, 24 Aug 42.

T.401 *Tral*: At Sevastopol, 22 Jun 41; participated in convoy escort and laid minefields; under repair, Sep–Oct 41; defence of Sevastopol, shelled German positions in Dec 41 and Jun 42; while escorting a convoy damaged by near-miss bombs, 27 Jun 42; entered the repair yard; renamed E*h*MTShch.401, 24 Aug 44; clearing minefields postwar; converted to target ship, 1 Sep 55.

T.402 *Minrep*: Under repair at Sevastopol, 22 Jun 41; repairs finished, Aug 1941; hit a mine near Feodosiya and broke in two, leaving only sixteen survivors, 12 Sep 41.

T.403 *Gruz*: At Sevastopol, 22 Jun 41; convoy operations; shelled advancing German troops at Perekop, 27 Sep 41; shelled German positions around Sevastopol, 26 Dec 41; while escorting a convoy, heavily damaged near Cape Kyz-Aul by a bomb hit on the stern, 16 Feb 42; made it back to base, but under repair for the rest of the year; sunk by German MTBs *S.28*, *S.51*, *S.72*, and *S.102* while delivering supplies to beachhead on Cape Myshako, 27 Feb 43 – the only ship of the *Fugas* class to be lost in a surface action.

T.404 *Shchit*: At Sevastopol, 22 Jun 41; convoy operations; laid a minefield with a total of 27 mines in the Dnepr-Bug estuary, 24 Oct 41; participated in amphibious landing on Feodosiya, Dec 41–Jan 42; while trying to supply Sevastopol, damaged by near-miss bombs, 2 Jul 42; returned to Novorossiisk; participated in amphibious landing at Yuzhnaya Ozereika, Feb 43; clearing minefields postwar; transferred to ship-raising division, 5 Mar 53; removed from the navy list and converted to a test ship, 7 Apr 56.

Project 53

T.1 *Strela*: With T.2, T.3, and T.4, left the Baltic Sea, 15 Jun 39; transferred to the Pacific Fleet via Panama

The typical anti-aircraft armament of the mid-war *Fugas*-class minesweeper comprised two or three 37mm 70-K cannon and several 12.7mm DshK machine guns. (Boris Lemachko Collection)

Canal, arriving 24 Aug 39; assigned to Vladivostok naval base, carried out routine tasks, 1941–45; clearing mines postwar; removed from the navy list, 1 Sep 55; converted to survey ship *Reduktor*.

T.2 *Tros*: With T.1, T.3, and T.4, left the Baltic Sea, 15 Jun 39; transferred to the Pacific Fleet via Panama Canal, arriving 24 Aug 39; assigned to Vladivostok naval base, carried out routine tasks, 1941–45; clearing mines postwar; transferred to the North Korean Navy, Dec 53.

T.3 *Podsekatel'*: With T.1, T.2, and T.4, left the Baltic Sea, 15 Jun 39; transferred to the Pacific Fleet via Panama Canal, arriving 24 Aug 39; assigned to Vladivostok naval base, carried out routine tasks, 1941-45; clearing mines postwar; removed from the navy list, 3 Jun 50; transferred to the North Korean Navy, 1950 (or after the Korean War, according to some sources).

T.4 *Provodnik*: With T.1, T.2, and T.3, left the Baltic Sea, 15 Jun 39; transferred to the Pacific Fleet via Panama Canal, arriving 24 Aug 39; assigned to Vladivostok naval base, carried out routine tasks, 1941–45; laid defensive minefield near Vladivostok, 9 Aug 45; wartime minesweeping operations; clearing mines postwar; removed from the navy list, 11 Oct 50; transferred to the border guards.

T.7 *Vekha*: With T.5, T.6, and T.8, left the Black Sea, 2 Apr 39; transferred to the Pacific Fleet via Suez Canal, arriving 17 May 39; assigned to Petropavlovsk-Kamchatsky, participated in amphibious landings on Kurile island of Shumshu, Aug 45; clearing mines postwar; no information on subsequent career.

T.8 *Cheka*: With T.5, T.6, and T.7, left the Black Sea, 2 Apr 39; transferred to the Pacific Fleet via Suez Canal, arriving 17 May 39; laid defensive minefield near Vladivostok, 9 Aug 45; wartime minesweeping operations; clearing mines postwar; transferred to the North Korean Navy, Dec 53.

T.405 *Vzryvatel'*: At Sevastopol, 22 Jun 41; mined the inner and outer ports at Odessa after the withdrawal of Soviet forces on 20-24 Oct 41; convoy escort duties; shelled German positions around Sevastopol, 27–28 Dec 41; while landing marines in Evpatoriya, damaged by German artillery, 5 Jan 42; thrown onto the beach by the subsequent storm.

T.406 *Iskatel'*: At Sevastopol, 22 Jun 41; laid defensive minefields, Jul 41; supported Soviet troops around Odessa, Sep 41; damaged by artillery near misses, 20 Sep 41; laid a minefield in the Gulf of Odessa, 14 Oct 41; shelled German troops in the Sevastopol area, 1 and 3 Dec 41; raided Romanian shipping but without sinking anything, ended up shelling Romanian coastal targets, 11–15 Dec 42; another unsuccessful raid, 26–29 Dec 42; laid several defensive minefields, Jul 44; clearing mines postwar; removed from the navy list and converted to a test ship, 18 Dec 54.

T.407 *Mina*: At Sevastopol, 22 Jun 41; convoy operations; shelled German positions around Sevastopol, 18, 28, and 29 Dec 41; while escorting convoy from Sevastopol, damaged by near-miss bombs, 27 Jun 42; several months of repair; raided Romanian shipping, 11–15 Dec 42; with T.412, found a Romanian convoy and shelled it for two hours, but unable to sink any ships, 13 Dec 42; another unsuccessful raid, 26–29 Dec 42; while escorting a convoy, hit by a bomb that penetrated the hull and exploded under the ship, 22 May 43; returned to Poti for extensive repairs; renamed *EhMTShch.407*, 24 Aug 44; clearing mines postwar; removed from the navy list and converted to a test ship, 18 Dec 54.

Table: Building Data

Type	Yard No.	Ship Name	Construction Dates (LD/LD/C)		
Zhdanov Shipyard No. 190 (Leningrad)					
Series I					
Project 3	459	T.201 *Zaryad*	12 Oct 33	10 Oct 34	26 Dec 36
	460	T.202 *Bui*	12 Dec 33	5 Nov 34	11 Aug 38
	461	T.203 *Patron*	28 Dec 33	30 Sep 34	4 Jul 38
	462	T.204 *Fugas*	5 Jan 34	25 Oct 34	10 Dec 36
Series II					
Project 53	491/111	T.1 *Strela*[1]	10 May 35	17 May 36	13 Aug 38
	492/112	T.2 *Tros*[1]	22 May 35	21 Jun 36	25 Sep 38
	493/113	T.3 *Podsekatel'*[1]	30 Jun 35	30 Oct 36	4 Dec 38
	494/114	T.4 *Provodnik*[1]	16 Sep 35	10 Dec 36	14 Nov 38
Petrozavod Shipyard No. 370 (Leningrad)					
Series III					
Project 53U	92	T.213 *Krambol*	26 Aug 38	30 Jan 39	30 Nov 39
	93	T.214 *Bugel'*	26 Aug 38	31 Jan 39	29 Jun 40
Ust'-Izhorsky Shipyard No. 363 (Leningrad Oblast)					
Series III					
Project 53U	14	T.205 *Gafel'*	12 Oct 37	29 Jul 38	21 Jul 39
	15	T.206 *Verp*	12 Oct 37	17 Jul 38	17 Jun 39
	16	T.207 *Shpil'*	17 Nov 37	18 Aug 38	23 Sep 39
	17/129	T.208 *Shkiv*[1]	18 Nov 37	31 Oct 38	12 Oct 39
	18	T.209 *Knekht*	16 Jun 38	17 Aug 39	3 Jun 40
	19	T.210 *Gak*	8 Aug 38	15 Apr 39	14 Nov 39
	20	T.211 *Rym*	21 Sep 38	5 May 39	25 Jun 40
	21	T.212 *Shtag*	6 Nov 38	2 Dec 39	26 Jul 40
	22	T.218	20 Mar 39	24 Nov 39	23 Dec 40
	24	T.215	23 Apr 39	17 Dec 39	30 Sep 40
	27	T.216	17 Sep 39	30 Apr 40	24 Dec 40
	28	T.217 *Kontradmiral Yurkovskii*	21 Sep 39	31 Jul 40	5 Aug 41
	34/34	T.219 *Kontradmiral Khoroshkhin*[2]	27 Apr 41	24 Sep 43	25 Sep 44
	35/36	T.220[2]	10 Apr 41	16 Nov 43	16 Oct 46
	36	T.221 *Dmitrii Lysov*	27 Jun 41	6 Jun 46	5 Nov 46
Ordzhonikidze Shipyard No. 201 (Sevastopol)					
Series I					
Project 3	67	T.401 *Tral*	5 Nov 33	23 Aug 34	23 Dec 36
	68	T.402 *Minrep*	5 Nov 33	14 Jan 35	28 Jan 37
	69	T.403 *Gruz*	20 Mar 34	21 Sep 35	25 Jul 37
	70	T.404 *Shchit*	17 Jan 34	17 Dec 35	19 Oct 37
Series II					
Project 53	175	T.405 *Vzryvatel'*	21 Jul 36	27 Apr 37	27 Apr 38
	176	T.406 *Iskatel'*	19 Sep 36	29 Aug 37	29 Apr 38
	177	T.411 *Zashchitnik*	10 Aug 36	31 Jul 37	31 Jul 38
	178	T.407 *Mina*	22 Dec 36	20 Aug 37	19 Aug 38
	179	T.7 *Vekha*	30 Dec 36	14 Mar 38	8 Sep 38
	180	T.8 *Cheka*	27 Dec 36	14 Mar 38	2 Nov 38
Series IV					
Project 58	187	T.5 *Paravan*	15 Mar 37	28 Jan 38	30 Dec 38
	188	T.6 *Kapsyul'*	21 Mar 37	28 Jan 38	3 Jan 39
	189	T.408 *Yakor'*	28 Mar 37	14 Jan 38	15 Feb 39
	190	T.409 *Garpun*	27 Apr 37	22 Mar 38	20 Feb 39
	191	T.410 *Vzryv*	29 Apr 37	6 Nov 38	9 Mar 39
	255	T.412	13 Apr 39	5 Nov 39	3 Mar 41
	256	T.413	29 Apr 39	12 Oct 40	21 Apr 41
	267	T.414	3 Jan 41	29 Apr 41	?
	268	T.415	20 Mar 41	5 Nov 47	?

[1] completed at Shipyard No. 196   [2] completed at Shipyard No. 370

T.407 *Mina* mooring in Sevastopol in 1946. (Boris Lemachko Collection)

T.411 *Zashchitnik:* At Sevastopol, 22 Jun 41; convoy escort duties; shelled German positions around Sevastopol, 10 May 42; damaged by German artillery during aborted amphibious landing near Stanichka, 7 Feb 43; sunk by German submarine *U.24* near Sukhumi, 15 Jun 43.

Project 53U

T.205 *Gafel':* At Tallinn, 22 Jun 41; laid a defensive minefield at Seskorsk roads, 16 Jul 41; damaged by Finnish artillery while at anchor near Hanko, 30 Jul 41; evacuation of Tallinn, 28–29 Aug 41; laid defensive minefields around Kronshtadt; evacuation of Hanko; during an amphibious landing on the island of Sommers, damaged by shells from a Finnish gunboat, two hits disabling the steering gear and flooding its compartment, 8–9 Jul 42; thrown onto the rocks near the island of Lavensaari during a severe storm, ultimately led to flooding of the machinery room and the steering compartment, 5 Aug 42; nine-day effort recovered the ship; towed to Kronshtadt for repairs; clearing mines postwar; removed from the navy list, 1 Sep 55; converted to non-self-propelled barge and renamed *BAMT.31250*, 17 Oct 55.

T.206 *Verp:* At Tallinn, 22 Jun 41; laid down a defensive minefield at the entrance to the Gulf of Finland, 6 Jul 41; laid a minefield between the islands of Bol'shoi and Malyi Tyuters, 16 Jul 41; evacuation of Tallinn, 28–29 Aug 41; slight damage from near-miss shells at Kronshtadt, 22 Sep 41; during the evacuation of Hanko, hit a mine that blew off its bow, causing the ship to sink, leaving only 21 survivors, 14 Nov 41.

T.207 *Shpil':* At Tallinn, 22 Jun 41; evacuation of Tallinn, 28–29 Aug 41; laid defensive minefields around Kronshtadt; routine wartime tasks (minesweeping, escorting submarines to and from base, minelaying); clearing mines postwar; removed from the navy list, 10 Sep 55; converted to a floating workshop and renamed *PM.93*, 17 Oct 55.

T.208 *Shkiv:* At Tallinn, 22 Jun 41; while escorting damaged cruiser *Maxim Gorky*, hit a mine near Nordvein shoal and sank, 24 Jun 41.

T.209 *Knekht:* At Tallinn, 22 Jun 41; laid a minefield near Ehre Island, 2 Aug 41; while steaming from Tallinn to Kronshtadt, hit a mine and sank near Keri Island, 24 Aug 41.

T.210 *Gak:* At Tallinn, 22 Jun 41; minelaying missions, Jul 41; evacuation of Tallinn, 28-29 Aug 41; laid defensive minefields around Kronshtadt, Sep 41; during the evacuation of Hanko, explosion of a mine in the minesweeping gear damaged the ship, 3 Nov 41; hit by an artillery shell in Kronshtadt, 13 Nov 41; routine wartime tasks (minesweeping, escorting submarines to and from base, minelaying); while escorting submarines, hit a mine near the island of Lavensaari, several compartments flooded but the ship remained afloat, 9 May 43; repairs took 46 days; hit a mine near Vyborg and sank in shallow water, 2 Jul 44; raised and repaired; sold for scrap, 13 Jan 61.

T.211 *Rym:* At Tallinn, 22 Jun 41; evacuation of Tallinn, 28-29 Aug; laid defensive minefields around Kronshtadt, Sep 41; collided with minesweeper *T.68*, suffering hole in the bow, 21 Sep 41; grounded near Tolbukhin lighthouse, damaging right propeller shaft and right engine, 28 Oct 41; during the evacuation of Hanko, explosion of a mine in the minesweeping gear damaged the stern, flooding the steering compartment, 10 Nov 41; routine wartime tasks (minesweeping, escorting submarines to and from base, minelaying); reclassified as a guardship, 19 Feb 59; sold for scrap, 12 Feb 60.

T.212 *Shtag:* At Tallinn, 22 Jun 41; several minelaying missions, Jul 41; while escorting submarine *Shch.322*, hit a mine and slowly sank near Soehlavejnu, 3 Aug 41.

T.213 *Krambol:* At Tallinn, 22 Jun 41; while escorting convoy from Tallinn to Kronshtadt, hit a mine near Cape Yumindanina and sank, 11 Aug 41.

T.214 *Bugel:* At Tallinn, 22 Jun 41; several minelaying missions, Jul 41; while heading from Tallinn to Kronshtadt, hit a mine near Keri Island and sank, 24 Aug 41.

T.215: At Tallinn, 22 Jun 41; evacuation of Tallinn, 28-29 Aug; evacuation of Hanko, Oct–Nov 41; routine wartime tasks (minesweeping, escorting submarines to and from base, minelaying); while escorting a submarine, damaged by several near-miss bombs, 10 Jun 43; removed from the navy list, 7 Jun 61; converted into floating barracks *PKZ.33*; sold for scrap, 9 Feb 62.

T.216: At Leningrad assigned to the naval academy, 22 Jun 41; while escorting submarines, hit a mine and sank near the Takhuna lighthouse, 6 Jul 41.

T.217 *Kontr-Admiral Yurkovskii:* Evacuation of Tallinn, 28–29 Aug; evacuation of Hanko, Oct–Nov 41; collided with tug in Kronshtadt, receiving hull damage, 21 Sep 41; routine wartime tasks (minesweeping, escorting submarines to and from base, minelaying); hit by artillery shell that damaged the port propeller shaft and knocked out electrical power, 6 Jun 43; sold for scrap, 6 Jul 61.

T.218: At Tallinn, 22 Jun 41; evacuation of Tallinn, 28–29 Aug 41; evacuation of Hanko, Oct–Nov 41; collided with T.204, bow damaged, 9 Nov 41; while attending the sinking passenger ship *Iosif Stalin*, hit a mine that damaged the stern, 4 Dec 41; routine wartime tasks (minesweeping, escorting submarines to and from base, minelaying); hit by bomb that penetrated the hull and exploded underneath, disabling engines and causing a fire, 10 Jun 43; towed back to base, two months of repair; in reserve, 30 Dec 55; reclassified as a test ship, 3 Apr 56; transferred to the Northern Fleet, 17 Nov 56; renamed *OS.9*, 27 Dec 57; removed from the navy list, 7 Mar 58.

T.219 *Kontr-Admiral Khoroshkhin*: Commissioned 1944; clearing mines postwar; placed in reserve and reclassified as a test ship, 30 Dec 55; transferred to the Northern Fleet, 17 Nov 56; renamed *OS.10*, 27 Dec 57; removed from the navy list, 7 Mar 58.

T.220: Commissioned postwar; removed from navy list, 13 Jan 61.

T.221 *Dmitrii Lysov*: Commissioned postwar; renamed *Dmitry Lysov*, 16 Nov 49; sold for scrap, 6 Jul 61.

Project 58

T.5 *Paravan*: With T.6, T.7, and T.8, left the Black Sea, 2 Apr 39; transferred to the Pacific Fleet via Suez Canal, arriving 17 May 39; clearing mines postwar; removed from the navy list, 11 Oct 50; transferred to the border guards.

T.6 *Kapsyul'*: With T.5, T.7, and T.8, left the Black Sea, 2 Apr 39; transferred to the Pacific Fleet via Suez Canal, arriving 17 May 39; no wartime combat, no information on postwar career.

T.408 *Yakor'*: At Sevastopol, 22 Jun 41; laid defensive minefields, Jul 41; mined Odessa harbor after the withdrawal of the Soviet troops, Oct 41; collided with transport ship *Kotovksy* exiting Gelenzhik harbor, 20 Nov 41; under repair until May 42; shelled German troops in the Sevastopol area, 10 May 42; damaged by near-miss bombs, 19 May 42; under repair until Dec 42; raided Romanian shipping but without sinking anything, ended up shelling Romanian coastal targets, 11–15 Dec 42; another unsuccessful raid, 26–29 Dec 42; clearing mines postwar; transferred to mine and torpedo research center, 17 Oct 53; removed from navy list, 7 Apr 56; converted to a test ship, renamed *Leshch*.

T.409 *Garpun*: At Sevastopol, 22 Jun 41; patrolled the area around Sevastopol and escorted ships; storm in Poti, thrown against the breakwater, bow damaged, 22 Jan 42; under repair until Mar 42; damaged by near-miss bombs, 18 Jun 42; damaged by near-miss bombs, 22 May 43; removed from active duty and converted into a training ship, 7 Apr 56.

T.410 *Vzryv*: At Sevastopol, 22 Jun 41; damaged by near-miss bombs, 30 Sep 41; shelled German positions around Sevastopol, 3 Dec 41 and 11 Jun 42; while steaming from Novorossiisk to Anapa, suffered direct hit by a bomb on forecastle, causing a serious fire and fears of a magazine explosion, deliberately run aground, but tug *Simeiz* arrived to help put out a fire, 14 Aug 42; refloated and towed to Novorossiisk, 15 Aug 42; under repair until Aug 43; struck a pier in Batumi and damaged, 10 Oct 43; a week before combat ended on the Black Sea, sunk by *U.19* near Constanza with 74 casualties, 2 Sep 44.

T.412: At Sevastopol, 22 Jun 41; laid defensive minefields, Jul 42; damaged by near-miss bombs during a resupply mission to Sevastopol, had to return to Novorossiisk, 2 Jul 42; raided Romanian shipping,

Tral-class minesweepers at Sevastopol during 1946–47. The final AA outfit of the ships can be seen in this photo: three 37mm 70-K and two 12.7mm DShK machine guns. (Boris Lemachko Collection)

Black Sea minesweepers post-war. From left to right: T.407 *Mina*, T.401 *Tral*, T.408 *Yakor'*, T.412 *Arseniy Rasskin*, and T.409 *Garpun*. (Boris Lemachko Collection)

11–15 Dec 42; with T.407, found a Romanian convoy and shelled it for two hours, but unable to sink any ships, 13 Dec 42; another unsuccessful raid, 26–29 Dec 42; named *Arsenii Rasskin*, 6 Jan 43; damaged by near-miss bombs while steaming between Batumi and Tuapse, 11 Nov 43; removed from active duty and converted into a training ship, 7 Apr 56.

T.413: At Sevastopol, 22 Jun 41; collided with tug *Simeiz* in Odessa and required repairs, 14 Oct 41; shelled German positions around Sevastopol, 1 Dec 41; hit by three bombs and capsized near Cape Fiolent, 13 Jun 42.

T.414: Under construction in Sevastopol at the beginning of the war; incomplete hull towed to Poti when the siege of Sevastopol began; postwar completion plans never carried out, hull scrapped.

T.415: Under construction in Sevastopol at the beginning of the war, abandoned on slipway when Sevastopol fell; hull damaged during the war, launched to clear the slip, scrapped.

Sources:

Bray, Jeffrey K, *Mine Warfare in the Russo-Soviet Navy*, Aegean Park Press, 1995.

Kostrichenko, V V and K L Kulagin, *Bystrokhodnye Tral'shchiki Tipa Fugas*, Modelist-Konstruktor, 2005.

Platonov, A V, *Ehntsiklopediya Sovetskih Nadvodnyh Korabley 1941–1945*, Poligon, 2002.

Platonov, A V, S V Aprelev, and D N Sinyaev, *Sovetskie Boevye Korabli 1941–1945: IV Vooruzhenie*, Tsitadel', 1997.

Rohwer, Jürgen and Mikhail Monakov, *Stalin's Ocean-Going Fleet: Soviet Naval Strategy and Shipbuilding Programmes 1935–1953*, Frank Cass Publishers, 2001.

Shirokorad, A V, *Entsiklopedia Otechestvennoj Artillerii*, Minsk, 2000.

Spasskiy, I D, *Istoriya Otechestvennogo Sudostreniya: Tom 3*, Sudostroenie, 1996.

Westwood, J N, *Russian Naval Construction 1905–1945*, Macmillan Press Ltd, 1994.

Three *Fugas*-class ships were transferred to North Korea, and at least one remained in service into the 2000s, although it is not clear which one. This 1993 photograph shows an 85mm tank gun mounted forward. (US Department of Defense)

DIVIDE AND CONQUER?
Divisional Tactics and the Battle of Jutland

Jellicoe's failure to achieve a decisive victory over the German High Seas Fleet at Jutland has often been attributed to his tactical system, described as centralised and rigid. The alternative usually suggested is 'divisional tactics', in which a fleet would be divided into independently-manoeuvring formations. But aside from vague generalities, little has been written about just how these tactics might have worked at Jutland. In this article **Stephen McLaughlin** seeks to rectify this gap in the historical literature.

In the decade or so before the First World War, there was an extended debate over tactics within the Royal Navy between 'single line' adherents and proponents of 'divisional tactics' or 'divided attack'. To understand the issues at stake, a summary of the opposing viewpoints is in order.

The Single Line and Fast Wings

By the 1880s the close-range ramming mêlée of the ironclad era was passing from the tactical scene as the gun once more became the primary weapon. In the late 1890s this trend led to a revival of the single line of battle – the line-ahead formation used during the sailing era. The advantages of the line were obvious: it left a battleship's guns free to fire over a wide arc on the broadside, and it could be controlled in battle by the follow-my-leader principle. By 1903 this development prompted the Admiralty to write to the Commander-in-Chief of the Mediterranean Fleet, Admiral Sir Compton Edward Domvile, that:

> Their Lordships... believe that the opinion of Flag Officers afloat – especially those who have the advantage of taking part in the 'Battle Exercises' of the past two years – trends on the whole towards the adoption of a single line or something approximating to a single line.[1]

Captain H J May, president of the Royal Naval College, was a strict adherent to the single-line idea, writing in 1902 that:

> All our encounters [during wargames] prove that the best

The 12in guns of the early British battlecruisers made them well-suited to lie in the line of battle, while their high speed enabled them to serve as a fast wing of the battle fleet. This is *Indomitable* in July 1908, less than a month after commissioning; she was taking HRH the Prince of Wales to Quebec for the tercentenary celebrations, and the Royal Standard can be seen flying from the mainmast. (Kingsway Real Photo, John Jordan collection)

place for a detached line is in prolongation of the main line, or in other words it is better not to detach it... From the tactical point of view pure and simple, *all* the battleships are better placed in one line than in any other way.[2]

Most officers, however, were not quite so doctrinaire. After the Mediterranean and Channel fleet combined manoeuvres of 1901, the umpire's report noted that:

> The general opinion, as regards numbers, is that 8 is the limit for placing in single line; if 12 or 14 Battleships compose the Fleet, the balance should be worked in a separate squadron and more especially for making flank attacks.[3]

At various times, other numbers were mentioned as the 'limit' for the single line: in 1903 Admiral Hedworth Lambton said ten ships, in 1909 Admiral Edward Bradford put it at twenty, while in 1915 Jellicoe considered 24 battleships the maximum.[4] Whatever the number in the line, the idea of employing 'wing' divisions for attacking the enemy's van and rear soon gained widespread acceptance among the Royal Navy's tacticians.[5] At the same time, it was recognised that these wing divisions would need to have a speed advantage over both their own main body and the enemy to reach a favourable position from which to threaten the enemy's flanks. As long as slower predreadnoughts made up a substantial part of the battle fleet, it was standard practice to place the dreadnoughts on the wings of the fleet, where they could make use of their superior speed, while the predreadnoughts formed the centre of the formation.

The idea of using cruisers as fast divisions was also discussed; as early as 1901 it was noted that:

> The opinion on the use of armoured Cruisers is practically unanimous, that with their speed and protection they should be utilised for attacking the van and rear of the enemy from the very commencement of the engagement.[6]

The idea of cruisers as a 'fast wing' was probably a factor in the decision to arm the new type of armoured cruiser (later rechristened 'battle cruiser') with 12in guns; during the design process it was argued that guns of this calibre would make them '*all the more qualified to lie in the line of battle if required to do so*'.[7]

But some doubted that the battlecruiser was suited to this role, among them Admiral Sir Francis Bridgeman, commander of the Home Fleet in 1907–09; after tactical exercises with the first battlecruisers, he expressed the view these ships 'must never be considered as dreadnoughts... [T]hey cannot work with battleships... [T]hey must be deployed as cruisers'.[8] Bridgeman's views were presumably based on the weak armour of the battlecruisers; they certainly did not arise from any antipathy to their speed, for two years later, when he was First Sea Lord, he pushed for the construction of fast *battleships*. These eventually materialised as the *Queen Elizabeth* class – ships specifically designed to act as a fast wing for the battle fleet.[9]

Whatever Bridgemen's views on the battlecruiser, however, his successor as C-in-C Home Fleet, Admiral Sir George Callaghan, was willing to use them in the role of a fast division:

> If the enemy has no battle-cruisers with his fleet... our battle-cruisers... may be employed as a fast division of the battle-fleet, or comparative freedom of action may be given to the Admiral commanding to attack the enemy in the manner... he may judge best.[10]

Rear-Admiral Sir David Beatty, who took command of the Home Fleet's Battle Cruiser Squadron in March 1913, agreed, writing in his orders to the squadron that:

> It may be expected that the Commander-in-Chief would use the fast division for attacking the enemy's flank and the [Battlecruiser] squadron must therefore be prepared to attack either van or rear. To attain quickly a good tactical position full speed will be needed, not only during the approach but perhaps also while engaged.[11]

But it was also recognised that speed would not be enough to keep a fast division out of trouble, as was shown by several cases during exercises:

> Battle cruisers acting as a fast division are especially vulnerable and if they lose their speed, they are immediately in a position of great inferiority to a battleship. They may then be cut off and destroyed, or left out of the action, as the *Princess Royal* and the *Indefatigable* were on 10th October 1913, and the three *Collingwood* class on 9th October 1913.[12]

In another instance, the battlecruisers were forced to deploy 'too near to the enemy Battle Fleet' during an exercise in July 1913 because a passing tramp steamer got mixed up in the action, with the result that Beatty's force 'came under heavy fire unsupported by our own Fleet which was not yet within gun range'.[13]

Whether composed of battlecruisers or fast battleships, however, by 1914 the concept of fast wing divisions was widely accepted in the Royal Navy, and provides a link to the ideas of the divisional tactics.

Divisional Tactics Before the War

Whether furnished with fast wings or not, there was a vocal minority of officers who believed that the single line of battle was a fundamentally misguided concept. These men were all members of the self-proclaimed 'historical school', who looked to the past, and particularly to the great wars of the sailing era for tactical lessons. As one of the proponents of divisional tactics, Admiral Sir Frederick Charles Doveton Sturdee, wrote: 'no battle in our history had ever been won' by the single line.[14] The line of battle in their eyes was a rigid, essentially defensive formation. To overcome its failings they looked to Nelson's victory at Trafalgar; instead of manoeuvring to form line of battle, he had charged headlong at the Franco-Spanish fleet in two separate divisions, each guided by an overall plan that emphasised individual initiative in the smoke and confusion of battle.

In addition to Sturdee, other critics of the single line included Admiral Sir Reginald Custance, and captains Herbert Richmond and William Reginald Hall (who later achieved fame as the head of the Naval Intelligence Division during the First World War). Mention must also be made of the Dewar brothers, Alfred and Kenneth, two young officers who would play a prominent part in the Jutland controversy after the war; both were already active participants in tactical debates before 1914, especially Kenneth.

But while all these men found their inspiration in Trafalgar, they disagreed on many details, so they never really coalesced into a unified 'school'. Another handicap they faced was the fact that just about all of them were regarded as 'difficult' personalities: Custance's 'inflexibility' and 'inability to recognise the full implications of technological progress' told against him; Sturdee was conceited and 'rubbed many people the wrong way'; and even Richmond's sympathetic biographer is forced to admit his subject's 'intolerance and intellectual arrogance'.[15] Thus, although there was a good deal of merit in many of their views, the leading proponents of divisional tactics were often their own worst enemies when it came to trying to get their points across.

Yet despite their personal shortcomings, they did manage to get their ideas tried at sea. Experiments with divisional tactics were carried out in the Channel Fleet under Admiral Lord Charles Beresford in 1907–08 – Custance was his second-in-command – but unfortunately almost no information about these exercises has come to light.[16] Far more is known about the tactical work carried out in the Home Fleet under the command of Admiral Sir William May (March 1909 – March 1911). Kenneth Dewar later described how Richmond, May's flag captain, after participating in many of these exercises,

> …wrote a paper for the Commander-in-Chief, suggesting various exercises, in which one and sometimes both fleets were to fight in widely separated divisions, instead of the stereotyped single line. May at first instinctively opposed any departure from the conventional idea and it is all the more to his credit that he accepted the suggestion and carried out a large number of appropriate exercises during the last year of his command.

However, these exercises soon revealed a fundamental problem:

> Instead of issuing brief general instructions beforehand and encouraging… subordinate Commanders to use their initiative and intelligence in forwarding the general plan of attack, Admirals tried to direct squadrons and flotillas by signal. The results were unsatisfactory and would probably have been chaotic in actual battle. The fault lay not so much in the actual tactics as in the system of command.
>
> *A dramatic improvement occurred during the last two months of May's command.* It was suggested, unofficially, to some of the junior admirals, who occasionally acted as Cs-in-C of the opposing fleets, that less centralised methods might give better results. The consequent improvement was almost miraculous. In an exercise carried out on 24th January, 1911, two divisions attacked their opponents [*sic*] van and three their rear. So far as could be judged from the exercises, the fleet in single line would have been paralysed and destroyed. Also, when the Home and Mediterranean Fleets met off the Spanish coast in February, 1911, Rear-

Analysis of Tactical Exercises 1909-11

Exercise	Total Gunfire, Divisional Squadron	Total Gunfire, Single Line Squadron	Difference	Advantage	Notes
I	1,600	944	656	Divisional	Gunfire assessed within 7,500 yds
II	4,410	4,890	-480	Single Line	
III	3,232	4,768	-1,536	Single Line	
IV	2,958	2,786	172	Divisional	Visibility only 6,000 yds
V	3,660	4,052	-392	Single Line	
VI	4,816	6,030	-1,214	Single Line	
VII	4,684	6,060	-1,376	Single Line	
VIII	10,154	10,768	-614	Single Line	
IX	N/A	N/A	N/A	N/A	No gunfire assessed
X	N/A	N/A	N/A	N/A	Both sides used divisional tactics
XI	4,578	4,586	-8	Single Line	
XII	2,636	3,146	-510	Single Line	
XIII	4,392	3,946	446	Divisional	
XIV	5,574	7,250	-1,676	Single Line	
XV	N/A	N/A	N/A	N/A	No gunfire assessed

Except in Exercise I gunfire was considered effective within 10,000 yds.
Except in Exercise IV visibility was in excess of 10,000 yds.
Exercises IX and XV took place at the end of strategic manoeuvres, and different rules were in force; hence no gunfire was assessed.

Admiral Sir George Warrender, in command of Red Fleet, assisted by Captain Reginald Hall, prepared brief instructions for divisional attacks on Blue Fleet. The plan was to isolate Blue's centre and concentrate against its van and rear from ahead and astern. The plan was a striking success for decentralisation of command.[17]

But it is noteworthy that Dewar does not actually say that *divisional tactics* were successful in these exercises – the phrases 'So far as could be judged from the exercises...' and 'The plan was a striking success for decentralisation of command' (but perhaps not for the tactics themselves) leave some room for doubt as to the actual outcome. In fact, we can analyse the results, for Admiral May compiled, and the Admiralty issued *Notes on Tactical Exercises, 1909–1911*, a massive volume detailing all the tactical work done during his period in command.[18] Divisional tactics were explored in fifteen exercises; of these, twelve involved actions between a fleet in single line and one attacking using independent divisions (see Figure 1 for an example). In each exercise, the gunfire the two sides would have brought to bear was tabulated, and in nine out of twelve cases the single line proved superior (see Table).[19] Nevertheless, May was converted to the cause of divisional tactics, writing that:

> Dividing the fleet at once gives freedom to subordinates and in so doing strikes at the root of the purely defensive formation of the single line, and leads to an offensive method of engaging.[20]

May also noted one of the lessons of these exercises:

> The speed of the Division is a most important factor, and unless the divisional attack has the superiority of speed, the difficulties of placing the Divisions in their correct position are considerable.[21]

Divisional tactics often took the form of fast wings attacking the van and rear of the enemy fleet; both Hall and Richmond recommended this technique.[22] There was in fact considerable overlap between the two tactical ideas, but they were not synonymous. For divisional proponents, the important element was always the independence of the detached force. A fast division that remained tied to the movements of the main body was merely a 'prolongation of the battle line'.[23]

According to Richmond, after May relinquished command of the Home Fleet in March 1911, 'the whole study [of divisional tactics] was dropped', and several historians have repeated this view.[24] But this is contradicted by Frederic Dreyer, who served in the Home Fleet throughout this period, first commanding the cruiser *Amphion* and then the super-dreadnought *Orion*; in his memoirs he wrote that divisional tactics were tried in exercises later in 1911 'without success', and that the following year 'Divided tactics were again put to the test in inter-Fleet manoeuvres, and again it was found that control of the situation was then easily lost'.[25]

The latter manoeuvres would have been conducted by Admiral Callaghan, who commanded the Home Fleet from December 1911 to August 1914. Richmond claimed that Callaghan 'did very little tactics in his time', but this seems grossly unfair.[26] Judging by the surviving documents, Callaghan devoted a great deal of thought to tactics, including divisional tactics. To take but one example, in a proposal for revising the 'Conduct of a Fleet in Action' (a very basic set of instructions included in the General Signal Book), submitted to the Admiralty in March 1913, Callaghan wrote:

> Commanders of divisions and squadrons detached from the main body of the fleet, and ordered to attack the enemy independently, should be careful not to expose their commands to the fire of a superior force by closing him prematurely, and should endeavour to bring their ships into action at a time and in a manner which will enable all portions of the fleet to give the most effective support to one another.[27]

This was very much in keeping with the basic tenets of the divisional school: detached units would be granted considerable independence, but must use that freedom to cooperate with the fleet as a whole. It is interesting to note that while many of Callaghan's other proposed revisions were rejected by the Admiralty, this paragraph was incorporated into the final version with only minor changes to the wording, which indicates that Their Lordships had no fundamental objection to the general trend of Callaghan's thinking, even though he might employ divisional tactics.[28] Unfortunately, we have very little information on actual exercises involving divisional tactics that Callaghan may have carried out. We do know that he commended Vice-Admiral Sir John Jellicoe when the latter carried out an independent attack on the 'enemy's' rear during the spring manoeuvres of 1912.[29] More intriguing is the fact that in May 1914 Dreyer – who painted such a negative view of divisional tactics in his memoirs – submitted a proposal for an instrument that could be used to calculate the best courses for use in divisional attacks, writing:

> Generally speaking it may be stated that when two unequal fleets engage, the Admiral of the stronger fleet while engaging his opponent with a main body of about equal strength, may endeavour to utilize his surplus ships, if of sufficient speed, in two separate squadrons. One to press on the van of the enemy's line; the other to prevent the enemy 'swinging his tail', i.e. to prevent him from altering course in succession away from the squadron attacking his van.[30]

Figure 2 shows one of the divisional formations Dreyer used as an example of such tactics. This strongly suggests that divisional attack had not been rejected prior to the war; certainly Callaghan kept an open mind on the subject, and it seems likely that exercises were carried out with them during his command of the Home Fleet.

DIVIDE AND CONQUER? DIVISIONAL TACTICS AND THE BATTLE OF JUTLAND

Fig 1: Prewar Exercise in Divisional Tactics

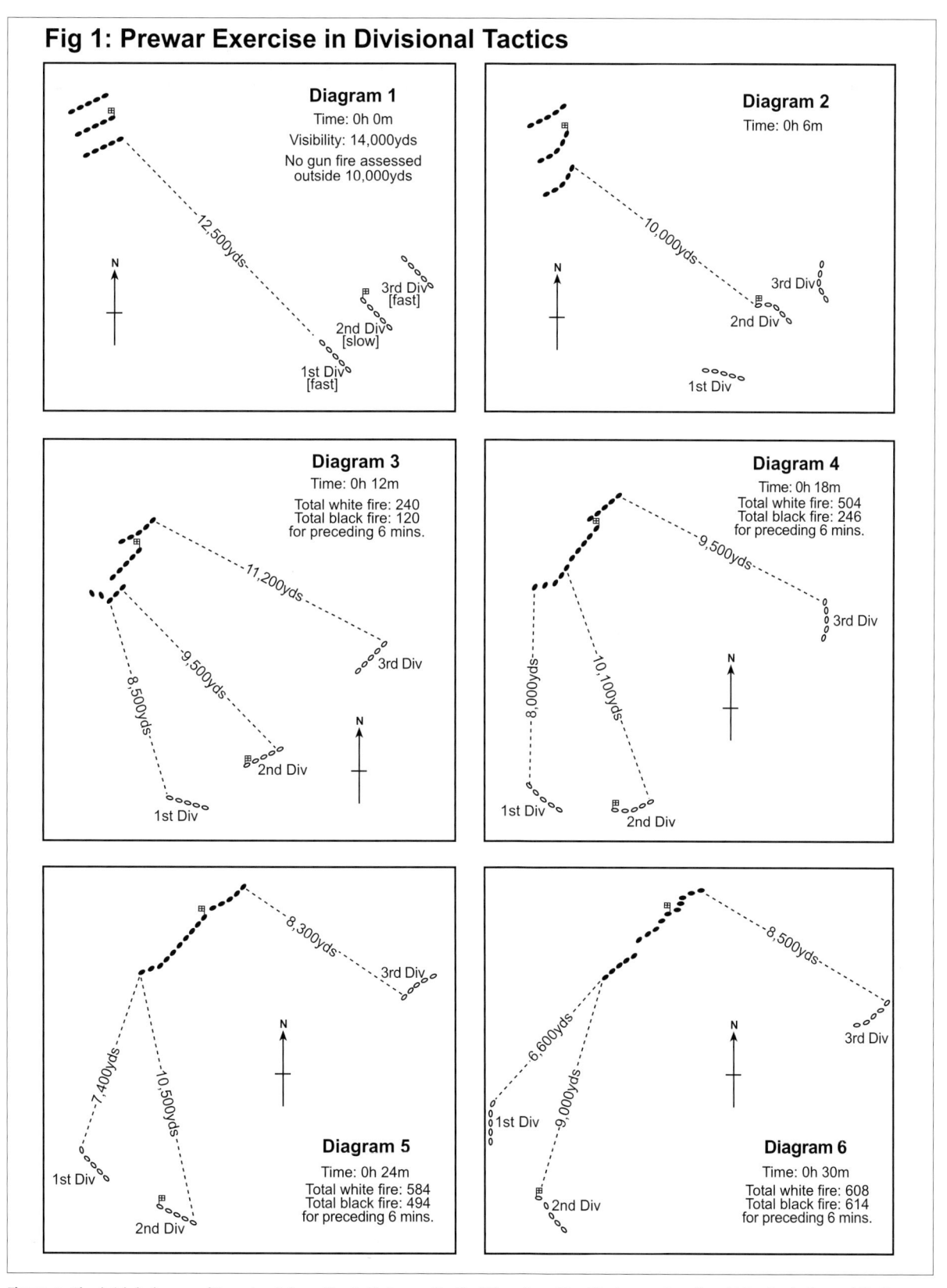

Diagram 1
Time: 0h 0m
Visibility: 14,000yds
No gun fire assessed outside 10,000yds

Diagram 2
Time: 0h 6m

Diagram 3
Time: 0h 12m
Total white fire: 240
Total black fire: 120
for preceding 6 mins.

Diagram 4
Time: 0h 18m
Total white fire: 504
Total black fire: 246
for preceding 6 mins.

Diagram 5
Time: 0h 24m
Total white fire: 584
Total black fire: 494
for preceding 6 mins.

Diagram 6
Time: 0h 30m
Total white fire: 608
Total black fire: 614
for preceding 6 mins.

Figure 1: The initial phases of Exercise II from May's *Notes on Tactical Exercises*. The Black squadron fought in single line, while the White squadron divided during deployment, two divisions moving to attack Black's rear while the third division steered to attack the enemy's van. In this exercise White had an initial advantage because Black's divisions overlapped, masking the fire of several ships, but ultimately Black was able to bring more guns to bear for a longer period of time. (© John Jordan 2015)

Fig 2: Dreyer's Diagram of a Divisional Attack by the Wing Divisions

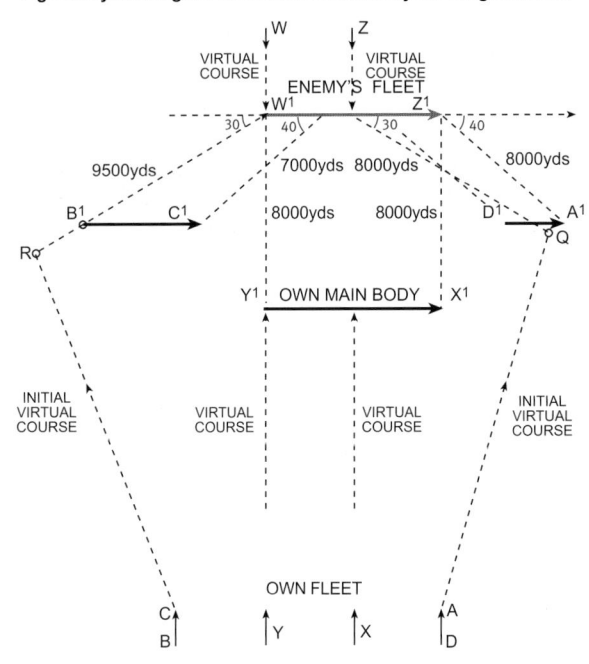

Figure 2: Dreyer's diagram showing a divisional attack. As the two fleets approach, 'Own Fleet' detaches the fast wing divisions (AD and CB) before the enemy has deployed; they steer for the virtual points Q and R, respectively, with the intention of reaching points 60° off the bow of each enemy column at a range of 11,000 yards. But when the enemy deploys to port, both detached divisions have to adjust course slightly to gain their position. The division C¹B¹, now in the rear, opens the distance between ships to 5 cables (1,000 yards) in order better to avoid any torpedoes fired by the enemy's battleships; the divisions of A¹D¹ and the main body X¹Y¹ maintain the standard interval between ships of 2½ cables (500 yards).
(© John Jordan 2015)

Divisional Tactics and Fast Wings in the Grand Fleet, 1914–16

The evidence for divisional tactics in the period between Jellicoe's replacement of Admiral Callaghan as C-in-C on 4 August 1914 and the Battle of Jutland on 31 May 1916 is minimal. John Irving, who was a midshipman in HMS *Ajax* at Jutland, wrote many years later:

> The possibility of divided-fleet tactics had been accepted in the Royal Navy since 1911: they had been practised by Jellicoe with the Grand Fleet in his 'PZ' exercises ... earlier in the war...[31]

This suggests that the Grand Fleet conducted at least a few exercises employing divisional tactics or detached wings. But while Jellicoe's initial Grand Fleet Battle Orders (GFBOs) did outline specific circumstances in which a group of ships might be detached from the line of battle, this was to be done only to counter an enemy move, such as an attempt to circle the rear of the battle line.[32] In other words, the purpose of detaching ships was entirely defensive, and no real tactical independence was granted to these ships.

The September 1914 revision of the GFBOs made some provision for detaching the wing divisions before deployment, but the circumstances under which this would be done are not mentioned.[33] The March 1915 revision outlined in more concrete terms the use of the battle-cruisers as fast divisions:

> If the enemy has (a) no battle-cruisers with his battlefleet, or (b) is much weaker in battle-cruisers, the function of our battle-cruisers may be an equally definite one: in the first case they would be best employed as fast divisions of the battlefleet, two squadrons in the van, one in the rear; in the second case a sufficient force should be retained by the Vice-Admiral commanding the Battle-cruiser Fleet to deal effectively *and with certainty* with the enemy battle-cruisers...[34]

The use of any battlecruisers left over after the 'sufficient force' had been detailed to deal with their enemy counterparts is not spelled out, but presumably they would have acted as a fast wing, albeit of much reduced force. With the arrival of the first fast battleships of the *Queen Elizabeth* class in the spring of 1915 – *Warspite* joined the Grand Fleet on 13 April and *Queen Elizabeth* herself arrived at Scapa on 26 May 1915 – Jellicoe had the nucleus of a true fast wing for the battle fleet, entirely separate from the battlecruisers, and this was soon reflected in a revision of the GFBOs issued in June 1915:

> ...'*Queen Elizabeth*' and '*Warspite*' will move to the van or rear of the main line of battle and will act there independently as a fast division, though conforming generally to its movements; unless there is special reason to the contrary they will be at the van...

> '*Queen Elizabeth*' and '*Warspite*' should, initially, concentrate on the leading ship of the enemy's line of battle, shifting their fire to the next ship if the leading ship hauls out of line, but this order does not tie the discretion of the captains if they consider it necessary or advantageous to direct their fire elsewhere.[35]

There are suggestions here of true divisional tactics, even though the fast wing was to 'conform generally' to the movements of the main body of the battle fleet. *Barham*, the third ship of the class, arrived in October 1915, and soon afterwards the 5th Battle Squadron was officially formed; *Valiant* and *Malaya* joined their sisters in the spring of 1916, giving Jellicoe a homogeneous force of five fast battleships. Under most circumstances they were to operate four to five miles ahead of the main battle line; but in the GFBOs in force at Jutland their role was more narrowly circumscribed, and nothing was said about them acting independently.[36]

If divisional tactics made only fleeting appearances in the GFBOs, there was at least one persistent proponent of them in the Grand Fleet: Vice-Admiral Sturdee, victor of the Falkland Islands, who took command of the 4th Battle Squadron on 7 February 1915. It would appear that soon after his arrival Sturdee began pressing the case for dividing the fleet while approaching the enemy; one of his memoranda survives – written, so he notes at the outset, 'In continuation of former papers', which clearly implies it was one of a series on the topic. His proposed method may be summarised as follows:

> An investigation has been made as to the desirability of dividing the Battle Fleet into two Divisions of about equal value. ... Diagrams have been prepared to show that there is no risk of one Division being overpowered, so long as the two Admirals work in co-operation, and that [sic – 'if'?] a clear plan as to what is intended to be done under certain circumstances is prepared beforehand. ... it would appear both wise to keep the fleet together until after the cruisers have got in touch, and separate them just before making the approach. This can be done in several ways, and, depending on the distance apart, need not take more than 15 or 20 minutes; during this tine [sic – 'time'] the divisions should cease approaching the enemy. ... The diagrams explain themselves. If either division is threatened, it avoids action until the other one is ready to support. If the enemy presses the attack for this length of time, i.e. less than half-an-hour, the first and most critical position in the action has been attained, and the enemy has to retire to avoid the annihilation of his van.[37]

Unfortunately, the self-explanatory diagrams are no longer with the paper, so it is unclear exactly how the separated divisions were to approach the enemy and act in battle. In any case, Jellicoe was not impressed with Sturdee's ideas; ten days before the Battle of Jutland, he wrote to the First Sea Lord:

> In regard to Sturdee, I should never feel safe with him in command of the most important squadron leading the van... I am very sorry to say that I do not trust his judgement in tactical questions. *I feel very strongly about this* and I know that other flag officers hold the same views...[38]

Nor did Sturdee help his own case; Rear-Admiral Alexander Duff, Sturdee's second-in-command, although he generally agreed with his admiral's tactical ideas,

Vice-Admiral Sir Frederick Charles Doveton Sturdee photographed on board his flagship *Hercules*, when he commanded the 4th Division of the Grand Fleet. Jellicoe, who disagreed strongly with Sturdee's tactical ideas, kept him on a short leash at Jutland, ensuring that he was under his direct command at the centre of the line – Sturdee's flagship *Benbow* was the 13th ship in a single line of 24 battleships. (Imperial War Museum Q 018063)

noted that Sturdee 'ruins his cause by impetuosity and tactlessness, and by an overweening belief in himself'.[39] The upshot is described by Sturdee himself:

> During the year [1916] I was ordered by the C in C never to discuss or try dividing the Fleet for the purpose of overpowering the enemy, & thus bringing a superiority of fire on part of the enemy's Fleet, so ceased having further Tactical Games on board my Flagship until there was a change in the Command.[40]

Jellicoe's fiat apparently came prior to Jutland, so it seems probable that, whatever experiments with divisional tactics may have taken place early in the war, by the time of Jutland they were no longer under consideration.

Jutland 1: Dividing before Deployment

And so we come at last to Jutland. Admiral May, in his *Notes on Tactical Exercises*, defined three distinct phases of any battle when divisional tactics might be used:

- Dividing and endeavouring to place the Divisions advantageously, while the Fleets are still well outside gun range.
- Withholding the Division until the enemy has committed himself to a movement, or till shortly before coming into gun range.
- Dividing at a still later stage after the action has actually begun.[41]

This scheme provides a convenient framework for discussing how divisional tactics might have been employed at Jutland, and so will be used here.

We can quickly rule out option number 1, since the discrepancies in reckoning between Beatty's Battle Cruiser Fleet (BCF) and the main body of the Grand Fleet, under Jellicoe, were so great that it would have been almost impossible to divide the fleet in a useful way before the two battle fleets came into contact. This highlights one problem with divisional tactics: the need for reliable information about the relative positions of the various forces. Lacking such information, the separated divisions would not be able to coordinate their actions. For the same reason, good visibility was also required; of the fifteen exercises in divisional tactics Admiral May included in his volume, only one took place where the visibility was less than 10,000 yards, and of this action May had noted that it was 'impossible to see the other divisions'.[42] It is also noteworthy that, despite the fact that he described line-of-battle tactics in misty weather, his table of contents mentioned only 'Divisional Attack in *Clear* Weather'.[43] This is a point worth bearing in mind as we proceed through the options available at Jutland.

Jutland 2: Dividing during Deployment

The second opportunity identified by Admiral May for divisional tactics was 'Withholding the Division until the enemy has committed himself to a movement, or till shortly before coming into gun range'. May considered this the most promising moment for dividing the fleet, since it would gain the advantage of surprise, leaving the enemy no time to devise a counter-move.[44]

At Jutland this would have been the moment of deployment at 6.15 pm, when the battle fleet changed from its cruising order in six columns, each of four ships, steaming on a broad front to the SSE, into the line of battle. Here we come to the first definite proposal for divisional tactics, put forward by Admiral Sturdee, who argued that this was a moment when 'either of the Admirals of the wing Divisions might have not obeyed the CinC's signal due to being in a better position for seeing how they could help the CinC better by so doing'.[45] Sturdee continued:

> …when the deployment was ordered to Port or away from the enemy, VA [Vice-Admiral] First Battle Squadron might have taken the great responsibility of deploying the other way? [sic] my [sic] hope at the time was that he would & I should have followed him. I do not say he should have done so, but if I had been in that position I should have been sorely tempted to do so.

Of course, the commander of the First Battle Squadron, Vice-Admiral Sir Cecil Burney – who has been described as 'orthodox, unimaginative, utterly lacking in initiative' – did no such thing, nor was he ever likely to.[46] But we can take a stab at analysing the possible results of such an action (see Figure 3).

To begin with, Sturdee's statement is somewhat ambiguous, since there are two possibilities: First, while the three divisions of the battle fleet on the port wing turned first NE, then turned SEbyE, the three starboard divisions could have deployed to the SW, opening an ever-widening gap between the two wings of the fleet. Probably the western half of the fleet, with the three *Queen Elizabeth* class ships of the 5th Battle Squadron (*Queen Elizabeth* herself missed Jutland because she was in dockyard hands, and *Warspite* had fallen out of line as a result of her steering difficulties), would have then shaped course more to the south, and then perhaps gradually turned more to the east, attacking the rear of the German line. This would have placed the German fleet between two fires, with Jellicoe's half of the fleet to the northeast and the other half 10,000-13,000 yards to the northwest. But this is where the visibility factor comes into play; Dreyer, now Jellicoe's flag-captain in *Iron Duke*, wrote that:

> …at about 5.40 pm Jellicoe had ordered me to have ranges taken on various bearings and report to him what would be the most favourable directions in which we could engage the enemy Fleet… As a result I reported to him that the most favourable direction was to the southward, and would draw westwards as the sun sank.[47]

As a result of these conditions, the western wing would have had to contend with a misty eastern horizon that

Jellicoe's flagship *Iron Duke* in 1914, shortly after completion. By placing himself at the head of the 4th Battle Squadron, the second of the three squadrons that made up the British line, Jellicoe hoped to be able to control the battle fleet during the decisive phase of the battle, and to be able to respond appropriately to the enemy's movements. Unfortunately, visibility during the evening of 31 May 1916 was poor, the patchy mist aggravated by the huge quantities of funnel smoke. As a result, he could never see more than three or four of the German battleships, and was unable to make out the enemy's movements. (Leo van Ginderen collection)

might have made effective gunnery very difficult, while at the same time the Germans might have seen them much more clearly to the westward. This is critical, since the western half of the fleet would have consisted for the most part of the older dreadnoughts with 12in guns, and would have been outnumbered by the German battleships, sixteen to fifteen. So if the High Seas Fleet turned westward to avoid Jellicoe's force, it might well have been able to engage the western half of the Grand Fleet on equal or better-than-equal terms. At this point, however, things become unpredictable. Would Admiral Reinhard Scheer, C-in-C of the High Seas Fleet, have made the same battle-turn-away he did? What would Jellicoe have done when he realised that half his fleet was going the other way? And so forth.

The second possibility would have been for Burney to steam straight ahead, then turn 90° to port and follow, at some distance, Jellicoe's half of the fleet. This is a bit more promising in some ways, since it preserves the better visibility while placing the trailing half of the fleet at a reasonable range for good gunnery – about 10,000 yards. The two halves of the fleet would have stayed within visual range and therefore could support one another effectively. By the same token, however, the trailing half would have come into action one ship at a time as they turned eastward, and so it would have been some while before they could develop their full volume of fire. Still, this course of action might have offered several advantages over the deployment as actually carried out.

Before moving on, we should note Sturdee's somewhat plaintive observation on his situation at Jutland:

> ....It was constantly present in my mind how any individual action of mine in the centre of the very long line could help the action, but I was painfully aware that I was powerless to move out of the line.[48]

This was exactly why Jellicoe had placed Sturdee's division in the middle of the battle fleet, rather than on one of the wings: he did not trust the renegade admiral not to try something along the lines of a split deployment. Sandwiched in the middle of the fleet, there was little Sturdee could do in the way of independent action.

Jutland 3: Dividing After Deployment

This brings us to May's third possibility for divisional tactics: 'Dividing… after the action has actually begun'. He treated this option somewhat dismissively, writing that:

> This method is more in the nature of a defensive than an offensive tactical operation. It implies that the Commander of a Fleet has begun his action by forming the line of battle, and that for some reason he finds it necessary to detach Divisions.[49]

However, the 'defensive' characterisation of this method

overlooks the ways divisional tactics might have been used in pursuing a retreating enemy, which is the situation that arose at Jutland.

Perhaps the most ardent advocates of this method were the Dewar brothers, Kenneth and Alfred, who wrote the controversial *Naval Staff Appreciation of Jutland*.[50] The Dewars were severe critics of Jellicoe's command style and his tactics, and they argued that the Grand Fleet would have achieved a decisive victory if only it had used divisional tactics. They pointed out two moments when this might have been done.

The first came at about 6.45 pm, half an hour after the British fleet had deployed. It was steering SEbyE, and the Germans were steaming right into its massed gunfire. Although Admiral Scheer could see little except the flashes of the British guns, he quickly realised that his fleet was in mortal danger, and at 6.35 he ordered a *Gefechtskehrtwendung*, a battle-turn-away, a not-quite-

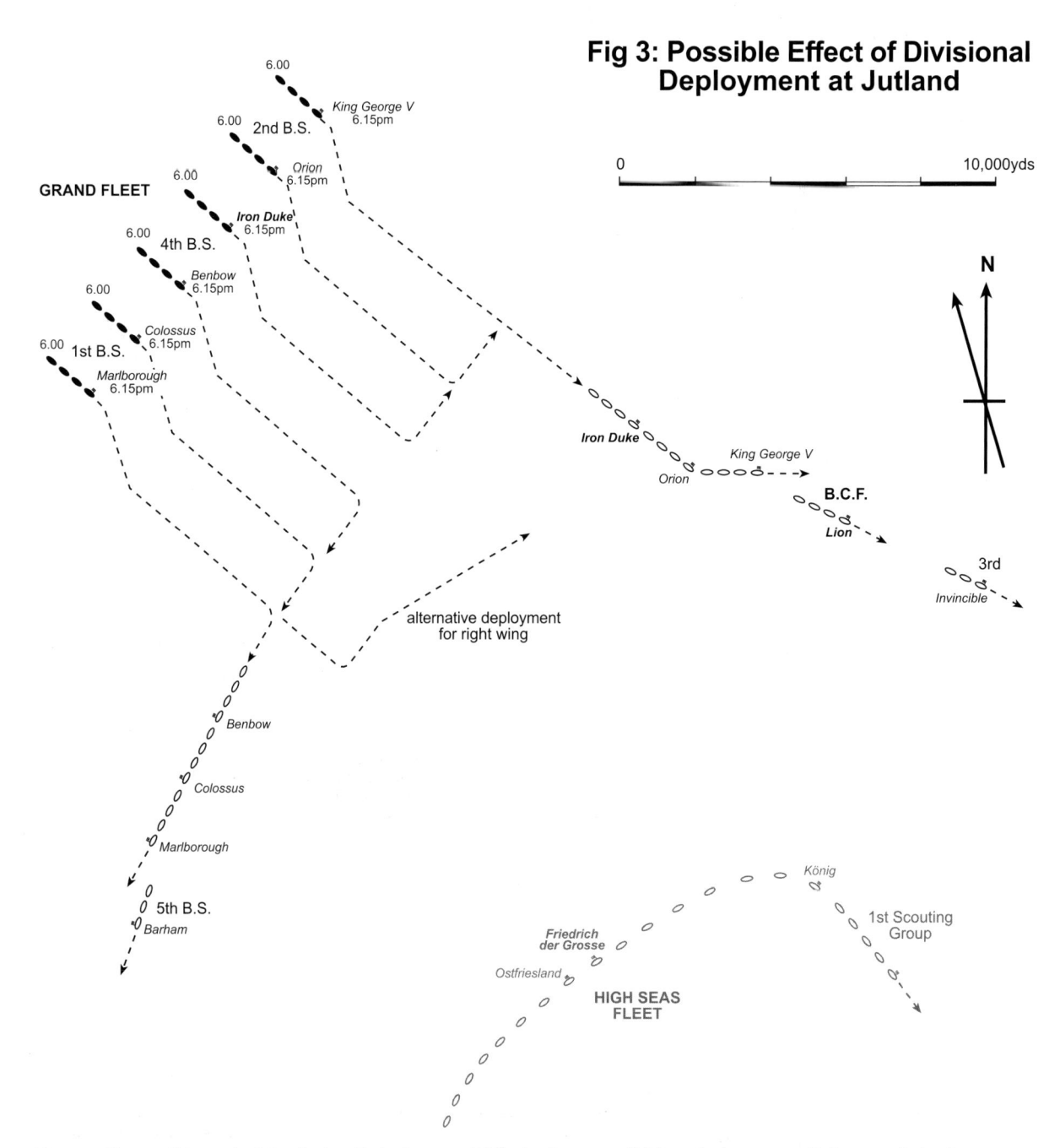

Fig 3: Possible Effect of Divisional Deployment at Jutland

Figure 3: The possible result of Sturdee's split deployment. While the three port divisions deploy as actually happened, the three starboard divisions deploy in the opposite direction, forming up astern of the 5th Battle Squadron. They either continue in a southwesterly direction, or turn to follow the port wing of the fleet at some distance. The High Seas Fleet is roughly equal to either wing, and probably has better visibility against the starboard wing than the latter would have against the German fleet. (© John Jordan 2015)

simultaneous 180° turn in which the last ship in the line turned first, the next ship turning as soon as it saw the ship astern start its turn, and so on up the line. Executed in some haste, the turn resulted in the Germans steering a roughly westerly course, moving away from the British. They soon disappeared into the murk, leaving Jellicoe somewhat perplexed as to where they had gone. So the British fleet continued at first on its SEbyE course, then as Jellicoe realised that the Germans had not simply been swallowed up by the mist he turned to SE at 6.44, and to South at 6.55. Here is where the Dewars saw an opportunity for a divisional attack. They wrote:

> ...if the 5th Battle Squadron, followed by the 6th, 5th and 4th Divisions had led round to the south-westward and proceeded at full speed to the northward of the enemy fleet, whilst the battle cruisers, followed by the 1st, 2nd and 3rd Divisions, had carried out a similar movement to the southward, the rear of the High Sea Fleet would have been enveloped and exposed to an overwhelming concentration.[51]

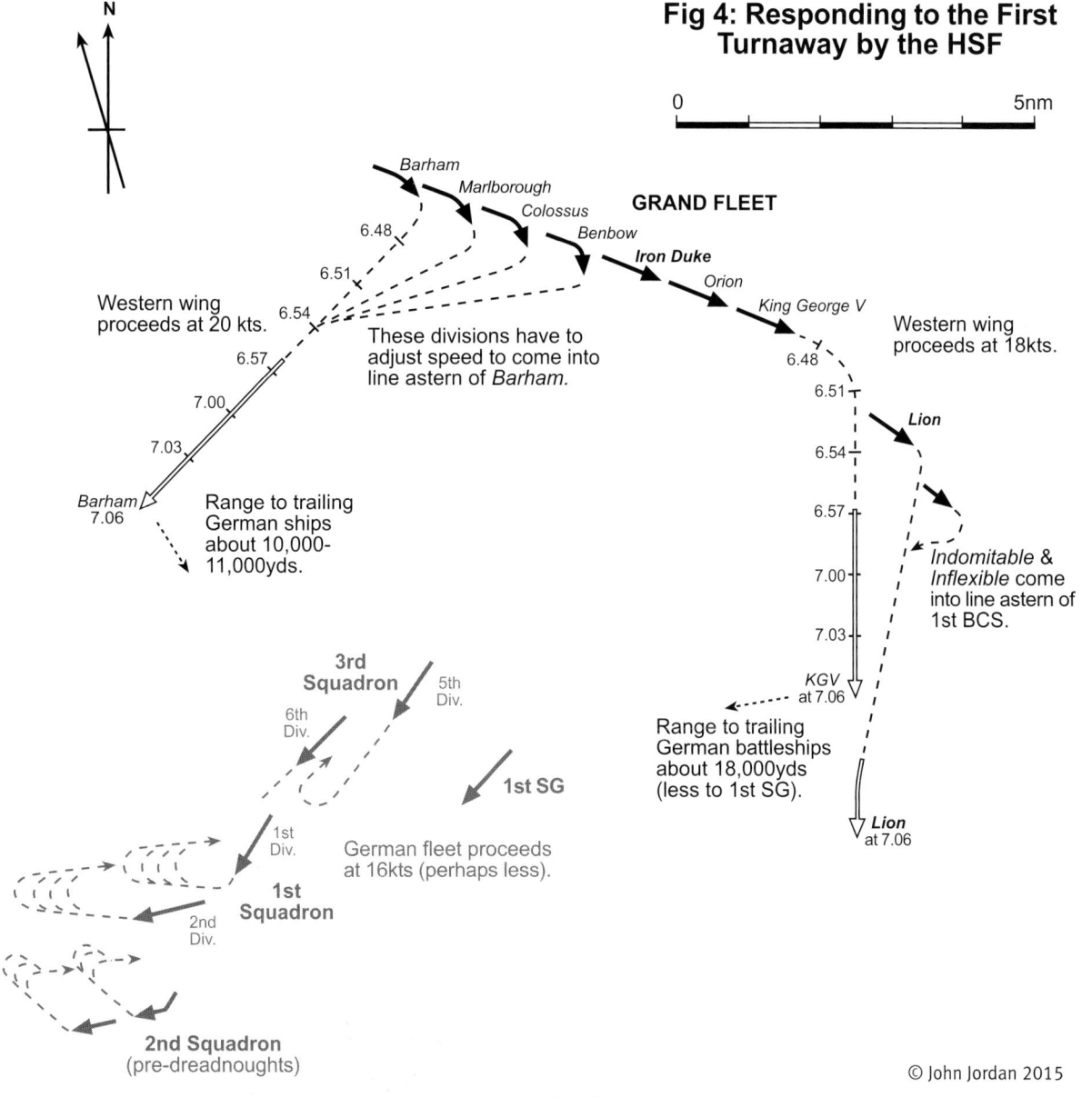

Figure 4: The Dewar brothers' proposed pursuit of the German fleet after the first *Gefechtskehrtwendung*. After Scheer turns away, the trailing divisions of the Grand Fleet, led by the 5th Battle Squadron, turn off to the southwestward, forming a west wing, while the east wing turns to the south, with the High Seas Fleet caught between. However, the two British wings are well out of sight of one another, and the Germans are again about equal to either of them; by turning to the northwest, Scheer can cross the 'T' of the west wing; by turning about (as he actually did) he can escape to the east between the two British wings. (Based on Diagram 33 of the *Naval Staff Appreciation of Jutland*.)

Figure 4 offers an approximation of what the Dewars were proposing – the rear half of the British battle line would split away, turning to the southwest, while the leading ships turned southwards, with the trailing divisions of the German fleet between them.

On paper, things look bad for the High Seas Fleet, but in fact there are several problems with this manoeuvre. First and foremost is the fact that the British didn't know exactly where the Germans had gone. Although *Falmouth* was able to discern the fact that the Germans had turned westward, to most observers it seemed likely that the enemy was simply concealed by a local thickening of the mist.[52] Even if the German turn-away had been understood, the poor visibility would have created other problems. The courses described by the Dewar brothers – southwesterly for the western wing and southerly for the eastern – lead to ranges on the order of 10,000–12,000 yards to the trailing German ships of the 5th Division and the 1st Scouting Group, and it is likely that the western wing would have been able to see the enemy only dimly at this range due to the mist. Moreover, the two wings of the British fleet would not have been able to see each other, making any sort of cooperation between them impossible. The divided Grand Fleet would also have offered two golden opportunities to the Germans. First, as Captain V S H Haggard, director of the Training and Staff Duties Division (the section of the Admiralty Staff charged with studying the lessons of the war) noted in 1922, this splitting of the British fleet would have 'resulted in the fleet being cut in half by Scheer's easterly movement' when the High Seas Fleet executed a second *Gefechtskehrtwendung* and turned back to the east at 6.55 pm.[53] In other words, Scheer would have had a wide-open pathway back home between the two wings. And secondly, if Scheer did not turn eastward at 6.55, he would have been in a very good position to cross the 'T' of the western wing.

The Dewar brothers believed that there was another opportunity for divisional tactics about half an hour later, after Scheer's third *Gefechtskehrtwendung* when he turned away from the massed firepower of the Grand Fleet for the second time under cover of a smoke screen and a torpedo attack by his destroyers; the British battle line first turned away from the torpedoes, then manoeuvred to reform. The Dewars asked:

> Was it not preferable to break up that long inarticulate line and for each division or sub-division to press forward independently at utmost speed to the westward, supporting each other in the envelopment and destruction of the German Fleet? If the Battle Fleet had proceeded in this direction, one half to the northward and the other half to the southward of the German line, led by the 5th Battle Squadron and Battle Cruiser Fleet respectively, the fate of the High Sea Fleet would probably have been sealed.[54]

An interpretation of this notion is given in Figure 5, although there are many uncertainties involved. The major interpretive issue is whether the divisions were supposed to turn west individually (as the first sentence quoted above seems to imply), or would they have formed northern and southern lines of battle (as the second sentence seems to indicate)? The best way to reconcile this contradiction is to assume that the most of the divisions first turn west, to comb the tracks of the approaching German torpedoes, then when the danger has passed each division turns to join either the 5th Battle Squadron or the Battle Cruiser Fleet. The exception is the van division led by *King George V*, which was least threatened by the torpedo attack, and so could steer directly to form up astern of the BCF. The overall result is somewhat messy, and not without dangers, as the German torpedoes were coming from somewhat different directions at high speed, so some dodging would have been necessary, and it is likely that some ships would have been hit.

But the real question is: What would the Germans have done? As can be seen from the diagram, the High Seas Fleet could turn either northwards or southwards and cross the 'T' of one or the other wing of the advancing British. In any case, there is little in the way of an 'envelopment' – the Germans simply had too much of a head start. Only if the action had developed into a long stern chase would there have been any chance of enveloping the enemy's trailing ships. But this would imply that the German fleet was fleeing, and despite the tactical blunders Scheer had committed during the battle, he and his fleet were never on the verge of being forced into a headlong flight.

Conclusion

Any review of divisional tactics must proceed on multiple levels. In the first place, the Royal Navy's attitude towards them before the war needs to be determined. Although several prominent historians flatly state that they were abandoned before the war, there is solid evidence in surviving tactical documents that they were still under consideration up to August 1914. It is particularly interesting that Frederic Dreyer, whose postwar comments about divisional tactics have often been quoted as evidence for their prewar rejection, developed a device to increase their effectiveness as late as May 1914. Moreover, the idea of using fast capital ships to attack the van and rear of an enemy's battle line was a widely accepted tactical concept, and paralleled one of the main lines of divisional thought. It seems likely, therefore, that the prewar tactical landscape of the Royal Navy was more complex and less dogmatic than has generally been asserted.

Another level of analysis concerns the general potential of divisional tactics. Their proponents insisted that the adoption of such tactics would have engendered greater initiative and aggressiveness in officers, and fostered a more flexible method of attack. Partially as a result of the inconclusive nature of Jutland, much of the Royal Navy's tactical thought between the wars focused on decentralisation, and this certainly paid handsome dividends in the Second World War. But it is by no means clear that the

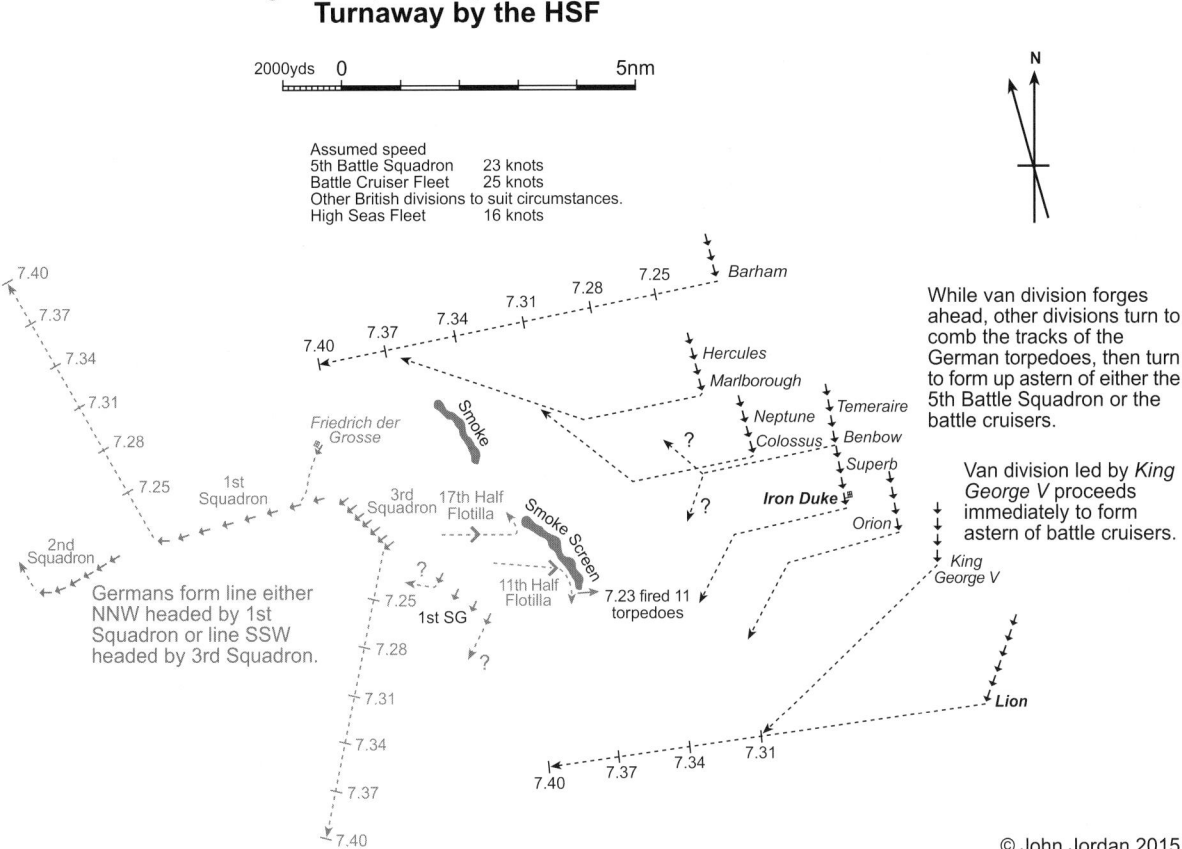

Fig 5: Responding to the Second Turnaway by the HSF

Figure 5: The Dewars' second suggestion for a divisional pursuit, after the third *Gefechtskehrtwendung*. In reality, as the Germans withdrew westwards under cover of smoke screens and a torpedo attack, the battle fleet turned away. In this hypothetical scheme, the leading division, led by *King George V*, instead turns to follow the Battle Cruiser Fleet (BCF). The other British divisions, threatened by the German torpedoes, turn to comb their tracks, then when the danger has passed turn to form a northern wing with the 5th Battle Squadron leading and a southern wing led by the BCF. But once again Scheer can turn to attack either wing on equal terms, while the two wings are out of sight of one another. (Based on Diagram 36 of the *Naval Staff Appreciation of Jutland*.)

later successes of tactical decentralisation can be read backwards to the conditions to the First World War; the battle fleets that confronted one another across the North Sea in 1914–18 were larger than any afloat in 1939–45, and they fought at much closer ranges. Despite the Dewar brothers' claim that divisional tactics were 'peculiarly applicable to a large fleet', at least one modern historian disagrees: Professor Jon Sumida has suggested that 'Divisional tactics were promoted by the fact that between the wars the active British battle fleet was less than half the size of the Grand Fleet at its wartime peak...'.[55] At Jutland there were 250 warships, from destroyers to battleships, manoeuvring in relatively narrow stretches of water; it seems probable that a certain amount of rigid choreography was necessary under such circumstances.

Before moving on, another factor should be noted. At the same time that the Royal Navy was experimenting with divisional tactics, the United States Navy was also debating the relative merits of the single line of battle and what it called 'group' – that is, divisional – tactics. One of the American critics of the single line of battle, a promising young officer, Lieutenant-Commander Dudley Knox, described the essential underpinnings of his preferred group tactics:

> The subordinate commanders must be students of tactics and thoroughly versed in its principles. They must be accustomed to working together in tactical situations. They should be governed by a system of command which fixes clearly the restrictions, duties and responsibilities of each in any situation that may arise. They must be uniformly indoctrinated in accordance with the views of their chief, and should have confidence in the soundness of such doctrine. They should be thoroughly acquainted with the particular plan of the commander-in-chief intended to govern that battle. Above all, far above all else, they must be loyal to the plan in every act.[56]

These were cogent observations, and point to one of the factors that would have crippled any attempt by the Royal Navy to develop divisional tactics before 1914: the lack of a strong staff system that could study the lessons of the various manoeuvres, digest them, formulate new tactics, and then promulgate the resulting doctrine throughout the fleet. Ultimately, the success of divisional tactics was dependent upon a common doctrine, as well as a common understanding of that doctrine and the communication of the commander's intentions as to how he intended to use that doctrine during battle. The administrative mechanisms needed to create this commonality were either weak or entirely lacking in the Royal Navy before the First World War.

Finally, we come to the question of the applicability of divisional tactics to the specific circumstances of Jutland. As is always the case with hypothetical situations, some aspects of the diagrams offered here may be questioned, but the overall concept is clear enough: Divide the Grand Fleet into two wings and catch some portion of the enemy fleet 'between two fires'. It sounds very dramatic, but those two wings would soon have been out of sight of one another in the prevailing visibility. If this divided force had not overwhelmed Scheer quickly, he might have had an opportunity to turn upon one wing and fight it on equal (or even superior) terms before the other wing could come to its aid.

These are problems that the more ardent proponents of divisional tactics ignored; to them, dividing the fleet was a universally applicable method. But the idea that the enemy would be 'paralysed and destroyed' by divisional tactics, as Kenneth Dewar put it, is based on the assumption that the enemy would not know how to cope with such an attack.[57] The German official history thought such an effect was unlikely:

> …in view of the low visibility then prevailing and the lateness of the hour, co-operation between [British] squadrons acting independently would have been so uncertain that the danger of the enemy overwhelming a part of the British Fleet with the whole of his force could not be overlooked.[58]

The proponents of divisional tactics, searching through history for the key to decisive victory, took Trafalgar as their model. But in so doing they overlooked a fundamental difference between Nelson's day and their own: Scheer was no Villeneuve, and the High Seas Fleet was not the brave but fundamentally incompetent Franco-Spanish fleet defeated by Nelson. In the Germans the British faced a skilled and well-trained opponent: twice Scheer extricated his fleet from extremely dangerous situations – albeit ones he had brought upon himself. Was such a man really likely to lose control of himself and his fleet because he was 'caught between two fires'? Or would he have grasped the opportunities offered by a divided enemy? We can never know for certain, and so the Jutland debate will continue long after the battle's centenary.

Acknowledgements:

I would like to express my gratitude to Simon Harley, who has generously shared the fruits of his own research into tactics in the dreadnought era; and to John Jordan, who took on the task of turning my crude sketches into superb diagrams of Jutland's 'might-have-beens'. John Brooks provided some valuable insights and corrections. Finally, and once again, the editorial skills of my wife, Jan Torbet, greatly improved this article. *[The Editor would also like to thank John Roberts for help with dating the photographs used to illustrate this article.]*

Notes:

1. Admiralty to Domvile, 26 October 1903; The National Archives (TNA), Kew: ADM 144/17. My thanks to Simon Harley for bringing this document to my attention.
2. H J May, 'Exercises Carried Out at the Royal Naval College, Greenwich', May 1902, pp.12–13; TNA: ADM 1/7597. Emphasis in the original.
3. Report of Umpire Committee on Fleet Action Off Cape St. Vincent on 6th September 1901, 'Combined Manoeuvres Mediterranean and Channel Fleets, 1901'; TNA: ADM 1/7506.
4. Nicholas Lambert, *Sir John Fisher's Naval Revolution* (Columbia, SC: University of South Carolina Press, 1999), p.212; Marder, *From the Dreadnought to Scapa Flow*, vol. III: *Jutland and After (May 1916–December 1916)* (second edition, London: Oxford University Press, 1978; hereafter *FDSF*), p.32 n.31.
5. See for example, 'Admiralty Policy in Battleship Design'; Peter Kemp (ed.), *The Papers of Admiral Sir John Fisher* (London: Navy Records Society, 1960), vol. I, p.325.
6. 'Combined Manoeuvres Mediterranean and Channel Fleets, 1901'; TNA: ADM 1/7506.
7. Kemp, *Papers of Admiral Sir John Fisher*, p.226; emphasis in the original.
8. Nicholas Lambert, 'Admiral Sir Francis Bridgeman-Bridgeman (1911–1912)'; Malcolm H. Murfett (ed.), *The First Sea Lords: From Fisher to Mountbatten* (Westport, CN: Praeger, 1995, pp.55–74), p.60.
9. Bridgeman to Churchill, 21 May 1912; Churchill Archive Centre (CAC), Cambridge: Chartwell Papers, CHAR 13/9/50-51. I am once again grateful to Simon Harley for directing me to this document. See also Winston Churchill, 'Oil Fuel Supply for His Majesty's Navy', 16 June 1913, pp.1–3; TNA: CAB 37/115/39.
10. Callaghan, 'Remarks on the Conduct of a Fleet in Action, Based on the Experience Gained in The Manoeuvres and Exercises of the Home Fleets During the Year 1913', 5 December 1913; Naval Historical Branch (NHB), Portsmouth: Backhouse Papers, Box 1, Folder T94608.
11. 'Confidential Battle Orders for 1st BCS', 17 July 1913; B McL Ranft (ed.), *The Beatty Papers: Selections from the Private and Official Correspondence of Admiral of the Fleet Earl Beatty*, vol. I (Aldershot: Published by Scholar Press for the Navy Records Society, 1989), p.74.
12. Captain A.E. Chatfield, 'Fast Division Work from a Gunnery Standpoint', October 1913; Ranft, *Beatty Papers*, p.91.
13. Chatfield, *The Navy and Defence* (London: William Heinemann Ltd, 1942), pp.106–107. Although Chatfield gives no date for this event, the fact that French Minister of Marine Pierre Baudin was present narrows down the time period; he observed the Home Fleet's annual manoeuvres in July 1913; Paul G. Halpern, *The Mediterranean Naval Situation 1908–1914* (Cambridge, MA: Harvard University Press, 1971), p.112.
14. Letter, Sturdee to Newbolt, March 1924; CAC: SDEE/3/5.
15. Matthew Allen, 'Rear Admiral Reginald Custance: Director

of Naval Intelligence 1899–1902' (*Mariner's Mirror*, vol. 78, no. 1 [February 1992], pp.61–75), p.73; Marder, *FDSF*, pp.42–3, 338; Barry D Hunt, *Sailor-Scholar: Admiral Sir Herbert Richmond, 1871–1946* (Waterloo, Ontario: Wilfrid Laurier University Press, 1982), p.25.

16. Reginald Custance, *The Ship of the Line in Battle* (Edinburgh: William Blackwood and Sons, 1912), p.103.

17. Kenneth Dewar, 'Battle of Jutland' (*Naval Review*, part 1: vol. XLVII, no. 4 [October 1959], pp.400–416), p.403; emphasis in the original. Warrender's instructions survive, pasted into Richmond's diary; National Maritime Museum (NMM), Greenwich: Richmond Papers, RIC/1/8, entry for 25 January 1911. My thanks to Simon Harley for a transcript.

18. Admiral Sir William May, *Notes on Tactical Exercises. Home Fleet. 1909–1911* (Admiralty, 19 September 1911); Admiralty Library, Portsmouth: Eb 012.

19. It is interesting to note that many years later Kenneth Dewar told Arthur Marder that *Notes on Tactical Exercises* 'consists almost entirely of diagrams and it would not be worth wasting your time in studying it'. Was he trying to dissuade Marder from consulting a volume that showed divisional tactics in a less-than-successful light? Letter, Dewar to Marder, 3 August 1963; University of California, Irvine: Marder Papers, Box 1.

20. May, *Notes on Tactical Exercises*, p.238.

21. May, *Notes on Tactical Exercises*, p.238.

22. Paper by William R Hall, 6 November 1907; NMM: Mercury Papers, MER/39; Hall, 'The Fleet in Action' (*Naval Review*, vol. II, no.1 [February 1914], 53–71); Arthur Marder (ed.), *Portrait of an Admiral: The Life and Papers of Sir Herbert Richmond* (Cambridge, MA: Harvard University Press, 1952), pp.170–2.

23. William Schleihauf (ed.), *Jutland: The Naval Staff Appreciation* (London: Seaforth Publishing, 2016), Chapter II, Sec. 11.

24. Marder, *Portrait of an Admiral*, p.263; Marder, *FDSF*, p.13; Jon Tetsuro Sumida, 'A Matter of Timing: The Royal Navy and the Tactics of Decisive Battle, 1912–1916' (*The Journal of Military History*, vol. 67, no.1 [January 2003], pp.85–136), p.91; see also p.104, where he states that the rejection of Arthur Pollen's fire-control system, which he argues took place in late 1912, was 'tantamount to the repudiation of divisional attack'.

25. Frederic C Dreyer, *The Sea Heritage: A Study of Maritime Warfare* (London: Museum Press Limited, 1955), pp.64, 68.

26. Marder, *Portrait of an Admiral*, p.263.

27. Callaghan to Admiralty, 5 March 1913, proposing 1) Instructions for the Conduct of a Fleet in Action, and 2) Training of the Fleet in Peace; NHB: Backhouse Papers, Box 1, Folder T94576.

28. H G Thursfield, 'Development of Tactics in the Grand Fleet'; NMM: Thursfield Papers, THU/107.

29. A Temple Patterson, *Jellicoe* (London: Macmillan and Co., Ltd., 1969), pp.48–49.

30. Memorandum, Dreyer to Vice-Admiral Commanding Second Battle Squadron, on Instruments for a) Plotting Approach and b) Divisional Attack, 4 May 1914; NHB, Backhouse Papers, Box 1, Folder T94578.

31. John Irving, *The Smoke Screen of Jutland* (New York: David McKay Company, Inc., 1966), p.144. 'PZ' exercises were tactical exercises that pitted two groups of ships against one another; the name derived from the signal used when initiating the exercise.

32. GFBOs, 18 August 1914, pph. 9; Addendum No. 3, 31 August 1914, pph. 9; TNA: ADM 116/1341.

33. GFBOs, Cruising Order No. 5, 12 September 1914; TNA: ADM 116/1341.

34. GFBOs, Cruiser Addendum, pph. 2, 3 March 1915; TNA: ADM 116/1341. Emphasis in the original.

35. GFBOs, section XXIV, 'Orders for "Queen Elizabeth" and "Warspite",' 3 June 1915; TNA: ADM 116/1341.

36. GFBOs, 'Orders for 5th Battle Squadron', December 1915; TNA: ADM 116/1343.

37. Sturdee, 'Dispositions for the "Approach"', 10 September 1915; British Library: Jellicoe Papers, Add. MS. 49012, ff. 56–60.

38. Jellicoe to Jackson, 21 May 1916; A Temple Patterson (ed.), *The Jellicoe Papers: Selections from the Private and Official Correspondence of Admiral of the Fleet Earl Jellicoe of Scapa*, vol. I (London: Navy Records Society, 1968), pp.241–2; emphasis in the original.

39. Marder, *FDSF*, p.33, quoting Duff's diary.

40. Sturdee, 'Summary of Proposals Made in Writing 1915 & 1916', undated (but postwar); CAC: Sturdee Papers, SDEE/3/6.

41. May, *Notes on Tactical Exercises*, p.238.

42. May, *Notes on Tactical Exercises*, p.293.

43. May, *Notes on Tactical Exercises*, pp.v–vi; emphasis added.

44. May, *Notes on Tactical Exercises*, p.240.

45. Letter, Sturdee to Sir Henry Newbolt, March 1924; CAC, Sturdee Papers, SDEE/3/5.

46. Marder, *FDSF*, p.42.

47. Dreyer, *Sea Heritage*, p.145.

48. Letter, Sturdee to Sir Henry Newbolt, March 1924; CAC: Sturdee Papers, SDEE/3/5.

49. *Notes on Tactical Exercises*, p.240.

50. The *Naval Staff Appreciation of Jutland* was printed in 1922 but never issued to the fleet due to its controversial nature; an annotated version has recently been published – see Schleihauf, *Jutland: The Naval Staff Appreciation*.

51. Schleihauf, *Jutland: The Naval Staff Appreciation of Jutland*, Chapter IX, Section 61, n.1.

52. See Cmd. 1068, *Battle of Jutland, 30th May to 1st June 1916. Official Despatches with Appendices* (London: HMSO, 1920; reprinted by the Naval & Military Press Ltd., 2006), pp.58, 71, 79, 89, 92, 99, 111, 120, 128, 151, 156, 161, 170, 186.

53. Memorandum, Haggard to First Sea Lord, 26 July 1922; University of California, Irvine, Langston Library: Microfilm M 000147. This microfilm was made for Arthur Marder and includes several documents related to the *Naval Staff Appreciation of Jutland*.

54. Schleihauf, *Naval Staff Appreciation*, Chapter X, Section 68.

55. Schleihauf, *Naval Staff Appreciation*, Chapter II, note 2; Jon Tetsuro Sumida, '"The Best Laid Plans": The Development of British Battle-Fleet Tactics, 1919–1942' (*International History Review*, vol. XIV, no. 4 [November 1992], pp.681–700), p.690.

56. Dudley W Knox, '"Column" as a Battle Formation' (*United States Naval Institute Proceedings*, vol. 39, no. 3 [September 1913], pp.949–958), p.957. Knox rose to the rank of commodore and became a prominent historian of the US Navy.

57. Dewar, 'The Battle of Jutland', p.403.

58. V E Tarrant, *Jutland: The German Perspective* (Annapolis: Naval Institute Press, 1995), p.147, quoting Groos, *Der Krieg in der Nordsee*, 5:306–8.

MODERN LITTORAL SURFACE COMBATANTS

In his latest article on modern warship developments, **Conrad Waters** looks at the new breed of surface ships designed specifically for littoral warfare.

The emphasis placed on the conduct of open-water, oceanic operations was a dominant influence on the structure of many leading fleets during the twentieth century. This reached a peak at the height of the Cold War. However, the priority for the vast majority of naval forces – lacking the resources to aspire to global reach – has always been the protection of national interests in the coastal and offshore waters of their littoral zones.[1] It is littoral operations in their various forms that have dominated the history of naval warfare. Today, economic and political developments mean the ability to control these waters is, again, assuming ever-increasing importance.

The growing importance of the littoral reflects a number of factors. An important starting point is the fact that almost three quarters of the world's population – and a similar proportion of capitals and major commercial centres – are located in littoral areas. The economic significance of these regions has been further enhanced by the expansion in national maritime rights provided for by the current United Nations Convention on the Law of the Sea (UNCLOS) regime, implemented from 1982 onwards.[2] Many fleets have undergone expansion to ensure these interests are properly protected. Equally, the reduced threat posed to the established 'western' maritime powers' command of the open seas since the end of the Cold War has allowed them to refocus in order to counter emerging threats. A renewed emphasis on expeditionary deployments to coastal waters to combat regional instability and, therefore, reduce hazards such as terrorism and piracy has been a conspicuous trend. More broadly, the ability to operate in – and to dominate – littoral waters increasingly has been seen as a key instrument of strategic leverage, not least by the United States.[3]

These trends have had a major influence on both the general types and, also, on the specific designs of warships being procured across the world. In terms of warship type, many smaller navies have, for example, stepped up investment in paramilitary patrol craft. Purpose-built offshore patrol vessels optimised for constabulary-type duties and capable of undertaking prolonged surveillance missions have assumed much greater popularity. A number of the larger fleets seeking to boost expeditionary capabilities have made the upgrade of amphibious capabilities a priority.[4] It should, of course, be evident that the requirements of those navies structured to protect littoral waters and those seeking to access them as part of an expeditionary deployment have tended to vary significantly.

Equally significant trends have become apparent in the design of warships intended to operate in offshore waters. This is particularly the case with respect to littoral surface combatants, ie those ships optimised for a warfighting role. These ships have had to adapt as a result of both operational experience and technological change. More specifically:

– The previous popularity of the specialised missile-armed fast attack craft (FAC) has struggled to survive the vulnerability of the type to air attack as revealed in the 1990–91 Gulf War. Whilst fast attack craft are still being constructed, their use is now largely concentrated on areas where the local topography is particularly well-suited to their use.[5]
– Replacement designs tend to be larger ships that incorporate a more balanced range of capabilities. In some cases, there has been an attempt to combine the roles previously performed by several types of littoral combatant in one class.
– The ability to support some form of helicopter operation has become increasingly common. This is an acknowledgement of the helicopter's utility in both antisubmarine and anti-surface operations.
– The trend towards larger ships has been reinforced by the desire of many of the more advanced navies to be able to deploy ships on expeditionary operations in littoral waters overseas.

This chapter examines some of the new littoral surface combatants that have been brought into service over the past decade, concentrating on US Navy and European designs. It aims to examine how differing national requirements have influenced these ships, and the extent to which designers have been successful in achieving their objectives.

United States: Littoral Combat Ship

An obvious starting point for any consideration of the current generation of littoral surface combatants is the US Navy's controversial littoral combat ship programme. Described as 'The Navy's most transformational effort' by the then-Chief of Naval Operations Admiral Vernon E

MODERN LITTORAL SURFACE COMBATANTS

The German fast attack craft *Frettchen*. Although specialised fast attack craft are increasingly confined to navies operating in geographical conditions that particularly favour their use, an increasing need for powerful littoral combatants is giving rise to a variety of alternative designs. (German Armed Forces)

Clark in 2003, the programme provided a clear indication of the US Navy's post-Cold War direction of travel from a focus on open-ocean warfare towards a greater interest in operations in coastal waters. Owing much to conceptual studies such as the 1990s 'Streetfighter', the new ships were intended to provide a relatively inexpensive 'seaframe' that could be rapidly reconfigured to undertake any one of a number of specific missions that might be required to defeat anti-access threats in the littoral.[6] Other key requirements included the ability to undertake sustained self-deployment to any part of the world, to combine a high sprint speed with good seakeeping and lower-speed stability, and to be interoperable with other US Navy combatants. Significant efforts were also made to limit the core crew size to the greatest extent possible.

Formally launched in November 2001, the littoral combat ship programme ultimately selected two very different design concepts. Contracts for detailed development work and the potential production of two pairs of prototype units were awarded in May 2004. One of these was awarded to a consortium headed by Lockheed Martin; the other to an alliance led by General Dynamics. Interestingly, the leaders of both groups allocated actual construction to other consortium members. The Lockheed Martin design is built by Marinette Marine (now part of Italy's Fincantieri) in Wisconsin whilst the General Dynamics variant is assembled by Austal USA in Mobile, Alabama.[7] It was intended that trials of the prototypes, alongside a competitive bidding process, would lead to the choice of a single design for series production of a class that would extend to over fifty ships. However, this down-selection process was never concluded as a result of implementation of an alternative strategy in 2010 that approved a block buy of a further ten units of each variant. It appears likely that construction will continue to be evenly split between the two types until at least twenty-eight littoral combat ships have been completed. It appears Saudi Arabia is also interested in acquiring four, much-modified Lockheed Martin variants.

The Lockheed Martin-designed littoral combat ship variant is named after lead ship *Freedom* (LCS-1). She was commissioned on 8 November 2008. Her design is based on the hull form of the Italian-built motor yacht *Destriero*, which gained the record for the fastest crossing of the Atlantic in 1992. Incorporating a semi-planing steel mono-hull and an aluminium superstructure, *Freedom* is a relatively large and heavy ship. Length – as originally designed – was 115m, beam around 17.5m and full load displacement in excess of 3,000 tons. The design fundamentals reflect the requirement to be able to deploy rapidly with meaningful payloads of mission equipment on a global basis; capabilities that necessitate a powerful propulsion plant. A combined diesel and gas (CODAG) arrangement is based around two Rolls-Royce MT-30 gas turbines and two Colt-Pielstick 16PA6B diesels. These provide a combined output of almost 85MW. This is sufficient to produce speeds in excess of 40 knots from four Rolls-Royce Kamewa waterjets. In

WARSHIP 2016

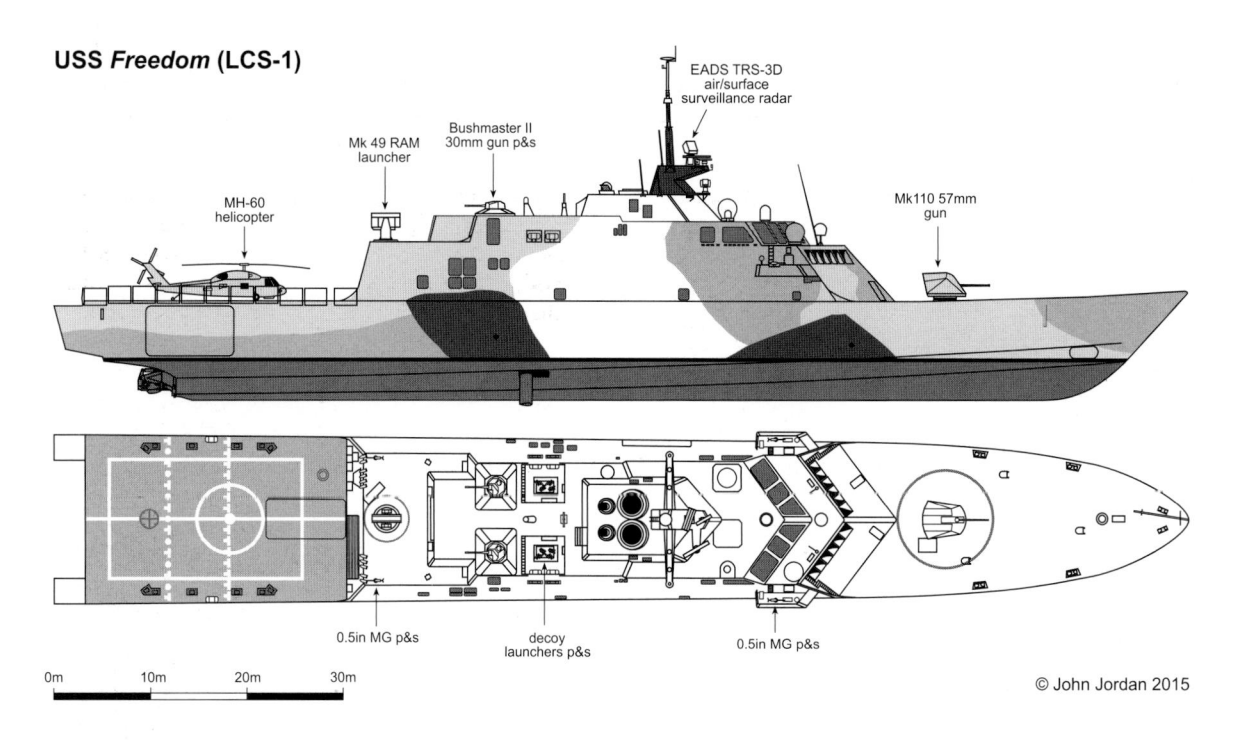

USS *Freedom* (LCS-1)

Freedom (LCS-1) is depicted as configured during her maiden deployment to Asia in 2013 with an interim surface warfare module embarked. The camouflage is derived from the wartime Measure 32 scheme and serves the additional purpose of hiding stains from the exhaust outlets in the hull. She had received a number of modifications since first delivery, notably the addition of two 3m-long extensions at either side of her stern. These 'water wings' are intended to increase reserve buoyancy and have been replaced by a full hull-extension in subsequent LCS-1 variant ships.

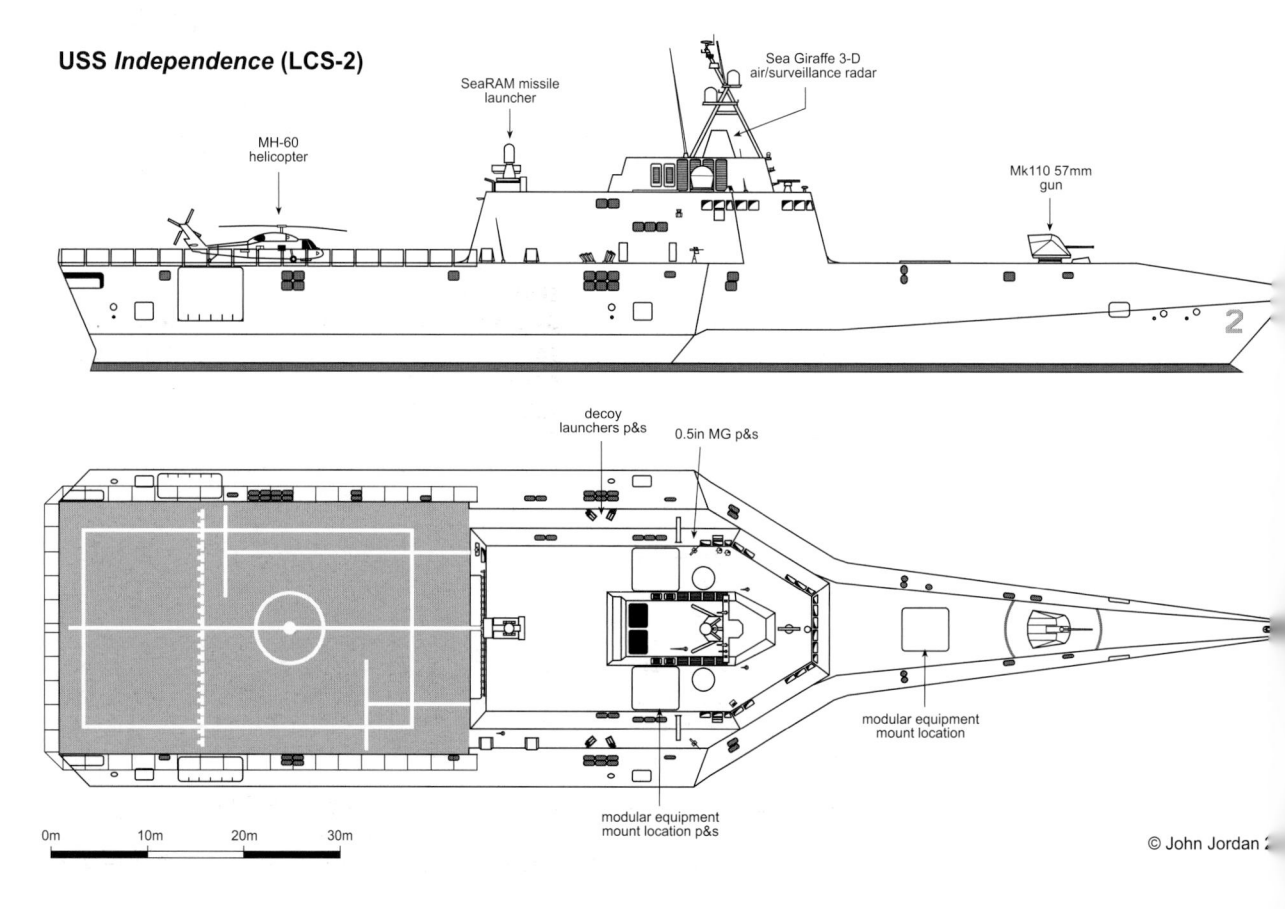

USS *Independence* (LCS-2)

MODERN LITTORAL SURFACE COMBATANTS

The US Navy LCS-1 *Freedom* type littoral combat ship *Fort Worth* (LCS-3) pictured during operations with the Royal Malaysian Navy in 2015. The littoral combat ship is being built to two distinct designs by different industrial consortia. The LCS-1 variant features a semi-planing steel hull derived from an Italian high-speed motor yacht. (US Navy)

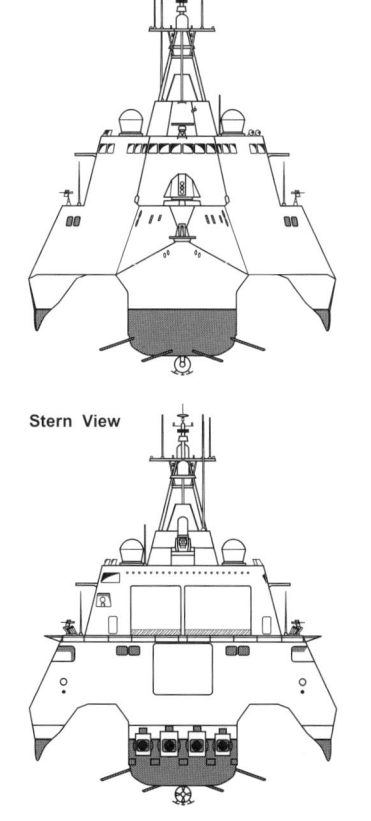

Bow View

Stern View

The profile and plan view of *Independence* (LCS-2) demonstrates the considerable volume provided by the variant's trimaran hull. This allows the incorporation of the largest flight deck found in any US Navy surface combatant, a large working/storage area beneath for mission equipment, and three mounting points for modular equipment.

comparative terms, it is interesting to note that two MT-30 turbines are able to produce nearly two thirds of the electrical output needed for the Royal Navy's new 65,000-ton *Queen Elizabeth* class aircraft carriers.

The General Dynamics/Austal *Independence* (LCS-2) variant has adopted a radically different configuration based on an all-aluminium trimaran design. Derived from the commercial fast ferry *Benchijigua Express*, the adoption of a lightweight trimaran hull provides much greater volume and deck area than a conventional mono-hull of similar displacement. Although displacing a little less than the LCS-1 type, *Independence* is a much larger ship. Overall length is c.127.5m and beam over 31.5m. This has permitted the incorporation of a flight deck that is larger than that of any other US Navy surface combatant and a mission bay more than twice as big as that found in *Freedom*. The lower hull resistance provided by the trimaran arrangement also allows achievement of a similar sprint speed to *Freedom* on lower installed power. Total output from two GE LM2500 gas turbines and two MTU 20V 8000 series engines is some 62MW, or around three quarters of that of the LCS-1 type. This has obvious benefits in terms of fuel efficiency and endurance.[8] As in the LCS-1 type propulsion is provided by four waterjets, in this case supplied by Finland's Wärtsilä. *Independence* was commissioned on 16 January 2010.

Although both are stealthy designs, *Freedom* and *Independence* are radically different in external appearance and incorporate significant variations in equipment fit. These include the use of distinct combat and platform management systems. There are also significant differences in sensor outfits. In spite of this, the two types offer fundamentally similar capabilities built around a core armament of a 57mm Mk 100 gun and a RIM-116 based point defence missile system (PDMS) supplemented by specific mission packages. The latter reflect an approach that originated with the Danish Stanflex 300 patrol vessels in the late 1980s. The Danish system is based on different combinations of weapons and equipment mounted in standardised containers that can be rapidly slotted into pre-prepared shipboard positions depending on the precise mission envisaged.[9]

The US Navy's modular mission packages have detailed differences from the Danish system. However, the overall 'plug and play' approach under which different equipment can be speedily integrated into a host ship and its combat management system is essentially similar. The LCS programme originally encompassed three specific packages aimed at providing capabilities particularly relevant to the littoral. These are antisubmarine warfare, mine warfare, and surface warfare. The packages – including mission modules and supporting aircraft and, importantly, mission crew – are common to both variants. They include emergent technologies such as autonomous mine-hunting systems and unmanned aerial vehicles (UAVs). A version of the surface warfare mission package has already been deployed operationally and the others will enter service shortly. Additional modular mission packages are under consideration. The modular system is arguably less expensive and more operationally effective than the use of a larger multi-mission combatant that would be a jack of all trades but master of none. It also facilitates upgrades as technologies develop.

The littoral combat ship programme has suffered from significant criticism almost from first inception. This has been based on concerns over both the underlying concept behind the project and its detailed implementation, and has not been altogether unjustified. Particular issues have included:

Construction problems: The two pairs of prototype ships were constructed under the then 'in vogue' concurrent design and engineering approach. Under this, fabrication commenced before the final design had been fully developed. Whilst allowing the prototype vessels to enter service more quickly than might otherwise be the case, the approach resulted in numerous construction defects and design deficiencies. This has required significant remedial work on the first vessels and redesign of the subsequent ships. For example, there has been a c.3m extension to the length of the *Freedom* variant to address reserve buoyancy issues.

Cost overruns: Partly as a result of the construction approach adopted, both variants of the littoral combat ship have cost much more to build than was initially expected. It was originally hoped that littoral combat ship seaframes would cost around US$220m in FY2005 terms. However, the first two ships cost around three times this amount. This resulted in orders for LCS-3 and LCS-4 – the final pair of prototypes – being temporarily shelved in 2007. Although construc-

The US Navy littoral combat ship *Independence* (LCS-2). An all-aluminium trimaran, she is the second of the littoral combat ship variants. (US Navy)

MODERN LITTORAL SURFACE COMBATANTS

The US Navy's littoral combat ships are built around a concept under which one of a range of specialised mission packages is embarked to support a particular operation. This 2012 photo shows *Independence* (LCS-2) testing a remote operating vehicle and AN/AQS-20 mine hunting sonar that forms part of the mine warfare mission package. (US Navy)

A MQ-8B Fire Scout unmanned aerial vehicle (UAV) being prepared for operation from the littoral combat ship *Fort Worth* (LCS-3) in August 2015. The operating potential of the US Navy's littoral combat ships is largely provided by the specialised equipment found in the various mission packages that can be embarked. (US Navy)

tion costs for the series-built ships are currently running around US$350m per unit (plus separately government-furnished equipment such as guns and other core equipment, which can add around US$100m per ship) this still makes them far from inexpensive. This is particularly the case when the additional cost of mission modules is factored in.

- **Mission module development:** Development and testing of specific mission modules has lagged behind both planned development timelines and the entry of the earlier littoral combat ships into service. There have also been significant changes in the specific components of the different modules. These have resulted in some reductions in performance.
- **Manning and support issues:** It was originally intended to operate each littoral combat ship with a core crew of around forty sailors. These would be supplemented by additional mission-related personnel and aircrew dependent on the requirements of a specific mission. In addition, much maintenance and administrative support traditionally carried on board ship would be carried out shore-side. When *Freedom* undertook her first operational deployment to South East Asia, the core crew was increased to fifty. However, official reports suggest this proved inadequate, with potential knock-on effects to operational availability.[10] There are concerns that the type will be significantly more expensive to operate and support than first envisaged.
- **Concept weaknesses:** Most significantly, there has been considerable concern that the littoral combat ship is not sufficiently well-armed or robust to survive in a higher threat environment. These concerns include the very limited potential for offensive warfare contained in the current designs; their relatively weak air-defence systems; and the acceptance of modest survivability standards. The littoral combat ships have been built in accordance with only limited US navy 'Level I+' survivability requirements. These are lower, for example, than those found in a legacy *Perry* (FFG-7) class frigate.[11] A number of commentators have suggested overseas multi-purpose frigate designs offer better capability for comparable cost.

The last-mentioned criticism, particularly, has been received with some sympathy by US Military leadership. This seems partly due to increased concerns over Chinese military capabilities, which have expanded significantly from the immediate post-Cold War environment in which the littoral combat ship concept was first conceived. In a February 2014 address, then-US Secretary of Defense Chick Hagel announced plans to truncate production after no more than thirty-two ships of a planned fifty-two unit programme had been ordered. Citing the need '…to closely examine whether the LCS has the independent protection and firepower to operate and survive against a more advanced military adversary and emerging technologies, especially in the Asia Pacific', he asked for proposals for a more lethal small surface combatant consistent with the capabilities of a frigate.

Many speculated the announcement marked a premature end to the littoral combat ship programme. However, in December 2014 it was decided that the new, multi-mission small surface combatant would be a modification of the existing designs. The upgraded ships – to be designated frigates – will essentially incorporate: permanently-installed anti-surface and antisubmarine capabilities; improvements to sensors and countermeasures; and additional protection at the expense of some loss to modularity. Some of the improvements will be back-fitted to earlier ships. In a further twist to the lengthy littoral combat ship saga, in December 2015 new Secretary of Defense Ashton Carter directed the Navy to reduce the overall programme to 40 ships to help fund more naval aircraft.

Norway: *Skjold* Class

The controversy impacting the US Navy littoral combat ship design is partly a reflection of the challenges and compromises inherent in designing a cost-effective warship that is intended to excel in performing a range of missions in coastal waters whilst being able to deploy globally. The LCS-1 and LCS-2 design solutions therefore make an interesting contrast with the Royal Norwegian Navy's *Skjold* class corvettes. These powerful, seaworthy vessels are much closer to the original FAC concept. They provide considerable anti-surface capabilities but have only limited potential in other roles. They are also arguably handicapped in terms of range and habitability. A variant of the *Skjold* – offered by Raytheon – was actually one of three concepts awarded a preliminary development contract for the US Navy programme in 2003. The design lost out to Lockheed Martin and General Dynamics when their competing proposals were selected for ultimate production the following year.

The *Skjold* class design was conceived towards the end of the Cold War as a replacement for earlier-generation fast attack craft. Although principally designed as a weapons platforms for high speed coastal operations, it was anticipated that the class would also be capable of wider deployment in support of NATO activities. Particular attention was paid to selecting the best hull form for this mission profile. A series of studies eventually resulted in the selection of an air-cushion catamaran/surface effect ship (ACS/SES) configuration.[12] As well as providing the best sea-keeping and shock resistance capabilities, this hull form offered a high speed-to-power ratio and maximised internal volume for any given displacement.

A contract for a prototype unit was awarded to what was to become Umoe Mandal in August 1996. The new ship was completed in April 1999 prior to commencing a four-year programme of trials. This included a year-long deployment to North American waters to showcase the concept to the US Navy. Throughout this period, *Skjold* operated with only a limited weapons and sensor fit, as the main aim was to demonstrate the design's overall functionality rather than to equip the ship for operations.

MODERN LITTORAL SURFACE COMBATANTS

The Norwegian *Skjold*-class corvette *Glimt* seen in February 2015. The class uses an air cushion catamaran (ACC) design and is capable of high speed and very shallow water operation when this is turned on. This picture shows the ship loitering at low speed with the air cushion turned off and supported only by the buoyancy of the twin hulls. Draught increases from 0.9m to 2.3m in this mode. (Henriette Daehli / Norwegian Armed Forces)

The lead *Skjold* class corvette pictured at speed. The operating concept behind the design is based on the rapid concentration of heavy firepower to overwhelm an attacking surface force. It relies on other elements of the Norwegian military handling air and underwater threats. (Torgeir Haugaard / Norwegian Armed Forces)

The end of the Cold War resulted in much debate as to whether the programme should be continued. Production contracts for a further five units were eventually signed in 2003. These included provision for *Skjold* to be upgraded to operational configuration. Completing the project proved to be more complex and took longer than first envisaged. However, the series production units started entering service from 2010 onwards. All six of the class are now operational.

Even taking into account their unusual hull form, the *Skjold* class are small, compact vessels. Length is 47.5m and full load displacement just 275 tonnes. Weight has been further reduced by the use of fibre-reinforced plastic as the main construction material. A lot of emphasis was placed on ensuring good stealth characteristics to assist overall survivability. As well as minimising radar, infrared and acoustic signatures this extended to the adoption of a specially-developed camouflage scheme based on the scientific study of Norwegian topography.

The principal role of the class is anti-surface warfare in the Norwegian littoral as a key part of broader sea control and anti-access operations. These would be carried out in conjunction with *Fridtjof Nansen* class multi-mission frigates, *Ula* class submarines and other elements of the armed forces. The operational concept involves the use of broadly distributed but fully networked firepower and sensors that can be rapidly but stealthily concentrated to overwhelm an enemy force. In addition to the stealth characteristics already referenced, this concept essentially requires the *Skjold* class to be capable of high-speed operation and carry a powerful surface-to-surface armament. The minimal resistance provided by the ACC hull allows speeds of 60 knots to be achieved from a combined gas and gas (COGAG) propulsion plant driving two Rolls-Royce Kamewa waterjets. This allows the entire Norwegian coast to be covered within six hours by just two units of the class. The ACC also facilitates operations in shallows of a metre or less when the cushion is in operation. Armament is focused on an enclosed battery of up to eight Kongsberg NSM Naval Strike Missiles housed in the after hull structure. They have a range in excess of 100 nautical miles and use an imaging infrared seeker in the final stages of an engagement to defeat electronic countermeasures. They are sufficiently powerful to disable a major surface combatant. The only other significant weapons system is a 76mm Oto Melara Super Rapid gun. In effect, stealth, speed and decoys provide the main defences against hostile counter-attack. The Thales MRR 3D NG radar which provides the main surveillance capability is specifically designed to identify incoming anti-ship missiles.

The *Skjold* class clearly provide considerable potential

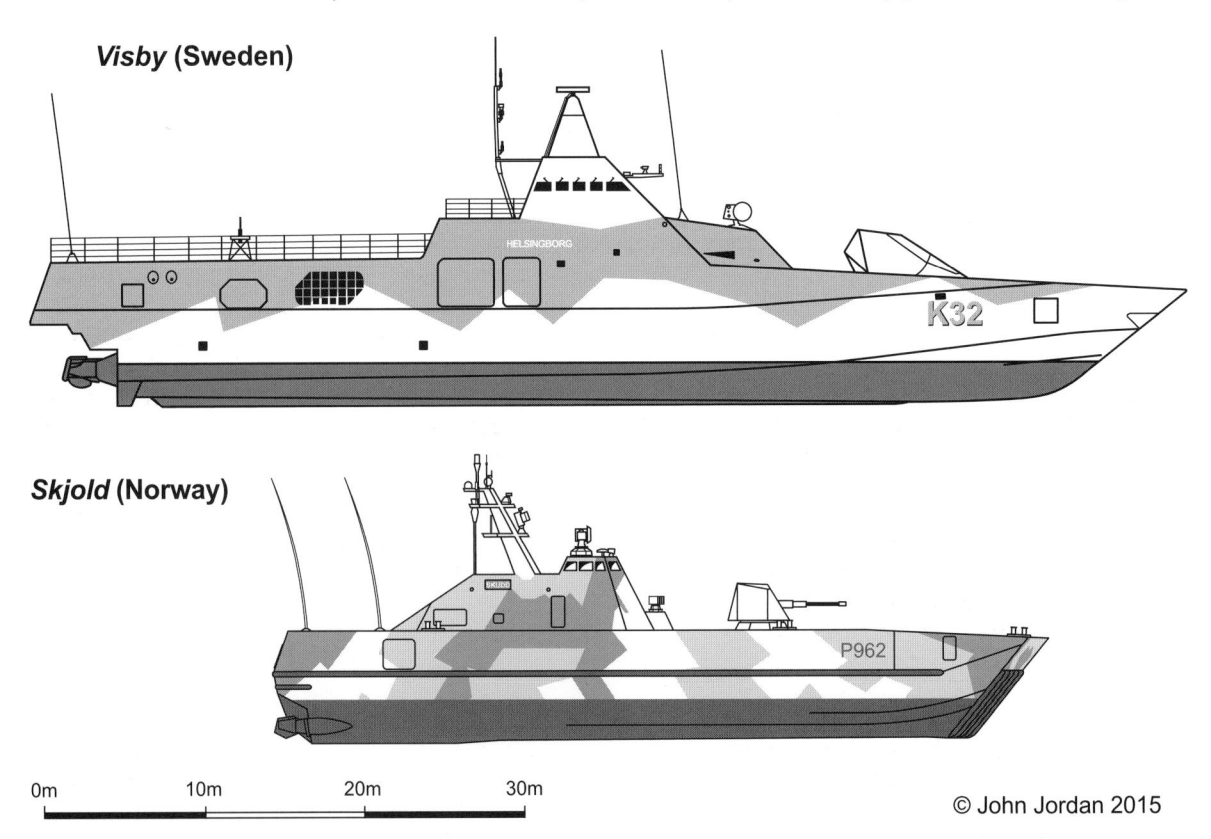

Sweden's *Visby* and Norway's *Skjold* classes – *Helsingborg* and *Skudd*, respectively, are shown here – represent different Scandinavian approaches to the design of a modern littoral warfare combatant. *Skjold* is a powerful but compact surface warfare combatant intended to operate with other units to ensure sea control in the littoral. The larger *Visby* has a much greater capacity for a range of operations, which extend to mine warfare and anti-submarine operations as well as the traditional fast attack craft's anti-surface role. Both ships rely heavily on stealth for survivability.

in their intended anti-surface, anti-access role. Norway also claims the basic design is sufficiently flexible to be adapted to support other missions. Certainly, the seaworthiness of the ACC hull facilitates open-water use. Moreover, their single-mission focus and modest crew size – they have a complement of just twenty – mean they are relatively cost-effective to acquire and support. However, lack of facilities such as a helicopter operating capability limits their potential for independent deployment. A range of 800 miles would also restrict the conduct of expeditionary operations, as would the minimal crew. Although arguably better-suited to high intensity warfare than the US Navy's littoral combat ships, they would inevitably be heavily reliant on the support of other units to counter air and underwater threats.

Sweden: *Visby* Class

An example of an approach somewhere between the modular and globally-deployable littoral combat ship concept and the potent but more fixed capabilities offered by the *Skjold* is provided by Sweden's *Visby* class corvettes. Also first conceived towards the end of the Cold War as a replacement for existing fast attack craft, the class shares the *Skjold* type's emphasis on stealth and has significant anti-surface warfare capabilities. However, possibly reflecting Sweden's previous use of FAC type vessels for antisubmarine operations, *Visby* and her sisters were also designed with significant antisubmarine and mine countermeasures potential. Their intended use therefore has similarities with the US Navy's littoral combatants, despite being much smaller ships.

The first pair of *Visby* class vessels was ordered in October 1995 following development work which began in the mid-1980s. Required class numbers fluctuated as a result of changing priorities and reduced funding. Ultimately, a class of only five vessels was completed. Similarly, plans to complete different variants with a focus either on mine countermeasures and antisubmarine operations (Series I) or anti-surface warfare (Series II) were also ultimately abandoned in favour of a single, multi-role configuration. Although *Visby* commenced trials towards the end of 2001, design changes, shipyard delays and cost overruns meant that deliveries were slow. It was only in December 2009 that the first ships were commissioned into frontline service. A series of incremental upgrades has been steadily implemented to enhance operational capabilities.

Incorporating a planing mono-hull design, the Swedish corvettes have an overall length of nearly 73m and are almost 10.5m broad. Full load displacement is c.650 tonnes. In similar fashion to *Skjold*, the hull structure utilises non-traditional materials comprising a double-skin, energy-absorbing carbon fibre-reinforced plastic sandwich. This reduces weight by as much as half of that of a conventional steel hull. *Visby* is also similar to the Norwegian ships in so far as the need to operate in confined and high-threat waters resulted in a premium

The *Visby* class corvette *Härnösand* seen operating in Norwegian waters in March 2012. The class emphasises stealth as a key means of maximising survivability during littoral operations.
(Torbjørn Kjosvold / Norwegian Armed Forces)

being placed on stealth characteristics. These are seen both as an aid towards maximising mission-performance and as a means of improving overall survivability. They are most evident in the shaping of the hull and superstructure to minimise radar cross-section. All major equipment is mounted internally, flush with the superstructure or subject to other concealment measures. Stealth also extends to techniques such as the use of radar-absorbent materials, infra-red reducing water-jet sprays and thermal-resistant camouflage paint. The main machinery is elastic-mounted to reduce radiated noise. Given a potential mine-warfare role, particular attention has been paid to demagnetising steel equipment and installing an advanced degaussing control system. Saab – current owners of the Kockums group that designed and built the class – describe their approach as Genuine Holistic Stealth or GHOST.

In terms of armament, the *Visby* class has greater conceptual similarity to the US Navy littoral combat ships in so far as a key design aim was to enable weapons fit to be tailored to a variety of specific missions. This essentially requires a combat management system – in this case the Saab 9LV CETRIS – which can readily integrate the various weapons and sensors that might be required, combined with sufficient space to accommodate a wide range of mission equipment. A voluminous equipment deck underneath the flight deck provides space for antisubmarine torpedoes, mines, depth charges and sonar. A working area immediately forward facilitates the deployment of remotely-operated Double Eagle

Mk3 and Seafox mine detection and mine disposal systems. Other weapons include a permanently embarked BAE Systems Bofors 57mm Mk3 gun (essentially the same weapon as that found on the littoral combat ships) and two quad RBS-15 surface-to-surface missile launchers. There is also provision for vertically-launched surface-to-air missiles, but funding constraints mean that these have yet to be authorised. Similarly, whilst there is a helicopter deck sufficient for light helicopter and UAV operations, proposals for a hangar were shelved on cost grounds. The principal surveillance and target indication radar is Saab's Sea Giraffe AMB; this is also used in the LCS-2 type littoral combat ship.

Possibly because the principal Baltic operating area is quite small, there is less emphasis on maximum speed in the *Visby* class than in some other littoral combatants. Nevertheless, the combined diesel or gas (CODOG) propulsion plant and Kamewa waterjets permit a sprint speed in excess of 35 knots. Equally, reported range of around 2,500 nautical miles is relatively respectable. This provides reasonable potential for out-of-area operations. Earlier-generation Swedish corvettes have already participated in the European Union's Operation Atalanta antipiracy mission off the Horn of Africa, and the *Visby* class would arguably be well-suited to this role.[13] However, it is questionable whether the type's small size and relatively limited maximum accommodation – space is provided for a maximum 53 personnel – would permit sustained, independent deployment on expeditionary duties. Indeed, design work on larger 'Visby Plus' variants with more internal volume and provision for a larger crew could be regarded as providing some acknowledgement of the limitations of the existing ships.

Germany: *Braunschweig* (K-130) Class

The growing importance of international missions was a key influence on Germany's *Deutsche Marine* when the time came to recapitalise its own fleet of littoral combatants. In a similar fashion to Sweden, it has a long history of fast attack operations in the Baltic. However, it had become increasingly aware of the weaknesses of its existing vessels as they were deployed more widely in support of NATO and United Nations stabilisation missions. The navy's involvement in the maritime component of the UNIFIL peacekeeping force in Lebanon was particularly influential in highlighting problems of endurance, sea-keeping and crew fatigue associated with the prolonged use of FAC-type vessels away from home. As a result, the user requirements for what was to become the K-130 *Braunschweig* class placed significant emphasis on suitability for sustained deployment in both oceanic and littoral waters. There was also a particular focus on anti-surface warfare and surveillance capabilities.

An order for five K-130 type corvettes was placed in December 2001 following the selection of the ARGE K130 Consortium as preferred supplier the previous year.[14] Each of the three members of the consortium was allocated fabrication of a specific part of the class; final assembly was split between all three yards. Construction commenced in mid-2004 and the first two ships were in commission by the end of 2008. However, a number of

The *Visby* class corvettes *Helsingborg* (front) and *Visby* (rear) pictured in trials in the Baltic in 2005. The design is based on a modular equipment concept that is not altogether dissimilar to that adopted by the US Navy's littoral combat ships. An extended development period has meant that it is only now that the class is reaching its full potential. (Peter Neumann/Copyright Saab AB)

MODERN LITTORAL SURFACE COMBATANTS

problems – particularly the discovery of defective gearboxes – significantly delayed the programme. This resulted in the temporary lay-up of the commissioned ships until new gearboxes were installed, and postponed the commissioning of the final three units until 2013.

Derived from the Blohm & Voss MEKO A-100 design, the *Braunschweig* class corvettes utilise a conventional hull form and displace around 1,850 tonnes. Overall length is just over 89m and beam is 13.3m. Although less obviously stealthy than the *Visby* class, significant attention was paid to reducing the overall signatures. This included the adoption of an X-form design, under which the ship's hull and superstructure are inclined at different angles. Thermal emissions are minimised by the injection of seawater into the exhaust piping and the use of jets to create a water mist around the outlets, which are positioned just above the waterline. The enhanced threat posed by mines in the littoral is reflected in measures to reduce acoustic signature and the use of an advanced degaussing system.

The class is built in accordance with the original MEKO concept. Under this, weapons and sensors are laid out as modules and the combat information centre and other key compartments feature modular foundation systems. However, actual equipment fit reflects the strong emphasis on anti-surface warfare. Two quad launchers for the latest Mk3 variant of the RBS-15 surface-to-surface missile provide a powerful strike capability against both ships and shore-based targets. These are supplemented by a 76mm Oto Melara Compact gun and two smaller Rheinmetall Mauser MLG 27mm weapons that have a secondary role against air threats. The main anti-air warfare system comprises two Mk49 launchers for RIM-116 Rolling Airframe Missiles (RAM). These are principally designed for point defence against anti-ship missiles but have been developed to provide a broader capability against air targets. An Airbus Defence & Space TRS-3D surveillance and target acquisition radar is the primary sensor. It is effective in identifying both surface and air threats and is optimised to deal with the 'cluttered' conditions found in the littoral. The same system is used in the LCS-1 littoral combat ship variant.

The *Braunschweig* design incorporates a flight deck that is sufficiently large to support operations by all helicopters in the *Deutsche Marine*'s inventory. However, storage facilities are only sufficient for UAVs. There is no inherent antisubmarine capability beyond that provided by any embarked helicopter, although installation of a torpedo defence system is one future enhancement that has been considered. There is also no installed mine detection or neutralisation equipment. It would seem that these design compromises reflect the German Navy's ability to use its Type 212A submarines and extensive mine countermeasures flotilla to undertake these missions, allowing the K-130 class to focus on their primary anti-surface role. This concept might, however, result in reliance being placed on allied forces to provide these capabilities in expeditionary operations if the complementary German units were not immediately to hand.

Another design compromise has been the class's relatively low speed. The diesel propulsion plant is linked to two conventional shaft lines by a cross-connected

The K-130 *Braunschweig* class corvette *Magdeburg* was delivered in 2008 but was subsequently laid up pending rectification of defective machinery. The class was developed to remedy limitations in habitability and endurance inherent in previous-generation fast attack craft as the German Navy became more involved in international stabilisation missions in the post-Cold War era. (German Armed Forces)

Magdeburg pictured in March 2014 during a littoral warfare exercise off the Norwegian coast. The K-130 class has strong anti-surface warfare potential in both littoral and oceanic waters but is more limited in other areas. (Morten Opedal / Norwegian Armed Forces)

MODERN LITTORAL SURFACE COMBATANTS: PRINCIPAL CHARACTERISTICS

Class:	*Freedom* (LCS-1)	*Independence* (LCS-2)	*Skjold*	*Visby*
Country:	United States	United States	Norway	Sweden
Number:[1]	3+11[2]	3+11[2]	6	5
Keel Laid:[4]	2 June 2005	19 January 2006	4 August 1997	17 December 1996
Launched:[4]	23 September 2006	26 April 2008	22 September 1998	8 June 2002
Commissioned:[4]	8 November 2008	16 January 2010	17 April 1999[5]	10 June 2002[6]
Full Load Displacement:	3,500 tons	3,000 tons	275 tonnes	650 tonnes
Principal Dimensions:	118.1m x 17.6m x 4.3m	127.6m x 31.6m x 4.3m	47.5m x 13.5m x 2.3m	72.7m x 10.4m x 2.5m
Propulsion:	CODAG + waterjets	CODAG + waterjets	COGAG + waterjets	CODOG + waterjets
	40+ knots top speed	40+ knots top speed	60+ knots top speed	35+ knots top speed
	3,500nm range	4,300nm range	800nm range	2,500nm range
Main Armament:	1 x 57mm Mk110 gun	1 x 57mm Mk110 gun	1 x 76mm Oto Melara	1 x 57mm Bofors Mk3
	1 x RAM Mk49 CIWS	1 x Sea RAM CIWS	2 x quad Kongsberg SSM	2 x quad RBS-15 Mk2 SSM
	Mission specific module[7]	Mission specific module[7]		4 x 400mm TT
				MCM ROVs
Aircraft:	2 x MH-60R Seahawk or	2 x MH-60R Seahawk	No provision	Flight deck for one light
	1 x MH-60R & UAVs	1 x MH-60R & UAVs		No hangar
Principal Radar:	1 x EADS TRS-3D	1 x Saab Sea Giraffe	1 x Thales MRR-3D NG	1 x Saab Sea Giraffe
Principal Sonar:	Optional mission module	Optional mission module	No provision	Hydra multi-sonar
Crew:	Core crew c.40	Core crew c.40	c.20	c.43
	Mission specific c.25–30	Mission specific c.25–30	Accommodation for 21	Accommodation for 53
	Accommodation for 75	Accommodation for 75		

Notes:
1. First number relates to ships delivered as of the autumn of 2015; second number relates to ships under construction or subject to firm plans.
2. A further twelve ships, based on either the LCS-1 or the LCS-2 design, are currently planned for a total of forty. It seems that only the last eight will be upgraded to the new frigate type standard.
3. Four additional units of a modified design are planned.
4. Dates relate to the first of class. Some launch dates relate to the formal naming date rather than physical launch.
5. Relates to first commissioning as a trials ship. Following rebuilding to production standard, she was recommissioned in April 2013. The first operational ship was commissioned in September 2010.

gearbox and is sufficient for speeds of up to 26 knots. The compensating benefit is a range of over 4,000 nautical miles at economical speeds of around 15 knots. This is bettered only by *Independence* amongst the other ships considered by this article. Accommodation for up to 65 personnel might also provide better potential for sustained deployment than the lower provision made in the Scandinavian ships.

Turkey: MILGEM ('Ada') Class

Broadly comparable in size to the German K-130 class are Turkey's MILGEM or 'Ada' class corvettes. The first of these, *Heybeliada*, entered service in September 2011. However, whilst the German ships have a strong focus on anti-surface warfare, the MILGEM class have been designed as general-purpose littoral combats with a particular emphasis on antisubmarine warfare. This probably reflects Turkey's continued investment in more traditional fast attack craft as the navy's primary anti-surface weapon. The waters of the Aegean archipelago remain well-suited for fast attack craft operations. This is reflected in the fact that Turkey's main regional rival, Greece, also retains significant numbers of the type. This combination of missile-armed fast attack craft and larger multi-mission corvettes for littoral defence is also found in other navies. For example, South Korea's new *Incheon*-class light frigates supplement its PKX missile-armed fast attack craft. Similarly, the Chinese People's Liberation Army Navy operates both Type 056 'Jiangdao' corvettes and Type 022 'Houbei' missile boats in the littoral role.

The MILGEM programme – taking its name from the Turkish *Milli Gemi* or 'national ship' – dates back to the 1990s. However, local conditions meant that it was to be a further decade before tangible progress was made. Design work was carried out by a dedicated project office within the Turkish Navy, with the keel of the lead ship being laid in 2005. From the start a key aim was to maximise the amount of indigenous equipment used in the ship. Turkish politics have had a significant impact on the programme, and construction of the type is likely to be curtailed at only four ships compared with eight originally planned. However, the final quartet are likely to be replaced by a larger, improved design with an enhanced equipment outfit.

Heybeliada and her sisters bear more than a passing resemblance to the LCS-1 type design. However, the class's conventional displacement-type mono-hull is not designed for high-speed operation and has more in common with the *Braunschweig* class. The MILGEM design is also somewhat smaller, with a full load displacement of around 2,300 tonnes and a length of just under 100m. The CODAG machinery arrangement – comprising two MTU 16V95 diesels and a single GE LM-2500 gas turbine – is capable of producing speeds of up to 30 knots through a cross-connected twin shaft arrangement. Range at cruising speed is a respectable 3,500 nautical miles.

The MILGEM class reflects the same attention to stealth seen in other current-generation littoral warships. This is most evident in the angled configuration of hull and superstructure and the enclosure of much radar-reflecting equipment. However, stealth characteristics are inherent throughout the design. The maintenance of minimal acoustic properties has been a priority given the antisubmarine role of the class, and the propulsion system and other key noise-generating machinery are raft-mounted. Infrared signature is computer-monitored and can be reduced by a wash-down system that also provides part of the ship's protection against nuclear, biological and chemical (NBC) attack.

The MILGEM type's general-purpose role is evidenced by an armament that includes a 76mm Oto Melara in Super Rapid configuration, Harpoon surface-to-surface missiles, a RAM point defence missile system, twin fixed torpedo tubes, and a flight deck and hangar for a medium-sized helicopter. The focus on indigenous equipment is reflected in the use of a locally-developed Havelsan G-MSYS combat management system, Aselsan ARES 2-N electronic support measures and the latter company's STAMP stabilised machine gun platforms. Aselsan also produces the MILGEM's Thales SMART-S Mk2 surveillance radar under licence. The bow-mounted sonar is a Turkish development of a US Navy design. An indigenous torpedo defence system is under development for installation in later units of the class.

Braunschweig (K-130)	*Heybeliada* (MILGEM)
Germany	Turkey
5	2+2[3]
December 2004	26 July 2005
19 April 2006	27 September 2008
16 April 2008	27 September 2011
1,850 tonnes	2,300 tonnes
89.1m x 13.3m x 3.4m	99.5m x 14.4m x 3.7m
2-shaft diesel	2-shaft CODAG
26 knots top speed	30 knots top speed
4,000nm range	3,500nm range
1 x 76mm Oto Melara	1 x 76mm Oto Melara
2 x quad RBS-15 Mk3 SSM	2 x quad Harpoon SSM
2 x RAM Mk49 CIWS	1 x RAM Mk49 CIWS
	2 x twin 324mm TT
Flight deck for one medium	1 x S-70B Seahawk
Hangar for UAVs only	
1 x EADS TRS-3D	1 x Thales SMART-S Mk2
No provision	Yakamoz bow-mounted
c.58	c.85
Accommodation for 65	Accommodation for 110

6. Relates to first delivery as a trials ship. The first two operational units were formally commissioned in December 2009.
7. The ships are equipped with mission bays and modular equipment mounts for specific (1) mine warfare (2) surface warfare or (3) anti-submarine warfare modules.

The lead Turkish MILGEM type corvette *Heybeliada* pictured in March 2014. A general-purpose littoral design with particular strengths in anti-submarine warfare, the type has been used as a means of developing the Turkish naval equipment sector. (Turkish Navy)

Heybeliada and her sister, *Büyükada*, have already undertaken extended deployments including operational service rescuing Turkish civilians from the civil war in Yemen and a circumnavigation of the African continent. The improved design – which will include a vertical launch system for surface-to-air missiles – will further improve the type's flexibility. Whilst not being able to match the speed and modularity inherent in the US Navy littoral combat ships, this is a balanced and practical design that marks a big step forward in the advancement of Turkish maritime capabilities.

Conclusion

The importance of the littoral has been reflected in considerable investment in surface combatants designed to control its waters. As evidenced by this chapter, the ships that have resulted from these efforts demonstrate a number of common features. These include trends towards greater size and flexibility, as well as the increasing use of stealth. Equally, however, the different designs have varied significantly in dimensions, capability and operational focus. This, in turn, reflects the differing requirements of the navies which have spon-

The MILGEM type corvette *Büyükada* conducting a replenishment at sea with the Royal Australian Navy's tanker *Success* in February 2015 during stabilisation operations in the Indian Ocean. The MILGEM design has proved capable of making lengthy deployments in support of Turkish interests. (Royal Australian Navy)

MODERN LITTORAL SURFACE COMBATANTS

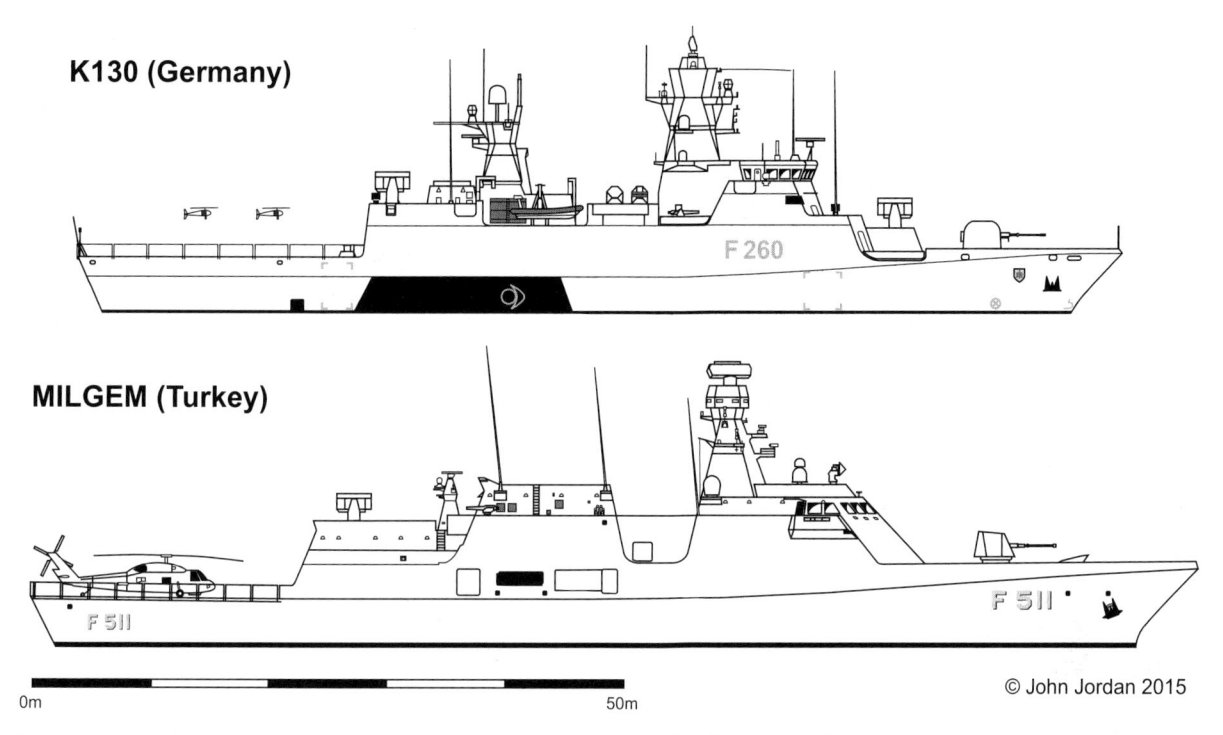

K130 (Germany)

MILGEM (Turkey)

© John Jordan 2015

Germany's K-130 *Braunschweig* class and Turkey's MILGEM are more traditionally-configured surface combatants that can be used in both littoral and oceanic roles. The K-130 is heavily orientated towards anti-surface warfare, whereas the MILGEM type has additional antisubmarine potential.

sored their construction. It is not surprising, for example, that a fleet with global responsibilities – such as the US Navy – has placed a premium on large and flexible warships with sufficient speed and endurance to deploy rapidly in order to cover a range of eventualities in any corner of the world. Equally, a ship intended primarily for operations in local waters – such as many of those operated by the Scandinavian fleets – is able to sacrifice some of these capabilities to maximise potency in a primary role. Consequently, direct comparisons between different designs have very limited value.

The limitations of such an exercise are further highlighted by the fact that warships increasingly undertake their missions as part of a wider network of systems rather than as discrete units. The *Skjold* class, for example, seems at first glance to retain much of the vulnerability to aerial attack that traditional fast attack craft demonstrated in the 1990–91 Gulf War. In practice, however, the corvettes would typically operate under an umbrella of air cover provided by the Royal Norwegian Air Force and the Aegis-equipped frigates of the *Fridtjof Nansen* class. This is even more the case with respect to the network-centric US Navy.[15] Although its littoral combat ships have been much-criticised for their potential vulnerability in a high-intensity area, these criticisms take no account of the fact that they would not be risked in such a situation without support from the very wide range of other ships and systems that the US military is able to deploy.

In conclusion, therefore, the success achieved by the designers of this new generation of 'western' littoral surface combatants can only be measured in terms of the objectives set by the fleets that have ordered them. This brief overview says most about how these navies envisage operating in the littoral, and the additional capabilities they see themselves requiring to achieve success.

Sources:

This chapter has been researched from contemporary industry and government literature, as well as press releases and news reports. The following sources provide more enduring reading material:

Clarke, Barry, Jurgen Fielitz & Malcolm Touchin (Editor Professor G Till), *Coastal Forces*, Brassey's (UK) Ltd, 1994.

O'Rourke, Ronald, *Navy Littoral Combat Ship (LCS)/Frigate Program: Background & Issues for Congress RL33741 – 12 June 2015*, Congressional Research Service, Washington DC, 2015.

Polmar, Norman, *The Naval Institute Guide to the Ships and Aircraft of the U.S. Fleet – Nineteenth Edition*, US Naval Institute, Annapolis, MD, 2013.

Slade, Stuart, 'Fast Attack Craft', *Conway's History of the Ship, Navies in the Nuclear Age*, pp.98–109, Conway Maritime Press (London, 1993).

Toremans, Guy, 'Sweden's *Visby* Class Corvettes – Stealth at All Levels', *Seaforth World Naval Review 2012*, pp.148–165, Seaforth Publishing (Barnsley, 2011).

Toremans, Guy, '*Braunschweig* Class Corvettes: Eagerly awaited by the German Navy', *Seaforth World Naval Review 2013*, pp.128–147, Seaforth Publishing (Barnsley 2012).

Toremans, Guy, '*Skjold* Class FACs: Norway's Fighting Cats –

Stealth reigns Supreme ', *Seaforth World Naval Review 2015*, pp.124–139, Seaforth Publishing (Barnsley, 2014).

Truver, Scott, 'USS *Freedom* Littoral Combat Ship – "Seaframe" for the future US Fleet', *Seaforth World Naval Review 2010*, pp.150–166, Seaforth Publishing (Barnsley, 2009).

Truver, Scott, 'USS *Independence* (LCS-2) Littoral Combat Ship – The "Wow!" Factor', *Seaforth World Naval Review 2011*, pp.136–153, Seaforth Publishing (Barnsley, 2010).

Footnotes:

1. The definition of the littoral varies widely. For the purpose of this chapter, the author has taken the broad delineation set out in the US Navy's operational concept, viz. 'The littoral is composed of two segments. The *seaward* portion is that area from the open ocean to the shore that must be controlled to support operations ashore. The *landward* portion is the area inland from the shore that can be supported and defended directly from the sea.' See *Naval Operations Concept 2010*, p 8, Department of the Navy, Washington DC, 2010.

2. Concluded in 1982 but not taking effect until November 1994, the current UNCLOS III regime formally adopted the concept of a national Exclusive Economic Zone (EEZ). Most significantly, this gives states sole rights to marine resources up to 200 nautical miles from their coasts. The convention also provides for more limited rights in respect of a country's continental shelf, which can extend national interests up to 350 nautical miles from its coastline.

3. Although somewhat outside the scope of this chapter, ensuring the US Navy's ability to operate securely in littoral waters in the face of the development of anti-access/area denial (A2/AD) strategies by potential adversaries such as China and Iran has become a key strategic objective for the US Military. One solution has been bolstering cooperation between the services, as evidenced by the development of the 2010 Air/Sea Battle doctrine, now renamed the less snappy Joint Concept for Access and Maneuver in the Global Commons (JAM-GC).

4. The author has examined recent offshore patrol vessel and amphibious assault ship design trends in previous editions of *Warship*.

5. The potential of the missile-armed fast attack craft was first demonstrated by the destruction of the Israeli destroyer *Eilat* by SS-N-2 Styx surface-to-surface missiles fired from Egyptian Project 183R 'Komar' class missile boats in October 1967. The engagement spurred many smaller navies to acquire similar weapons as a perceived equaliser to the major surface combatants deployed by the larger fleets. The experience of the Gulf War suggested that these craft were themselves highly vulnerable to counter-attack from missile-armed helicopters and other aircraft operating outside the envelope of their defensive systems.

6. The Streetfighter was part of a concept that envisaged the acquisition of large numbers of powerfully-armed, fast coastal warships that would be sufficiently cheap and numerous to be regarded as expendable. The US Navy littoral combat ships, as built, are far larger and more complex vessels.

7. Austal USA has subsequently assumed the role of primary contractor for the LCS-2 programme. It is also responsible for the US Navy's JHSV joint high speed vessel, which is built in the same Mobile yard.

8. This argument is a simplification, as it relates to operation at full power. *Freedom*'s diesels are reportedly more economical than those installed in *Independence,* suggesting she should be more efficient to operate at normal speeds. Similarly, whilst the LCS-2 design provides more space, supporters of the LCS-1 variant suggest it is more manoeuvrable in the littoral waters for which the type has been designed. It is also interesting to note that *Freedom* and her sister, *Fort Worth* (LCS-3), were selected over the LCS-2 variant for initial overseas deployments of the LCS type.

9. The experience of the Stanflex approach was not entirely positive with regard to the Stanflex 300 class, largely because their crews lacked the experience to be fully effective in all the envisaged roles. As a result, members of the class were allocated specific Stanflex outfits and their crews trained accordingly. The modules have, however, been successful and many have been transferred to new ships as the Stanflex 300 vessels were phased out due to the Royal Danish Navy's post-Cold War re-orientation to more blue-water roles.

10. A widely-reported Government Accountability Office report on *Freedom's* inaugural overseas mission suggested that the enlarged crew struggled to get adequate sleep in spite of assistance from the specialised mission module crew and outside contractors. Defects also prevented the ship spending as much tine underway as planned. See *Deployment of USS Freedom Revealed Risks in Implementing Operational Concepts and Uncertain Costs GAO-14-447* published by the Government Accountability Office, Washington DC in July 2014. In fairness, the experiences of a pilot deployment of a new ship are unlikely to be typical of ongoing operations once initial snags have been ironed out.

11. Until the adoption of new standards in 2012, US Navy warships were designed to one of three survivability standards, viz. Level I (low), Level II (moderate) and Level III (high). Aircraft carriers and major surface combatants have been designed to Level III standards, with frigates and amphibious vessels to Level II. The Level I standard selected for the littoral combat ships also applies to mine warfare vessels, patrol vessels and support ships. See *OPNAV Instruction 9070.1* issued by the Office of Chief of Naval Operations, Department of the Navy on 23 September 1988.

12. A surface effect ship combines the air cushion of a hovercraft with the rigid twin hulls of a catamaran. The ship is supported by the buoyancy of the catamaran hulls when the cushion is turned off, rising so that less of the fixed hull area remains in the water when the air cushion is turned on.

13. The deployment of the older corvettes *Stockholm* and *Malmö* to the Indian Ocean in support of Operation Atalanta in 2009 was regarded as a success, with the speed and manoeuvrability of the ships being an asset in anti-piracy duties. However, the deployment required the use both of a heavy lift ship to transport the corvettes to and from the area of operation and ongoing deployment of a support ship. The experiment has yet to be repeated.

14. The ARGE K-130 Consortium comprised Blohm & Voss, Nordseewerke Emden and Friedrich Lürssen Werft and was selected over the HDW Project Group K-130. The first two yards were later to merge with HDW to form ThyssenKrupp Marine Systems. The latter's direct involvement in surface warship construction has subsequently been scaled back through the disposal of many shipbuilding facilities.

15. A good overview of network-centric warfare is contained in Norman Friedman's *Network-centric Warfare: How Navies Learned to Fight Smarter Through Three World Wars*, US Naval Institute, Annapolis, MD, 2009.

THE CHINESE FLAGSHIP *HAI CHI* AND THE REVOLUTION OF 1911

Richard Wright, author of *The Chinese Steam Navy*, attempts to resolve a mystery concerning the activities of the Chinese flagship *Hai Chi* during 1911–12.

To the casual naval historian a scenario whereby the Chinese cruiser *Hai Chi* was sent to represent her country at the Coronation Fleet Review of King George V at Portsmouth (UK) in June 1911, followed by a visit to New York (and possibly Mexico) in September, and was then hurriedly recalled to defend Shanghai at the outbreak of the Chinese Revolution in October, might seem to provide a satisfactory précis of her movements for that year. The ship's presence at the Spithead Fleet Review is well documented, as is her visit to New York, and her triumphant arrival (*sic*) back in Shanghai straight from New York is written into a history of the Revolution first published in 1913.[1]

Yet there are unexplained gaps in her progress through western waters, and garbled mentions of her various visits, such as 'the Chinese cruiser *Hai Chi* took part in New York's Spithead Naval Review', render the official version suspect. Since no single concise account of her wanderings in western waters appears to exist, the following outline, resurrected in the main from contemporary reports but utilising various points brought out in an article on the *Hai Chi* in the Chinese Wikipedia dated 2013,[2] is presented as a reasonably accurate reconstruction of the lead-up to her deployment to the West in 1911 and of the deployment itself; the account which results contains some surprises.

The story begins in 1895 at Wei-Hai-Wei at the conclusion of the Sino-Japanese War. The remnants of the Peiyang (or Northern) Fleet had surrendered to the Japanese. There were, amongst others, three Chinese naval officers of later note in the fleet. One was Liu Kuan-hsiung (pinyin: Liu Guanxiong),[3] the second in command of the cruiser *Ching Yuen*. His political sympathies appear to have been firmly with North China. Another was Cheng Pi-kuan (also Ching or Chin; pinyin: Cheng Biguang), in command of the last of the three ships of the (attached) Canton contingent, the torpedo gunboat *Kuang Ping*. His political sympathies were so emphatically Cantonese (or Southern) that he even petitioned the Japanese with the request that his ship not be surrendered, as Southern China was not really at war with the Japanese – his request was ignored. The third was Sah Cheng-ping (pinyin: Sa Zhenbing), in command of the training ship *Kang-Chi*. He had no political affiliations, but served his current naval masters faithfully throughout his long career.

Rebuilding the Chinese Fleet

Following the total loss of all the ships of the Peiyang Fleet, replacements were sought in European dockyards. From a thin budget these included four German-built torpedo boat destroyers, three small German-built cruisers of 2,950 tons (*Hai Yung*, *Hai Chou* and *Hai Chen*), and two larger protected cruisers of 4,300 tons named *Hai Tien* (pinyin: *Hai Tian*) and *Hai Chi* (*Hai Qi*). The latter two ships were built in the UK at Armstrong Elswick on the Tyne to a fairly standard design of export cruiser. Armed with two 8in guns and ten 4.7in quick-firers (QF), their near-sister was the Argentinian *Buenos Aires*. The foremasts of these ships, fitted unusually ahead of the bridge, were a distinguishing feature. Cheng Pi-kuan was one of the Chinese naval officers sent to oversee their construction 1896-1899.[4]

Both ships were completed during the first half of 1899, and were then steamed out to China by contract crews. Some attempt was apparently made by Italy to buy *Hai Chi* that year, but this came to nothing. Ironically, Italy was pressurising China into giving it a lease over Sanmen Bay (just south of the Chusan Islands)

Characteristics

Built:	Armstrong Elswick
Laid down:	1896
Launched:	24 Jan 1898
Completed:	1899
Displacement:	4,300 tons
Dimensions:	396ft pp x 46ft 8in x 16ft 9in (mean)
Machinery:	2-shaft TE, 17,000ihp = 24 knots
Main guns:	2 – 8in (2xI), 10 – 4.7in QF (10xI)
ATB guns:	16 – 3pdr (16xI)
Torpedoes:	5 – 18in TT
Protection:	protective deck 1½in with 3in slopes; 4½in gunshields, 4in hoists; CT 6in
Complement:	350

Source: *Conway's All the World's Fighting Ships 1860–1905*.

Hai Chi sailing for China from England in 1899 with a contract crew and flying the British Red Ensign. Her paint scheme is the 'Victorian Livery' of black hull, white upperworks and buff funnels. (Newcastle City Library)

at the time, following similar deals made by Britain, Russia, Germany and France in 1898, but the approach was somewhat half-hearted. The only sabre-rattling undertaken by Italy was by a small naval force in the autumn of 1899 which penetrated north as far as Chefoo (Yentai), before retiring south after achieving nothing.[5] Had it gone farther west, into the Gulf of Pechihli (Bohai), there could have been a surprise awaiting it in the shape of a small fleet of five modern Chinese cruisers, four torpedo boat destroyers, plus two torpedo gunboats, which were all reported assembled in the Taku (Tagu) roadstead by the second week of August[6] – albeit not necessarily fully manned or worked up. (The naval ports of Port Arthur, Wei-Hai-Wei and Tsingtao [Qingdao] had been leased to the Russians, the British and the Germans respectively the previous year.)

A year later (1900) the Boxer Rebellion took place, centred in and around Peking, and the international legations were besieged. A large multi-national naval force assembled off Taku, initially coexisting uneasily with the strange Chinese squadron on the far side of the anchorage, which showed no sign of hostility. In the end the Chinese flagship, *Hai Yung*, was invited to anchor in the centre of the international armada, which she did, and the remainder of the Chinese warships in the roadstead, led by *Hai Tien*, were ordered south to Shanghai by the local viceroy. However, *Hai Chi*, now commanded by Commodore Sah Cheng-ping, elected to remain in the gulf for a while, in the process providing valuable aid for stranded missionaries along the coast and standing by the US battleship *Oregon* which had run aground. In the end only the torpedo boat destroyers proved a problem for the Allies, as they were berthed in Taku dockyard and were assessed as a possible threat to the forthcoming assault on the Taku forts. They were quietly and efficiently boarded and captured.

Following the relief of the Peking legations and the suppression of the Boxers, things settled down to normal on the China coast; that is until the cruiser *Hai Tien*, coming south to Shanghai in thick fog, overshot the entrance to the Yangtze and impaled herself at an angle of 15 degrees on a pinnacle rock just off Elliot Island in the approaches to Shanghai. She proved incapable of salvage.

Hai Tien hard aground on rocks in the approaches to the Yangtze, 1904. (MPL)

The name of her commanding officer seems to have been air-brushed out of history, but it is likely to have been that of Liu Kuan-hsiung, of whom Professor Rawlinson wrote: '1904, cashiered for grounding his ship'.[7]

This tragedy left the *Hai Chi*, at 4,300 tons, the largest and finest ship of the Chinese Navy for the next thirty-odd years. Under the command of Sah Cheng-ping she had already earned a name as the ship with the highest standard of cleanliness and efficiency in the service, and this tradition was to continue. And in 1905, on his predecessor's death, Admiral Sah was to take over the job of Commander-in-Chief of the newly combined Chinese Navy, amalgamating the former Peiyang, Nanyang, and Kwangtung (North, Central and Southern) squadrons, with a new base at the Kiangnan dockyard in Shanghai. In this role, in 1908 and from the deck of *Hai Chi*, he was to welcome to the port of Amoy the visit of a squadron (two divisions of four battleships each) from the American Great White Fleet during its world tour.

Hai Chi, in company with the smaller cruiser and former flagship *Hai Yung*, was kept busy during 1907 and 1909 with flag-showing missions around South East Asia. Also in 1909 Rear-Admiral Cheng Pi-kuan reappeared, being appointed Admiral of the Cruiser Squadron. Then the death of the Emperor Kuang-hsu and the appointment of his three-year-old nephew Pu Yi in his place, together with the death of the implacable old Empress Dowager who had held all the strings, began to threaten the whole fabric of the Empire. Revolution was in the air; and when it came the action would be, to put it simply, the revolutionary South against the Imperialist North, with the dividing line formed by the Yangtze river.

The Coronation Review of 1911

In 1910 came the death of the British King Edward VII, and in the following year the coronation of his successor, King George V. One of the major events in the forthcoming pageantry was a Fleet Review, to be held in June 1911 in the traditional area for these events, the sheltered waters of the Eastern Solent, with the main anchorage of Spithead marking the entrance to the impressive naval dockyard of Portsmouth. Invitations were sent out to the seafaring nations of the world to send a warship to represent their countries at this event. Sixteen accepted, including China; Greece was a latecomer.

China's selection of a warship for the review was the logical one, the flagship *Hai Chi*, the largest ship in the fleet. Little appears to have been previously published about her role in this and other events, but the Chinese Wikipedia of 2013 did feature an article on the service life of *Hai Chi* which supplies some clues as to her comings and goings, although the chronology is woefully weak. (It would also be an understatement to say that the article suffered somewhat in its translation into English.)

Broadly speaking, there were a number of different purposes behind the deployment of *Hai Chi* to the west in 1911, which would be under the overall command of Rear-Admiral Cheng Pi-kuan:

- She would represent China at the Fleet Review in England.
- She would be carrying part crews for the two small training cruisers which were being built for China at the shipyards of Armstrong Elswick (Newcastle) and Vickers (Barrow-in-Furness).
- She would remain in western waters until the end of the year, when the latter cruisers were expected to be completed, and would then escort them back to China.
- In the intervening period she would 'have a month of repair' at Newcastle, and would also take the opportunity of paying an official visit to New York. (This was to be in return for the hospitality offered to ships of the Great White Fleet at Amoy in 1908.)

Looking at this programme, which involved the Fleet Flagship sailing from China in April 1911 and not returning to China until February 1912 at the earliest, one might be permitted a thought about the imminent Revolution. *Hai Chi* could have been back in Chinese waters after the Coronation Review by September if this had been thought necessary. Was something else being planned?

In a book about China's Revolution published as early as 1913,[8] the author surmises that some of China's provincial viceroys, who were in secret sympathy with the reformers or would-be revolutionaries, had actually urged the Court to send the *Hai Chi* to King George's Fleet Review, simply as a means of removing China's's most prestigious naval unit from the forthcoming defence of the Manchu Dragon Throne. This is a thought-provoking assertion; however, in fact it would make equal sense if it was the Dragon Throne and Admiral Sah that considered it would be better if Rear-Admiral Cheng Pi-kuan and *Hai Chi* were well out of the way in the probable event of a Revolution breaking out in 1911, in case the Fleet Flagship defected at an early stage, thereby striking an enormous blow against the Navy's somewhat tenuous loyalties to the Throne.

What actually happened when the Revolution finally broke out can in fact be deduced from *Hai Chi*'s movements at the time, and the author hopes that the following reconstruction will shed a new light on the subject.

Deployment to the West, April to October 1911

Certain points regarding *Hai Chi*'s deployment are culled from the Chinese Wikipedia, while most of the detail about her arrivals and departures are taken from contemporary newspapers and journals as noted. Estimates of her dates of arrival and departure where these are not known are calculated using distance tables and an economical speed of 12.5 knots, which was conveniently disclosed during an interview at New York.[9]

Shanghai: 11 April 1911 is quoted as the date on which the *Hai Chi* was detailed to attend the Coronation Review. The ship was given a brief refit at the Kiangnan

Hai Chi at the Coronation Review of June 2011, flying the yellow dragon ensign. (CPL)

naval dockyard, embarked part crews for the new cruisers together with a band, and sailed for Europe on 21 April. This date fits with her published date of arrival at Singapore.

Singapore: Arrived 28 April; sailed 1 May.[10] Thereafter she visited briefly (for coaling purposes) Colombo, Aden, Port Said (Suez Canal), and Gibraltar. While on passage Admiral Cheng somewhat surprisingly ordered the removal of all pigtails or queues, the traditional symbol of servitude to the Imperial Throne. The few crew members who demurred were put off the ship at the next port of call.

Plymouth (England): Arrived 4 June.[11] This was her host port in the UK. She spent a fortnight there, and would have been busy cleaning ship after five weeks at sea, as well as with drills in preparation for the Review. The steaming time was half a day between Plymouth and Portsmouth, where the foreign visitors were expected to assemble in the Solent on Monday 19th.

Portsmouth 19 June: The Fleet Review anchorage was an area of the East Solent six miles long by two miles wide, bordering Spithead. The Royal Navy had assembled a formidable array of warships ranging from dreadnoughts to submarines, and the area echoed to the sound of gun salutes as the visiting ships arrived. *Hai Chi* was in good company with two elderly near-sisters from Argentina and Chile, but both Germany and the United States had their very latest dreadnoughts representing their countries, and Greece's contribution, the armoured cruiser *Giorgios Averof*, built in Italy, was so new that she joined the review only a month after her first commissioning and without ammunition, which was to be embarked in England.[12]

19 June was also the date of what appears to have been the first official notification by the Chinese to the US Authorities of the contemplated visit by *Hai Chi* to New York.[13]

On Wednesday 21 June all the flag officers or captains of the foreign ships were escorted by train to London in order to attend the coronation of the new king, which was held on Thursday 22nd. Two days later, on Saturday 24th, the Fleet Review took place. King George V, in the Royal Yacht *Victoria and Albert*, steamed between the lines of assembled ships to the cheers of the ships' companies, finally coming to anchor in the centre of the armada. Once again, all flag officers or captains of the visiting ships were invited to congregate, this time on board the Royal Yacht in order to meet the King. The day ended with a full-scale illumination of the fleet.

Monday 26 and Tuesday 27 were given over to mass entertainments, including a party for officers at Whale Island (the gunnery school) for 3,500 guests; this was unfortunately blighted by rain showers. The somewhat crowded delights of Portsmouth were open to all and sundry, and for the adventurous London lay at the end of the railway line from Portsmouth Harbour to Waterloo.

Portsmouth to Plymouth: The fleet started to disperse on 28 June. *Hai Chi* sailed from Portsmouth back to Plymouth on 29 June.[14] (She was followed four days later by *Averof*, presumably to collect her outfit of ammunition, but in her unworked-up state the Greek ship unfortunately clipped a rock while entering Plymouth Sound and ended up in dry dock for a month or more.)

On 4 July it was reported that Admiral Cheng had sent a telegram to Peking about a projected visit to Mexico which was to be combined with the ship's visit to New York.[15] Relations between China and Mexico had been strained as the result of the massacre of several hundred Chinese immigrants in May during the Mexican Revolution, and reparations had been demanded.

Plymouth 12 July: *Hai Chi* sailed for Newcastle and refit.[16]

Newcastle: Estimated date of arrival is 14 July. A journal reported later: 'For some weeks the *Hai Chi* was in the Tyne pending repairs and overhaul to her machinery, and attracted considerable attention on account of the condition of peculiar cleanliness in which she was kept. In fact the crew displayed considerably greater ability and discipline than has often been attributed to minor navies…'.[17]

The cruiser building for China on the Tyne was the *Chao Ho*, a training ship of 2,750 tons; she was due to be launched in October. *Hai Chi* would have transferred embarked crew members to her during this period at Newcastle. The second cruiser, *Ying Swei*, a near-sister ship of 2,460 tons, was building at the Vickers shipyard at Barrow-in-Furness, on the other side of the country,

Newcastle to Plymouth: *Hai Chi* probably sailed back from Newcastle mid-August for her host port. She was later reported embarking 600 tons of coal at Plymouth on 21 August.[18]

Plymouth 31 August: 'The Chinese cruiser *Hai Chi* sailed from Plymouth yesterday [31st] for New York, Mexico and Cuba after three months in British waters … The vessel is to return to England to escort a new warship to China'.[19] This is the first time that there were reports of only one new cruiser to be escorted to China, and of a visit to Cuba.

New York: *Hai Chi* arrived at Sandy Hook in the afternoon of Sunday 10 September after a ten-day voyage, and proceeded up harbour the following morning on a reported ten-day visit, the first warship flying the Chinese dragon flag ever to enter American waters. Units of the US Fleet were anchored in the Hudson river and she received a thunderous welcome. Her stay is extensively reported in the pages of the *New York Times*. Only the relevant highlights of a very successful fortnight are recorded below.[20]

The Press noted that there was not a single pigtail or queue amongst the ship's company; the high standard of discipline and the overall state of cleanliness of the vessel were also remarked upon. Captain Tang (Tong) Ting Kwang was noted as the commanding officer (with Rear-Admiral Cheng in overall command). Amongst other features mentioned was the Marconi wireless set with one operator, one of apparently only four sets fitted in Chinese warships at the time – doubtless in the largest units, the cruisers. Her economical cruising speed of 12.5 knots was also reported.

Rear-Admiral Cheng, presumably taking the train, paid official visits to Secretaries of State in Washington, and then to President Taft at his holiday home at Beverly, Boston. Then on 18 September, together with the Mayor of New York, he led a parade of Chinese sailors up Riverside Drive, to lay a wreath at the tomb of General Grant. There was also a visit to the newly-laid hull of the third of the Chinese training cruisers, *Fei Hung*, which was being built at New York. Finally, there was the city's China Town, which took the ship's company to its heart,

The training cruiser *Chao Ho* immediately after launch at Armstrong Elswick just after the start of the Chinese Revolution in October 1911, and on the day that *Hai Chi* sailed from Bermuda. She is flying no fewer than three 'dragon' flags of the Imperial Chinese Navy, but probably for the last time. (Newcastle City Library)

and 'open days' when crowds turned up in their thousands to visit the cruiser. The press stated that after leaving New York the ship would call at several other American ports as well as Mexico (Vera Cruz) and Cuba, after which she would return to New York for a short visit. She would then return to the UK in order to escort (one) new cruiser back to China.

Details of *Hai Chi*'s departure from New York are difficult to find. She may have been delayed a few days while awaiting clearance for the visit to Mexico. However, it is estimated that she left on Monday 25th, for Charleston, possibly for coal, arriving 29 September, as there was a succinct paragraph about her in a South Carolina newspaper: 'Charleston SC 29th September: The Chinese cruiser *Hai Ch*i has received orders from Peking and will sail Monday [2 October] for Havana, thence through the Mediterranean...'.[21]

The visit to Mexico had been called off at the last minute following a reconciliation between the two governments. The visit to Cuba, reported earlier, appears to have been inserted into the ship's programme as a time filler, an opportunity to visit the local Chinese population of an isolated country as she had done in 1907 and 1909. The reference to returning 'thence through the Mediterranean' is unclear, unless it meant 'after collecting the newly built cruiser'.

Havana: It is estimated that *Hai Chi* arrived from Charleston on 5 October. Her time in Cuba is under-reported, except that the visit was planned for ten days and the band was very popular.

Meanwhile, in China the thunder clouds of revolution had been gathering for weeks. The contest when it came would be between Dr Sun Yat-sen and his revolutionaries in the south, and General Yuan Shih-kai, commanding the Peiyang army, defending the dragon throne and the Manchus, in the North. Admiral Sah, newly-appointed C-in-C of the Yangtze, was to have an unenviable task when the time came.

Then, unexpectedly, as apparently the start of the revolution was planned for a week later,[22] fighting broke out out at Wuhan (Wuchang/Hankow/Hanyang) on 10 October, and spread rapidly. Admiral Sah started moving his ships up river, and over the next few days the media swung into action. That most prestigious newspaper the *North China Herald* assured its readers that the flagship was already returning from Europe via Mexico to station itself off Shanghai (as a defensive measure). This was not quite true, but the book on the Revolution published in 1913 corrupted the story with a remarkably graphic account of the cruiser *Hai Chi* leaving New York in October, arriving in Shanghai in about December; according to this account she promptly hauled down the dragon flag and became the flagship of the rebel navy. A month later this was contradicted by a report in a Japanese newspaper stating that the Chinese cruiser *Hai Chi* was 'missing', and that since leaving Mexico nothing further had been heard of her and grave fears, etc... The fact of the matter was that the ship had been quietly tied up alongside at Barrow-in-Furness in the northwest of England since November 1911.[23]

What happened on board the flagship at Havana during the days which followed the outbreak of the Revolution has to be regarded as pure speculation. Admiral Cheng, a southerner, was an admirer of Sun Yat-sen, and agitators amongst his pigtailless crew may well have advocated a quick defection and a speedy return to China to aid the rebels. However, whether it was by accident or by design, it would have quickly become apparent that the ship was just about as far from China as it was possible to be; there was also the question of who would pay for all the coal and sort out the complexities of a passage of the Suez Canal. Meanwhile, it seems likely that signals from Peking were ordering a neutral course of action.

In the event, the ship is estimated to have sailed from Havana on the weekend of Saturday 14 or Sunday 15, as she arrived at Bermuda on 20 October, presumably to coal. This is known from British Foreign Office correspondence with Bermuda, relating to the discovery of the corpse of a drowned Chinese member of crew which was found after her departure. The same correspondence also records the fact the cruiser *Hai Chi*, of the Imperial Chinese Navy (*sic*), had sailed from Bermuda on 23 October 'for European waters'.[24] The distance from Bermuda to Barrow-in-Furness is about 3,060 nautical miles, or just over ten days at 12.5 knots, so her arrival at Barrow on 3 November (11 days) appears to bear out the report by the Foreign Office.[25] It looks as if *Hai Chi* had been effectively sidelined from any participation of any sort in the Revolution of 1911.

In the meantime Chinese naval forces had been moved up the Yangtze to the centre point of the Revolution, at Wuhan, and Peiyang military forces under General Yuan Shih-kai moved south and recovered part of Wuhan by the end of October. However, the flag of revolution was spreading fast. Shanghai and the Kiangnan dockyard suddenly fell to the insurgents 3/4 November. Admiral Sah and his warships found themselves without a base, and units started defecting to the rebels. Then, on about 12 November (references vary), Admiral Sah himself was deposed by the commanding officer of the flagship *Hai Yung*. The 'white flag' of the naval rebels (see below) spread rapidly throughout the fleet thereafter.

November 1911 to June 1912

The arrival of 'His Imperial Chinese Majesty's cruiser *Hai Chi*' at Barrow on 3 November was low-key. She was berthed in the Ramsden dock. Further instructions for her were being awaited, but it was anticipated that she would escort the new cruiser *Ying Swei* back to China in due course, using one third of her own crew. The ship would be open to visitors, and on two nights a large party of her officers and crew was entertained at the local Hippodrome music hall.[26] It was a far cry from the chaos taking place on the Yangtze. Subsequently the training cruiser *Ying Swei*, which had been launched at Vickers shipyard in July, carried out her sea trials successfully on 2–4 December, no doubt with observers from *Hai Chi* on board.

By December the political climate had changed considerably. General Yuan Shih-kai had suddenly allied himself and his armies with the rebels, and had arranged for the abdication of the Manchu Emperor, Pu Yi. After much bargaining the new First Provisional Government of China was announced for 1 January 1912, with Dr Sun Yat-sen as Provisional President at Nanking, and with Yuan Shih-kai holding most of the strings. The five-barred horizontal red, yellow, blue, white and black flag, prominent during the revolution (see illustration), was chosen as the national flag. Thus, on 1 January 1912 the crew of the *Hai Chi*, under the watchful eye of Admiral Cheng, would have marched from one side of the ship to the other, thereby transferring their loyalty from the Manchu dynasty to the Republic, and the five-barred flag would have been raised for the first time[27] in place of the former dragon flag. However, there was not much money available to the new Republic.

In Barrow, the month of January was quiet for *Hai Chi* while Admiral Cheng awaited orders from Peking about his future movements. *Ying Swei* was presumably ready to sail, once payment had been received. The only news concerning *Hai Chi* was a funeral for one of her ship's company, a stoker named Chang Ho Tong, aged 57, who died on 19 January and was interred in Barrow cemetery the following day with full military honours.[28]

On 12 February the Emperor abdicated as arranged, transferring all powers to Yuan Shih-kai; and Sun Yat-sen resigned the Presidency in Yuan's favour. On 10 March Yuan Shih-kai was installed as Provisional President in Peking.[29]

It must have been of the greatest interest to Admiral Cheng, in Barrow, when he read on 11 March that 'Admiral Chen Pu-kwang (commander of the cruiser *Hai Chi*)' had been appointed Minister of Navy in Yuan Shih-kai's new inter-party Cabinet.[30] There was no further commitment to stop him sailing for home at once, as arrangements had evidently been made for a contract crew to bring out *Ying Swei* if necessary . The main stumbling block was, unhappily, a national coal strike that had started in Britain on 26 February, and which was to continue until 5 April. Furthermore it would take a week or two to build up supplies. Until then *Hai Chi* was trapped.

Hai Chi sailed eventually from Barrow on 17 April 1912 flying the five-barred flag of the new Republic, and carrying the new Minister of the Navy. She was bound for Cardiff, in Wales, to be docked briefly for small repairs, and then to neighbouring Barry to embark the coal for passage to China. She was finally reported sailing from Cardiff for Port Said on 25 April,[31] after many months of frustrations. Normally there would have been a valedictory paragraph or two under Naval and Military Intelligence in *The Times* newspaper about her prolonged stay in western waters since the Coronation

Hai Chi sailing from Barrow-in-Furness in April 1912 and flying the five-barred flag of the new Republic of China. (Courtesy of the *North West Evening Mail*).

Review, but the great liner RMS *Titanic* had sunk in the North Atlantic the previous week with horrendous loss of life and diverted all attention.

Hai Chi's passage east was the reverse of her passage out without any time being wasted, that is until she arrived at Singapore from Colombo on 30 May.[32] There was then a definite pause in her progress, with a report that Admiral Cheng and officers had entertained onboard for tea the Chinese consul and his family, on 5 June.[33] It is probable that by this stage Admiral Cheng had received the news of a Cabinet reshuffle, and that Mr Liu Kuan-hsiung (Admiral, ex-*Hai Tien*) was now Navy Minister in his place.[34]

In the end *Hai Chi* sailed from Singapore for Shanghai on 7 June after a full week in port.[35] It is estimated that she arrived at Shanghai, her final destination, on 15 June, slightly later than various other predictions, and after a round trip of some 31,000 nautical miles. She had been overseas for fourteen months and, apart from attending the Coronation Review and visiting New York, had achieved remarkably little, all of which seems to reinforce the author's theory that the prime reason for her deployment to the West had actually been to remove her from the country ahead of the impending Revolution, given that her loyalties were decidedly suspect.

The Training Cruisers

The finances of the new Republic of China were found to be such that no money was available to pay for the completion of the two British-built training cruisers early in 1912 , and they were both eventually put up for sale. *Chao Ho* was nearly sold to Greece in October 1912,[36] but clearly that deal fell through. However, after finance again became available both she and *Ying Swei* were withdrawn from sale, repurchased by China and brought out to Shanghai by contract crews, arriving March and May 1913 repectively.[37] However, naval funding was still insufficient for the purchase of *Fei*

Hung, building at New York. She was instead sold on to Greece, in May 1914, and renamed *Helle*.

Flags and Ensigns

During the research for this article it became evident that the present listings of the naval ensigns of China are suspect for the years 1911/1912. During the last three months of 1911, between the outbreak of the revolution on the Yangtze and the promulgation of the First Provisional Government of the Republic, there were a variety of rebel flags. The most prominent was the five-barred horizontal red, yellow, blue, white and black flag representing the five races of the union (Chinese Han, Manchu, Mongols, Mohammedans and Tibetans), while the army had quickly produced its own Wuhan 'iron blood' red flag with a black star and eighteen gold balls for military use. Meanwhile there were frequent allusions to warships on the Yangtze hoisting the 'white flag' when changing their allegiances in November 1911. This appears to have been the 'white ensign', with a black star with eighteen gold balls on a red field in the upper canton; this ensign features in some modern references and was a simple adaptation of the military flag for temporary naval use. Then, on 1 January 1912 the five-barred flag became the national flag, and *Hai Chi* was seen flying it as an ensign on her departure from Barrow (see photo p.149). Finally, the new official naval ensign for the Republic, red with a white sun on a blue background in the upper canton, together with other flags, was being debated by the National Council on 6 May 1912 and may have been adopted for use as soon as August.[38]

Chinese Naval Ensigns of the Period

'Dragon' flag of the Qing Dynasty; originally triangular in configuration, this was rectangular from 1889–1911.
Colours: blue dragon with white horns and dorsal scales, green spurs and a red tongue.

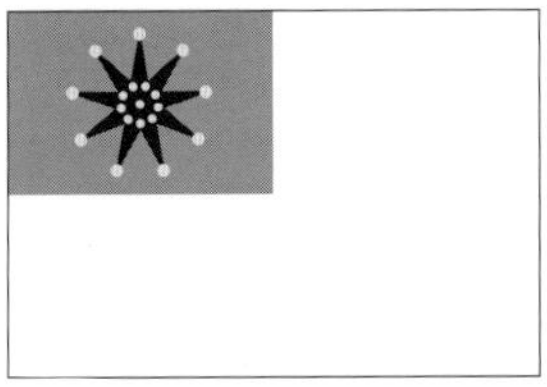

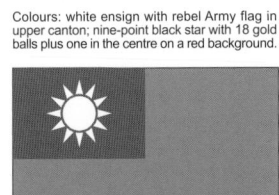

'White Ensign' (conjectural) flown by Chinese naval ships when they changed their allegiances on the Yangtse in November 1911. (There are other known variants of this ensign.)
Colours: white ensign with rebel Army flag in upper canton; nine-point black star with 18 gold balls plus one in the centre on a red background.

First flag of the Republic of China, flown 1912. The five coloured bands represented the five great ethnic groups in Chinese history.
Colours (top to bottom): red (Han Chinese), yellow (Manchus), blue (Mongols), white (Chinese Muslims) and black (Tibetans).

Naval ensign of the Republic of China, flown from late 1912.
Colours: red flag, white sun on blue background in upper canton.

Hai Chi: The Later Years

Thereafter, from 1913, China was rent by civil wars which lasted up until the Second World War. Admiral Cheng achieved a short spell as Navy Minister 1916–17, and then defected with the flagship and half the small fleet (including the cruiser *Chao Ho*) to Canton in 1917, where he was assassinated the following year. In 1923 *Hai Chi* and a small number of warships again defected, this time to the north where a warlord had reputedly purchased the entire squadron to defend Tsingtao, his recent acquisition. This team, known as the North-East Squadron, following a much-needed refit in Japanese-owned Port Arthur in 1927 and with the addition of small auxiliaries carrying seaplanes, became a surprisingly effective seaborne opposition to Chiang Kai-shek's forces as they fought their way north by land in 1926–28 for the eventual reunification of the country.

Subsequently the North-East Squadron, now known as the Chinese Navy's 4th Squadron, remained in the north and based on Tsingtao but in 1933, following a dispute about conditions, the three cruisers (*Hai Chi*, *Chao Ho* and *Hai Chen*) defected once again and sought sanctuary back south in Canton. Once again the conditions proved unsatisfactory, and *Hai Chi* and *Hai Chen* broke out two years later from the Pearl River, leaving *Chao Ho* stuck on a mudbank. This time the central government forces succeeded in rounding up the two older cruisers and decommissioned them. Together with most of the remaining ships of the 1895–1902 era, they would subsequently be used as blockships at Kiangyin on the Yangtze, thereby denying access to Nanking to the Japanese forces in 1937. Both *Chao Ho* and *Ying Swei* were sunk in action that year, the former defending Canton, the latter on the Yangtze river.

The Defection of the Flagship *Chongqing* in 1949

An indication of what might have happened had the flagship *Hai Chi* remained in China in 1911 can be found in the example of the later *Chongqing*. After the war in 1945 China found herself with very little in the way of warships and the beginnings of a new Civil War fomented by the People's Liberation Army (PLA) from a power base in Manchuria. In part payment for Chinese vessels appropriated by Britain in 1941 a small 6in-gun cruiser was offered to China as a flagship. This was HMS *Aurora*, a vessel with a fine war record but remarkably inauspiciously named in the circumstances, as it was the Russian cruiser *Aurora* which had fired the first shot of the Russian Revolution. Renamed *Chongqing*, she was refitted in Britain and sailed out to China in May 1948, manned by the Chinese Navy. She had political agitators amongst her crew and conditions were described as poor.

Between August and September *Chongqing* was operating off the Chihli and Shantung coasts, providing gunfire support for China's Nationalist armies as they were steadily forced south by the People's Liberation Army; then from November she was back in the Shanghai area for the defence of the Yangtze, as the PLA was converging on the north bank. The remainder of Chinese Navy, mainly ex-Japanese and ex-US Navy

The Chinese cruiser flagship *Chongqing* (ex-HMS *Aurora*) in UK waters in 1948. (Kennedy, courtesy of the World Ship Society)

destroyers and frigates, was strung out along the river. Then, on 25 February 1949 there was a mutiny on board the *Chongqing*, and she absconded that night for the north and the PLA-held port of Yentai (formerly Chefoo). Pursued and bombed by the National Air Force she was moved further north, to Hulutao in the Bohai. The PLA actually had no use for her at that time, as it had no naval ships, little oil fuel and none of the appropriate spare ammunition. So, after the ship had sustained a bomb hit at Huiutao, she was deliberately scuttled on her side alongside her berth at the end of March, for future recovery when the time was right.

The PLA may not have wanted a cruiser just yet, but the effect on morale of the defection of the flagship was obviously immense amongst the naval forces of the Yangtze. And then, when the PLA fired on and disabled the neutral British warship HMS *Amethyst* on 20 April, that incident assisted in the complete collapse of morale in the Chinese Nationalist warships on the river; eighteen in number, they all defected a few days later.[39]

Sources:

Books and Papers:
Bell, Christopher & Elleman, Bruce, *Naval Mutinies of the Twentieth Century*, Routledge (Oxford 2003).
Bell, Montague & Woodhead, *China Year Book 1913*, Routledge (Oxford 1913).
Clubb, O Edmund, *Twentieth Century China*, Columbia UP (New York 1964).
Foreign Office, London, Papers.
Lanxin Xiang, *The Origins of the Boxer War: A Multi-National Study*, Cambridge UP (UK 2003).
Rawlinson, J L, *China's Struggle for Naval Development 1839–1895*, Harvard UP (USA 1967).
Swanson, Bruce, *Eighth Voyage of the Dragon: A History of China's Quest for Seapower*, USNIP (Annapolis 1982).
Thomson, J S, *China Revolutionized*, Bobbs-Merrill (Indianapolis 1913).
US State Papers.
Wright, Richard N J, *The Chinese Steam Navy 1862–1945*, Chatham Publishing (London 2000).

Newspapers and Journals:
The Advertiser, Adelaide, Australia.
Barrow News, *Barrow Guardian*, *Barrow Herald*, UK.
Charlotte News, South Carolina, USA.
Chinese Wikipedia 2013.
The Engineer, UK.
Evening Telegram, New York.
New York Times.
North China Herald (NCH), Shanghai.
Peking and Tientsin Times (PTT), Tientsin.
Straits Times, Singapore.
Singapore Free Press and Advertiser.
The Times, London.
Weekly Sun, Singapore.

Footnotes:
1. Thomson, pp.69–70.
2. Presently off-line.
3. As the action takes place in the early 20th century, the Wade-Giles spelling of Chinese names has been used, with Pinyin in brackets as necessary on the first occasion of use.
4. Rawlinson, p.267.
5. Lanxin Xiang, p.99.
6. *PTT*, 29 July, 12 August 1899.
7. Rawlinson, p.279.
8. Thomson, p.4.
9. *Evening Telegram*, New York, 17 September 1911.
10. *Singapore Free Press & Advertiser*, 29 April 1911, p.12.
11. *The Times*, 5 June 1911, p.10.
12. Wikipedia, *Giorgios Averof*.
13. US State Papers relating to Foreign Relations, XXXIV, p.84.
14. *The Times*, 30 June 1911.
15. *Straits Times*, 4 July 1911.
16. *The Times*, 11 July 1911, p.15.
17. *The Engineer*, 17 November 1911, p.524.
18. An unidentified newspaper cutting.
19. *The Times*, 1 September 1911, p.2.
20. *New York Times*, 12, 13, 19, 25 September; *Evening Telegram*, 17 September, 1911.
21. *The Charlotte News*, 29 September 1911, p.2.
22. Clubb, p.40.
23. *NCH*, 21 October 1911; Thomson, pp.69–70 ; *Straits Times*, 31 January 1912; *The Times*, 6 November 1911, p.24.
24. The National Archives, Kew, FO371–1099, China series 1, 1906–1919, part 2 1909–1911, p.263.
25. *The Times*, 6 November 1911, p.25.
26. *Barrow Guardian*, 4 and 11 November 1911; *Barrow News*, 4 November 1911; *Barrow Herald*, 11 November 1911 (this report contains a typographical error giving the departure date from Bermuda as 2 October, instead of the 23rd).
27. Chinese Wikipedia 1913.
28. *Barrow News*, 27 January 1912.
29. Clubb, p.44.
30. *The Advertiser*, Adelaide, 12 March 1912, pp.9–10.
31. *Barrow News*, 20 April 1912; *The Times*, 27 April 1912, p.6; *Straits Times*, 11 May 1912.
32. *Straits Times*, 30 May 1912.
33. *Weekly Sun*, 8 June 1912.
34. *China Year Book 1913*, pp.506–507.
35. *Singapore Free Press*, 6 June 1912.
36. *The Times*, 12 October 1912, p.6.
37. *NCH*, 29 March, 3 May 1913.
38. *China Year Book 1913*, p.506; Wikipedia: 'Republic of China – Naval Ensigns'.
39. Bell and Elleman, Ch.11 ; Swanson, pp.180–182.

THE BATTLESHIP *COURBET* AND OPERATION 'SUBSTANCE'

Stephen Dent investigates an almost forgotten episode in the long and varied career of one of France's first dreadnoughts.

In histories of the Second World War the careers of the French battleships *Courbet* and *Paris* are often passed over in a couple of sentences (that is if they get mentioned at all). One recent study simply states that after their seizure in British ports in the summer of 1940 they were used 'only for training and administration until 1944'. While *Paris* survived the war and was eventually returned to France in August 1945, *Courbet* is at least rather better known for her final fate, as one of the blockships sunk off the Normandy beaches in June 1944. Yet in fact for a part of the four years that she was based in Britain the venerable battleship played an important, if almost entirely forgotten, role in the development of what was one of the most unusual weapons devised by any of the belligerents.

At the start of the conflict both *Courbet* and *Paris* were serving as training ships, a role they had carried out for most of the previous decade, being too old and outdated to be regarded any longer as front-line fighting vessels. However the invasion of France in May 1940 resulted in both seeing action in support of retreating French land forces (*Paris* being damaged by German bombing) before withdrawing across the Channel. *Courbet* was one of a number of French warships that began arriving and anchoring in Spithead on 19 June. A brief period of uncertainty and no little tension followed; when the French gave the impression that they might sail again, Admiral Sir William James, the combative C-in-C Portsmouth, closed the port due to fictitious enemy minelaying and even issued orders that if the ships tried to sail without permission they were to be stopped by force. After a bit of verbal sparring the French eventually agreed to berth their ships in Portsmouth Harbour itself (James claiming that the British required the main anchorage for other purposes) and it was here that they were seized on 3 July. As with the demilitarisation of 'Force X' at Alexandria, the combination of the demonstrable threat of overwhelming armed force plus an appeal to common sense was successful, and unlike at Devonport the operation was carried out without any serious violence.

Of the ships seized in British ports, many were elderly, obsolescent units, and the only ones of much real use were four destroyers, some sloops (subsequently reactivated and employed for convoy escort duties) and half a dozen submarines. (James rather dismissively describing the rest as 'rubbish'.) Although at that point in the war Britain was seriously short of battleships (*Royal Oak* having been sunk and none of the new *King George V*s as yet in service), *Courbet* and *Paris*, easily the largest units taken, really represented more of a burden than an asset, being of negligible fighting value but still requiring

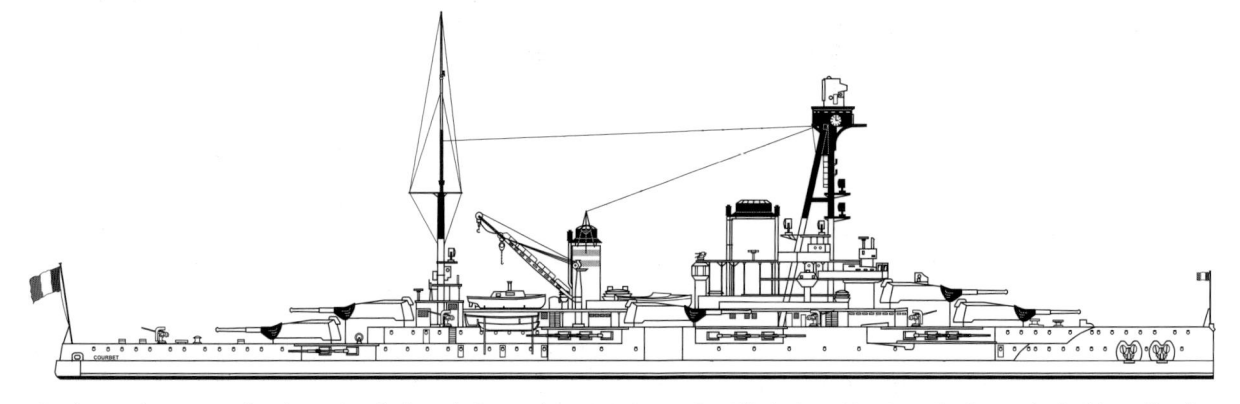

Courbet as she appeared at the outbreak of war in September 1939. In a major refit during 1927–31 at La Seyne she had been fitted with seven 75mm Mle 1922 HA guns, and two years later a quad 13.2mm Hotchkiss machine gun had been added abaft the forefunnel. When she was reactivated to provide fire support for the Allied armies in Northern France in May 1940, *Courbet* and her sister *Paris* received additional light AA guns: six 13.2mm twin Hotchkiss MG (turrets 2 & 5, quarterdeck) and two single 13.2mm Browning MG (quarterdeck). *Courbet* was also fitted with a single 37mm Mle 1925 gun on turret 5. This led to the use of *Courbet* as a floating anti-aircraft battery at Portsmouth during 1940–41. (Drawing © John Jordan 2015)

152

Courbet at Portsmouth in late 1940. The forward group of 14cm casemate guns has already been removed; the ship is otherwise in the condition she arrived in Portsmouth in June, shortly before the Armistice. (Courtesy of Philippe Caresse)

sizeable crews, stores, ammunition and so on. The tragic events at Mers el-Kebir only exacerbated the problem of manning such ships with French personnel.[1]

The Forces Navales Françaises Libres (FNFL), operating under de Gaulle's French National Committee based in London, had already been founded by Vice-Admiral Emile Muselier on 30 June. *Courbet* was returned to the FNFL on 10 July, becoming an administrative unit a few days later, after which she was used as their depot and accommodation ship at Portsmouth.[2] By March 1941 her secondary guns had been removed (subsequently to be used for coast-defence purposes – see *Warship 2013*, pp.168–71), but her main and AA armaments were retained, the latter proving its worth during the Luftwaffe raids on Portsmouth in March 1941 (during which she was also near-missed). On 19 March, though still a FNFL ship, she was reduced to a Care & Maintenance status, and there her war might have ended. Indeed at some point during the next eighteen months, as the course of the conflict ebbed and flowed, the British actually suggested that she be broken up for scrap. Fortunately, as things turned out, this proposal was not taken up, instead remaining under discussion (and no doubt providing plenty of scope for political wrangling between the slightly uneasy wartime allies) while the ship herself lay quietly in Portsmouth harbour.[3]

'Bouncing Bombs'

The story of the invention and development of the 'bouncing bombs' – the famous 'Upkeep' used by the Dam-busters of 617 Squadron and its lesser-known cousins 'Highball' and 'Baseball' – has been told in detail elsewhere, and only needs to be summarised here. The British aviation designer Dr Barnes Wallis had come up with the idea of skipping a bomb across the surface of water in the manner of skimming a pebble from a beach, but attempts to interest the Air Ministry in this unconventional scheme instead attracted the attention of the Admiralty's Department of Miscellaneous Weapons Development (DMWD, known as the 'Wheezers and Dodgers'), who were never shy of looking into such things. The Royal Navy was keen to find ways of eliminating Germany's remaining major surface warships, which had such an impact on its own dispositions and operational planning, and here was something that offered the chance to do just that. During the summer and autumn of 1942 initial model tests in the National Physical Laboratory tank in Teddington and then full-scale trials at Chesil Beach in Dorset convinced both the Admiralty and the Air Ministry of the viability of the concept as a weapon of war, and after further successful trials in January and February 1943 development diverged onto two different paths, with the larger version eventually resulting in 'Upkeep', used to destroy the Ruhr dams in May 1943, the smaller one becoming the aircraft-launched 'Highball', and its surface-launched cousin 'Baseball'.[4]

The RN, impressed by the weapon's potential, began also to think of it as a way of attacking Italy's capital ships, German U-boat pens and possibly other targets. Two new RAF squadrons were formed for the development and operational use of the new weapons: Bomber Command's 617 Squadron and its much less famous Coastal Command counterpart, 618 Squadron. The latter was intended to use the high-performance twin-engined Mosquito bomber to launch 'Highball', its first target being the battleship *Tirpitz*, and overall planning envisaged this attack and that on the dams happening simultaneously, since otherwise once one had taken place the enemy would almost certainly implement countermeasures at other potential targets and much of the weapon's potency would be lost. Unfortunately inter-service rivalry brought about a continual changing of priorities during the winter and early spring of 1943, hindering development of both weapons. In an indication of the attitudes that developed, Air Chief Marshall Sir Arthur Harris said 'Highball' was 'just about the maddest proposition as a weapon we have yet come across and that is saying something', while in contrast Admiral Sir Dudley Pound, First Sea Lord, described it as 'the most promising secret weapon yet produced by any belligerent.' The truth, as events were to prove, was somewhere in between.

The Requirement for a Target Ship

618 Squadron was initially tasked with 'preparatory work for Highball operations', while the target date for Operation 'Servant', the attack on *Tirpitz*, was set for 15 May 1943. Ongoing trials at Reculver in Kent were resulting only in disintegration of the weapon (usually referred to as a 'store') on impact with the sea, and each time this occurred new, modified versions had to be made. In the meantime the squadron's aircrew, after initially practising low-level navigation over the North Sea, began carrying out high-speed dummy runs at low level in a confined location against the submarine depot ship *Bonaventure*, anchored in Loch a' Chàirn Bhàin, south of Cape Wrath. However, as modified stores being

dropped off Reculver were more successful, the requirement was now for a ship that could actually be 'attacked' (albeit using inert practice bombs). *Bonaventure*, as an operational unit, was clearly unsuitable, both because of the risk of damage to the ship and because security would inevitably be compromised.

The need had already been anticipated back in early February for 'full scale trials to ascertain the effect of (a) Highball hitting the side of a ship at high speed, (b) subsequent trajectory and behaviour of hydrostatic valve setting and operation'. There was a belief that the weapon would tend to 'cling' to a ship's side, but there was no certainty as to its precise path once under water, so the idea was to fit a ship with underwater sensors to get some sort of picture of this. The Admiralty began considering a number of old vessels that were potentially available, to see which might be most suitable, including several of the older battleships still in active service, the target ship *Centurion* (then in the Middle East), *Iron Duke* (aground in Scapa Flow), the Netherlands cruiser *Sumatra* (at Portsmouth awaiting modernisation), and *Courbet* and *Paris*. *Sumatra* had the advantage that she could steam under her own power, while *Paris*'s use as a depot ship meant that a replacement would be needed. *Courbet*, at Portsmouth, was manned by 'at present a bare minimum maintenance party of Free French ratings' and was 'not in use, cannot steam, but could probably be towed. Is so old and shaky that it is probable that if she received a heavy bang she would disintegrate.' Despite the rather dismissive (and, fortunately as it turned out, quite inaccurate) tone of this final comment, it was clear that 'taking all factors into consideration… *Courbet* is the most suitable vessel' due to her size and availability. A small maintenance party would be needed, also pumping arrangements – 'it is not anticipated that the damage to the ship would involve more than local distortion and possible leakage' – while a suitable tug could be made available in a fortnight. A report was quickly compiled on the current state of the ship and what would need doing to prepare her for her new role:

- Watertightness. Hatch and armoured hoods on forecastle and quarterdeck to be made watertight. All doors and hatches between decks to be closed and made watertight. Openings left after removal of guns to be blanked off.
- Towing. Method required by tow master or towing company. Additional clenches to be fitted for blake stoppers. Personnel for tow.
- Machinery. No steam available. Lister petrol dynamos required to supply power for lighting and 10kw for degaussing arrangements.
- Fuel. 128 tons of coal on board. Petrol for listers to be stowed on deck in drums.
- Messing arrangements. Galleys and bakery in working order. FW[fresh water] tanks require overhaul. Two pumps to be repaired. Cabins to be used by towing personnel. Personnel for trials – French or British.
- Armament. Approximately 1,000 12" fuzed shells on board. Estimated time for removal six months. DAS to say whether he is prepared to take shell and where it is to be stowed. Who is to remove shell? Any AA guns (Oerlikons to be fitted for towing)?
- Free French stores. Approximate time four weeks to de-store.
- Life-saving appliances. Propose to use Carley Floats. Power required to work cranes if ordinary boats are used.
- Labour requirements. Three gangs for four weeks subject to men being available or higher priority given.

In addition to the above work, there were also concerns as to whether or not the ship needed to be degaussed, since she would be passing through waters that were known to have been mined by the enemy. An additional complication was that *Courbet* was part of the FNFL, meaning that everything was 'subject to the concurrence of the Free French Authorities.' On 20 February it was reported that the ammunition on board did not need to be disembarked(!);[5] otherwise, three to four weeks preparation time was required. Six Oerlikons were needed for AA defence during the ship's passage, for which two tugs would be necessary plus one in reserve, as well as fighter cover and minesweeping where needed. Something between seventy and a hundred men would be required to crew the ship, of which twelve were to man the Oerlikons, while six were for hand steering. Around fifty would have to remain on board for the period of the trials themselves. (Such was the shortage of trained manpower at the time that it was noted that this could impact on forthcoming commissioning or re-commissioning of ships.) Vice-Admiral E L S King, the Principal Naval Liaison Officer (PNLO), approached Rear-Admiral Philippe Auboyneau, commander of the FNFL,[6] to request agreement, saying that the British would make good any damage that was done, though this occurrence was unlikely since only 'dummy weapons' were to be used. He also pointed out that the possibility of scrapping was still under consideration, and the trials would take precedence and so serve to put this off. The time allowed for the ship's passage was between one and three weeks; the trials themselves could last for up to three months. The cover story was simply that the ship was to be used to give aircrew training against a full-sized target.

By 24 February the necessary agreement had been given and, with a crew consisting of Cdr. H F Nash (rtd.) as CO, five other officers and 119 ratings having been arrived at, it was decided that *Courbet* would sail at high water on 30 March, with three tugs until she passed the Needles, two at the bow and one at the stern; the latter would then leave the ship. Arrival would be on 4 April. However, perhaps inevitably, a problem now arose. The ship was supposed to fly both the White Ensign and the French ensign (*Paris*, at Devonport, at that time flew both), but on 29 March Auboyneau objected to this, and wanted her to fly just the French ensign, since she was in effect only being lent to the RN. The Admiralty, believing that the British crew would in turn object to sailing a ship under the French ensign, regarded the French attitude as variously 'pernickety' and 'somewhat unreasonable'. The French threatened to withdraw permission entirely, at which point the Admiralty decided that the sensible approach was not to press the matter any further, and that the ship would be commissioned to sail under the French ensign, though a large White Ensign would be kept available on board if needed, and once the operation in question was completed she would fly the White Ensign thereafter.

Courbet went into Portsmouth dockyard in the middle of March 1943 for what were termed 'temporary repairs'; she was commissioned into the Royal Navy on 22 March, and by the end of the month, now out of dockyard hands and attached to HMS *Victory* II for accounting purposes, she had officially become a unit of Portsmouth Command. Her original mission had expanded considerably, and she was to be used not only for the trials and for training the airmen but also for looking into counter-measures. From early on in the weapon's development it had been felt that the enemy would copy the principle more or less straight away, particularly if they got to capture an example (or, it was noted, if any of them played billiards!) and Britain, as a maritime power, had a great deal to lose by this.[7]

'Port HHY', Loch Striven, Argyll

Loch Striven, on the western side of the Firth of Clyde, was chosen on 23 March as the location for the trials – Operation 'Substance' – as offering a good balance between security and accessibility. While further up the Clyde, above the anti-submarine boom that ran from Dunoon to the Cloch Lighthouse near Gourock, would have been safer, it was also far too busy, being the location of all manner of merchant shipping activity, assorted naval establishments, shipbuilding and so on. Security was an absolutely paramount consideration, and the loch, sheltered from the outside world, fitted the bill well. It was already being used for training with 'human torpedoes' and 'X-craft' and now, closed to traffic and in due course also protected with smoke screens, it was given the code name of 'Port HHY', while *Courbet* became 'BV7'. As part of the security arrangements the considerable number of senior officers due to observe the trials were told to come in plain clothes, so as not to draw attention to themselves. Later, in June 1943, a memo was issued by Greenock sub-command to the effect that there was going to be a large amount of torpedo bomber practice in the area, thus helping to disguise the purpose of all the low-level aerial activity taking place. Local inhabitants around the loch were either evacuated or else ordered to stay indoors with their curtains drawn and guards outside. However on a wider level this same secrecy could also cause unexpected problems: with numerous meetings about the project taking place in various locations across Britain, and representatives of so many different organisations involved (including the RAF, RN, DMWD, HMS *Vernon*, DSR (Admiralty), Vickers, SMD, MAP and others), people sometimes turned up to meetings not knowing what it was that they were there to discuss! This did nothing for the speed of development of the weapon, while there were also problems with the conversion of the Mk.IV Mosquitos to carry it, thanks in part to the earlier inter-service disagreements about priorities.

There seem to have been various delays in *Courbet* making passage up the west coast of Britain. Her initial departure from Portsmouth was delayed due to concerns about the arrangements made for the storage of items usually kept on board. Arrival date was put back from 4 to 8 April; eventually *Courbet* made it to the Clyde by 15 April, which had been the intended start date for the trials. While she remained officially part of Portsmouth Command, the following month she became attached to HMS *Orlando*, the wartime base at Greenock (a sub-command of Western Approaches).

Loch Striven had much to commend it, not only in terms of its relative seclusion but also that its geography bore some resemblance to the sort of area where the *Tirpitz* might be found. However, one major drawback was its depth: the trials necessitated the laying of three first class moorings by the Boom Defence Service in 40 fathoms and 33 fathoms (240ft & 198ft / 75m & 60m), in this remote location. Then there was the actual securing of *Courbet* and other craft involved, berthing and maintenance, rigging and recovering nets, laying and maintenance of range buoys, plus the accommodation, transport and welfare of personnel. At least visiting senior officers could enjoy the accommodation of the Glen Burn Hotel in the little seaside town of Rothesay (a popular peacetime destination) at the end of the loch.

Operation 'Substance': April–December 1943

The trials had been given 'A*' priority; however, by this time the delays in getting the Mosquitos converted meant that the end of May was looking a more likely target date, meaning simultaneous attacks on the dams and *Tirpitz* were out of the question. In fact 618 Squadron transferred to the nearby Coastal Command training station at Turnberry[8] between 8 and 19 May 1943, so by the time they had all arrived the dams raid had actually taken place. While the aircrew themselves knew what the 'stores' that were now being delivered were for, the cover story was that they were a new type of mine for use overseas, and units using them were to be known as 'Special Mining Squadrons'.

The crucial issue of the depth setting for detonation would depend on the underwater trajectory of the 'Highball' bomb, and so determining this was one of the main aims of the trials against *Courbet*. The belief was that imparted backspin would cause the store to 'cling' to the ship's hull, at least initially, but this needed to be proved, and if so for how long before the store lost momentum and sank. Six booms were fitted projecting from the ship's sides, with a piezo-electric pressure gauge

Courbet shortly after arrival in Loch Striven. Note the painted on bow wave, left over from 1940 when it was a standard feature of French warships operating in the Atlantic. (*Paris* sported the same feature, though it was painted out at some point after her arrival at Devonport.) The generally rather dishevelled state of the ship is readily apparent. (ADM277/46)

suspended from each, 40ft below the surface of the water. The detonation of a small, ¼lb charge would be picked up by these, giving an indication of underwater trajectory. It soon became apparent that while this was good for determining position, it was less so for depth, and so four hydrophones were also suspended, three at 120ft and one at 40ft. Later a fifth was added, as a spare.

In terms of the accuracy of the actual impact, it was not anticipated that the influence of tides would be significant, however rough sea conditions might well be, so this too was to be carefully measured. There was also the matter of dropping position for the stores, the speed and height of the aircraft, and the rotation speed of the stores.

Then there were the defensive counter-measures to be tested in anticipation of a German copy of 'Highball'. Two possible approaches were to be investigated: a row of LCTs moored about 45ft from *Courbet*'s side with a wire 'fence' strung between them; and gunfire from Oerlikons mounted on the battleship's deck and also on the LCTs. In due course the ship acquired yet another role as well, as a conventional practice target for FAA pilots, although this would only be after the conclusion of Operation 'Substance'. This would both strengthen the existing cover story and also further postpone the tricky question of the battleship's ultimate fate. This latter was now a serious issue, for breaking up a ship the size of

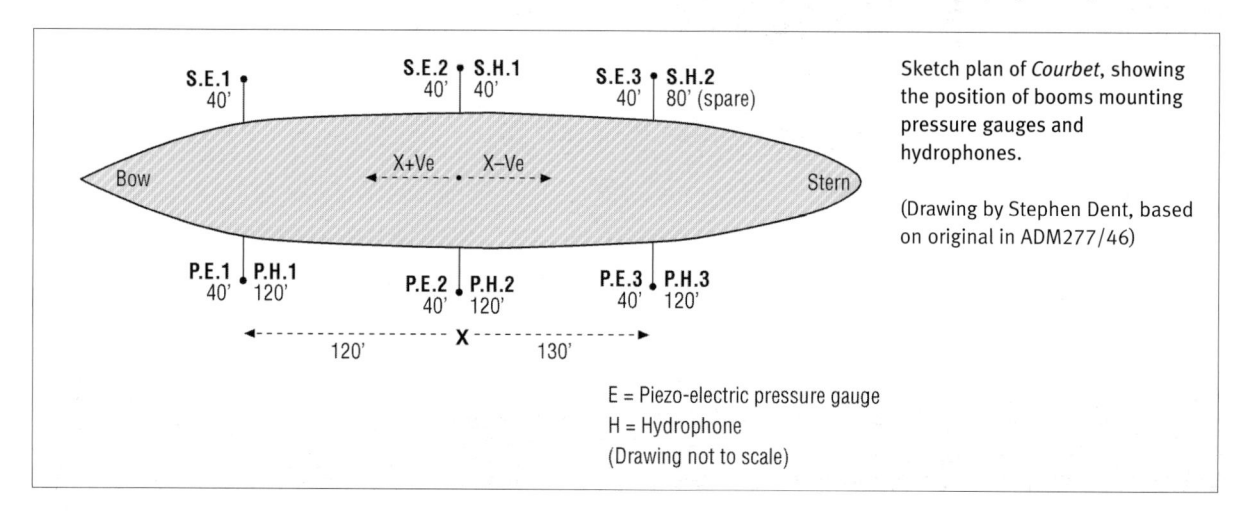

Sketch plan of *Courbet*, showing the position of booms mounting pressure gauges and hydrophones.

(Drawing by Stephen Dent, based on original in ADM277/46)

E = Piezo-electric pressure gauge
H = Hydrophone
(Drawing not to scale)

Courbet would provide five weeks' worth of steel for Britain's hard-pressed wartime industry, at that time largely dependant on scrap. However, for Admiral Auboyneau the scuttling of most of the remainder of France's fleet at Toulon the previous November gave *Courbet*'s retention 'as a depot or training ship' great importance compared to the 'inglorious fate' of scrapping – in effect political and symbolic considerations had come to outweigh simple war requirements.

The first batch of trials on 9–10 May were failures, due to problems with the release gear on the aircraft. The trials recommenced a week later, lasting until 22 May. One problem that quickly became apparent was that the marker buoy for dropping had been incorrectly positioned, being some 800yds from the ship instead of 1,200yds, resulting in an impact velocity far greater than intended, causing damage to the stores themselves (despite their being made of ⅜in steel plating), to the hydrostatic pistols, and on one occasion to *Courbet* herself. Observers were stationed on board the ship, in the attacking aircraft, and elsewhere around the loch; in this instance a number of senior RN and RAF officers and civil servants were on board. The store, instead of hitting the armour belt and then disappearing below the surface of the loch, went slightly off line and tore a hole in the bow. As the observers scattered in panic, red Very Lights were instantly fired off as a signal to stop any further drops.

Other factors that contributed to the very mixed results of these initial trials were the poor weather, and incorrect painting of the stores. These were painted half in black and half in white so as to enable the observers to gauge the rotation; unfortunately gloss black paint had been used instead of matt, resulting in light reflections which meant it couldn't be distinguished from the white, thus making it hard to measure RPM.

On 26 May the trials were 'temporarily suspended', before recommencing on 5 June. Between then and 2 July some eleven successful runs took place: of these, eight of the stores sank more or less vertically after striking the side of the ship; two progressed some 20ft (just under half way) towards the keel; and one ran off course and impacted near the bow, passing right under the ship and out the other side. With these trials the dropping range was varied, starting at 1,600yds and then shortening in progressive steps of 200yds until damage to the stores resulted in failure of the firing pistols. Then the reverse was done, eventually arriving at an optimum range of 1,000–1,400yds with the aircraft flying at 360mph at 30–50ft above the surface of the loch. However there were still problems with accuracy and with the release gear, and it became evident that revised target dates of 12 July and then 12 August were still impractical.

Meanwhile other, more fundamental difficulties were becoming clear, mostly to do with *Tirpitz*'s location at the very north of Norway. Quite apart from the obvious problems of attacking a heavily defended target at low level in a confined airspace, accurate long-distance navigation over sea in poor visibility would be a considerable challenge, and reliable intelligence about the ship's precise whereabouts was another potential problem. Most seriously, the Mosquito's range was inadequate. Various solutions were considered, including flying from carriers, ditching in the sea, attempting to crash-land in neutral Sweden, or operating from the Soviet Union, but none were very satisfactory. With 'Highball' continuing

The 'Highball' that ran off course and smashed into the unarmoured part of *Courbet*'s side, giving an indication of the amount of damage that even an inert practice round could do. (ADM277/46)

Courbet during one of the 'Substance' trials, pictured just as a 'Highball' impacted with her side. The false bow wave is still present, however the battleship has now acquired a modified paint scheme with a broad white band on the hull amidships, and the sides of both funnels also painted white (as well as the top of the fore-funnel), all presumably as aiming markers for the Mosquitos. One of the booms mounting measuring instruments can be seen projecting from the ship's side. Note also the LCT in the foreground. (ADM277/46)

not to perform as hoped, the decision was taken to attack *Tirpitz* using X-craft. The success of this, Operation 'Source', in September, and the near-simultaneous ending of hostilities with Italy, removed the two most likely targets from the list, leaving only the U-boat pens and other 'Atlantic wall' defences, plus the major units of the Japanese navy. It made sense for the majority of 618 Squadron to be dispersed elsewhere, leaving just a nucleus to continue 'Substance', albeit now with rather less urgency.

Lt.Cdr. F J Boswell replaced Nash as *Courbet*'s CO during August, and at the start of the following month the trials recommenced. It had become clear that the hydrophones positioned at 120ft were leaking as a result of water pressure, thus nullifying their readings, and so on 20 September it was decided that a new type would be used henceforth. It was also felt that the single small explosive charge did not give enough information to accurately gauge underwater trajectory, and so this was replaced with a series of six 'cracker' charges set to detonate at two second intervals. Another problem was that it had not been possible to recover any of the stores used in the most recent set of trials. Even so, it was evident that impact with the battleship's armoured side was still causing damage to the store, even with a wooden layer to absorb some of the shock. While some 80% of the hydrostatic pistols worked satisfactorily, it was now clear that underwater trajectory was not at all what had been hoped for. Very few of the stores ended up right beneath *Courbet*, while some still simply rebounded from the ship's side and then sank. Further modifications to the design were required. On 23 October it was finally admitted that 'the project has not proved the success anticipated'.

During November nets were strung below the ship to catch stores as they sank, rather than them being lost in the depths of the loch. A further series of twenty-five trial drops with a modified 'Highball' took place during 26 November-2 December, and while these were more successful the stores still tended not always to screw underneath the ship, due, it was now realised, to insufficient back-spin. Once more Operation 'Substance' was suspended. There were now more pressing matters to attend to.

'Gooseberry' Blockship

On 28 January 1944 *Courbet* was again reduced to Care and Maintenance status, though still on the Clyde. There she stayed for the next couple of months, but in late March she was settled on by those planning the invasion of occupied France to be one of the 'Gooseberry' blockships sunk off the Normandy beaches as part of the artificial ports that were an essential part of the operation. Ironically, of the other three major naval vessels chosen for this important if slightly unglamorous fate, two were *Sumatra* and *Centurion*, both of which had a year before been other candidates for the 'Highball' trials; the fourth was the old British cruiser *Durban*. While these other three were all capable of steaming to France themselves, *Courbet* was not. So, while they would require proper crews (albeit much reduced in overall numbers) to man their machinery, as well as defensive armament (plus the necessary stores and ships' boats), in the case of *Courbet* it was 'considered that sufficient officers and men to handle the tow, and to provide such damage control as is practicable, plus close-range weapons, crews and lighting and domestic facilities are all that are required. An entirely French complement is acceptable but a liaison party will be necessary. It is suggested that Free French headquarters be asked to provide a complement.' Rear-Admiral William Tennant, the officer responsible for the entire 'Mulberry' harbour operation, ordered that *Courbet* be ready for this new service by 15 May.

In mid-April she officially returned to the strength of the French fleet,[9] although she was still RN manned, and by the end of the month she was back on the south coast, undergoing repair and refit at Devonport. Work was completed by 12 May, by which time she had also acquired a French crew. On 21 May (now with her British liaison officer on board) she left Devonport, under tow of the tugs *Growler* and *Samsonite*, arriving at Portland the following day. A fortnight later the same two tugs towed her across the Channel, where she was finally sunk off Ouistreham on 9 June 1944, thus ending surely one of the most unusual wartime careers of any battleship.[10]

Conclusion

The trials of 'Highball' continued on Loch Striven after *Courbet*'s return south, with her place taken by the newly decommissioned HMS *Malaya*. Four years of hard wartime service had left the old British battleship seriously in need of a major refit, and it made more sense

Courbet sunk as a blockship off Ouistreham during the Normandy Landings. The crews for the anti-aircraft guns (35 men) remained on board for the first 24 hours, before being evacuated on 10 June. Note that the colour scheme remains the same as during the 'Substance' trials, though the fore-funnel has lost its cap. (Courtesy of Philippe Caresse)

instead to release her crew for new vessels, in particular those required for the imminent Operation 'Neptune'. The trials took place during May and September 1944, but continued to be dogged by problems, including the unsuitability of the heavily bulged hull of *Malaya*, further instances of damage to the stores on impact or rebounding from the ship's side, and 'extremely severe' weather conditions which resulted in damaged aircraft, some injured aircrew, and a large number of misses. 618 Squadron was eventually sent to Australia, with the intention of using the weapon against Japanese shipping,

Courbet being broken up *in situ* after the end of the war. (Courtesy of Philippe Caresse)

but a combination of lack of targets and a distinctly unenthusiastic attitude on the part of the Americans put paid to this too. Although a small detachment was used for anti-U-boat operations, using cannon-armed Mosquitos, for a unit formed with elite personnel theirs had been a distinctly underwhelming wartime career, and one punctuated with a number of fatal crashes, a particularly sad irony given that 'Highball' never made it to operational use.

The wreck of *Courbet* was broken up where it lay after the end of the war. Loch Striven has intermittently been used since to lay-up large merchant ships during times of economic downturn such as the recent global recession, while in 2010 a team of divers led by Dr Iain Murray from Dundee University located between eight and fourteen 'Highballs' on the bed of the loch, along with what appeared to be part of the moorings laid in 1943. Plans remain to return at some point and recover at least one of these; however, at the time of writing, although a number of 'bouncing bombs' recovered from other test sites exist in various museums across England and Wales, all the intact examples represent the larger and more famous 'Upkeep', and only pieces of 'Highball' can be seen.

Acknowledgements:
The author wishes to thank the staff at the National Archives (formerly the Public Record Office), Kew, London, in particular Nick Barrett and Hugh Alexander, as well as John Jordan and Ian Johnston, for help with this article.

Principal sources:
Primary:
The National Archives (formerly the Public Record Office), Kew:
ADM1/12924 French Battleship *Courbet*: French concurrence in use as target ship.
ADM1/14827 Consideration of naval targets for operation 'Highball' and appreciation of tactics and force involved.
ADM1/16009 Results of 'Substance' trials.
ADM1/17053 Blockships. Requirements for HM Ships 'Durban', 'Centurion', Dutch 'Sumatra' and French 'Courbet' ie complements, stores armament, etc.
ADM116/4539 Greenock Memoranda: information and instructions affecting ships and authorities in the Greenock Sub-Command.
ADM116/4843 'Rota Mine' 'Highball' anti-ship trials Operation 'Substance'.
ADM187/23-37 Lists of stations and movements of Allied and Royal Naval Ships ('Pink Lists'), January 1943–June 1944.
ADM277/46 Lt.Cdr. L Lane papers: bouncing bombs, Highball and Upkeep; development, tests, results and German versions.
AIR6/63 Air Board, and Air Ministry, Air Council: Minutes and Memoranda, Precis Nos.1–89.
AIR20/1000 Operation 'Highball', 618 squadron: operational planning.
AIR20/2617 Operations 'Highball' and 'Upkeep': reports and papers.
Secondary:
Brickhill, Paul, *The Dam Busters* (Evans Bros, London, 1951).
Clayton, Anthony, *Three Republics, One Navy: a Naval History of France 1870–1999* (Helion & Co. Ltd, Solihull, 2014).
Curtis, Des, DFC, *A Most Secret Squadron* (Skitten Books, Wimborne, Dorset, 1995, republished, Grub Street, London, 2009).
James, Admiral Sir William M, *The Portsmouth Letters* (Macmillan & Co, London, 1946).
Pawle, Gerald, *The Secret War* (George G Harrap & Co, London, 1956).
Warlow, Lt.Cdr. B, RN, *Shore Establishments of the Royal Navy* (Maritime Books, Liskeard, Cornwall, 2000).

On-line:
http://news.bbc.co.uk/1/hi/scotland/glasgow_and_west/8127423.stm

http://www.heraldscotland.com/news/home-news/estate-where-dambusters-tested-bombs-up-for-sale.13822135

https://secretscotland.wordpress.com/2013/05/17/after-the-dambusters-raid/

http://forargyll.com/2010/07/the-day-of-the-highball-discovery-in-loch-striven/

Footnotes:
[1] In all some 9,000 French personnel were eventually repatriated, via Southampton and Liverpool.

[2] In contrast *Paris*, similarly seized at Devonport, was commissioned into the RN on 12 July, being used initially as an accommodation ship and later as a depot ship for trawlers and other vessels of the Auxiliary Patrol.

[3] Some dates are uncertain, since sources vary.

[4] Work on 'Baseball' was carried out at DMWD's station at Weston-super-Mare in Somerset, but changing operational priorities meant that in the end it never progressed beyond a lengthy and varied series of trials.

[5] Presumably because it already had, probably back during 1940–41. However, no definite records seem to exist.

[6] The first commander of the FNFL, Vice-Admiral Muselier, the only flag officer to rally to de Gaulle, was a controversial and individual figure who proceeded to antagonise the British, the Americans and eventually (perhaps not entirely surprisingly) de Gaulle as well, not only losing his job but even spending time in prison! He was replaced in March 1942 by Captain Auboyneau of the destroyer *Le Triomphant*, at that point the major active unit of the FNFL, who was recalled to London and promoted to Rear-Admiral. Auboyneau not only managed to run the small but growing naval force effectively, but unlike his predecessor also steered a much more successful path through the tangled politics of the period, in due course having a notable active career during the latter years of the war and on into the post-war era.

[7] In fact the Germans were already considering something of this nature, and the dams raid gave the project, code-named 'Kurt', a new impetus. However, due to disappointing initial results which replicated the experience of the British, it was then assigned a low priority, and development did not proceed at any pace.

[8] The site, both before and since, of the renowned golf course.

[9] The same was the case with *Paris*, still at Devonport. The French Navy at this point therefore officially boasted an impressive five battleships on its strength, though in reality only two (*Richelieu* and *Lorraine*) were actually operational.

[10] In an indication of the widespread secrecy of the time, her official status on 9 June was 'at sea', and even at the end of June Plymouth Command still had her down as being 'at Portsmouth'.

THE 'FLAT-IRON':
The Coast Defence Battleship *Tempête*

Philippe Caresse tells the story of this unusual vessel, one of a number of coast defence monitors built during the period which followed the American Civil War.

Under the influence of the 'Jeune Ecole', which made the reduction in the displacement of larger warships a virtue, France laid down from 1865 a series of coast defence battleships. The latter, whose inspiration was the American monitors, were designed to defend the ports and the French littoral. The fleet role of these ram vessels equipped with gun turrets was poorly defined, and most of them would be stricken within a few years of their completion. The few survivors took little active part in the First World War. The coast defence battleship *Tempête*, whose brief service is recounted in this article, boasted one of the most unusual silhouettes of the end of the 19th century.

In its meeting of 10 November 1871, the Council of Works (*Conseil des travaux*) was tasked with drawing up a programme of 1st and 2nd class coast defence ships. Following lengthy discussions, proposals were submitted by the engineers Woirhaye, Selleron, Clément and de Bussy and were formally considered on 30 July 1872. The proposals were purely for 2nd class ships, and staff requirements were as follows:

– draught: not to exceed 5.20m
– speed: 10 knots
– radius of action: 750nm
– armament: two 27cm (10.8in) guns in a twin trainable turret
– hull: to be fitted with a ram

The Woirhaye proposal was rejected on the grounds that the figures for weights were arbitrary and the machinery too heavy. Selleron's plans were likewise rejected, because his distribution of armour flew in the face of the requirements. The Clément design had insufficient speed and endurance, excessive stability, and hull lines which did not find favour. Louis de Bussy, on the other hand, had modelled his proposal on the British breastwork monitors *Glatton* and *Rupert*. The light upper hull allowed a small forecastle with a breakwater to be fitted, and superstructures were reduced to a minimum to permit maximum firing arcs for the turret. Dimensions and displacement were also within the required limits. On 19 August of the same year the project was presented to the Navy Minister and approved. Two new coast defence battleships with the names *Tempête* and *Vengeur* would be laid down to de Bussy's plans at Brest. The order for *Tempête* was placed on 26 December 1872; the ship would be the first of her size to be built almost exclusively of iron and steel in the naval dockyard.

The Coast Defence Battleship *Tempête*

The steel hull was built on the bracket frame system developed in the United Kingdom with a double bottom. Nine main watertight transverse bulkheads divided the hull into ten compartments which were numbered from forward to aft. The hull was further subdivided into sixty watertight compartments.

The armour belt enclosed the entire hull; it was 250mm thick at the forward and after ends, and 330mm amidships. The armoured breastwork was 40 metres long and rose to a height of 2.60m above the waterline; it too had a thickness of 330mm. The main deck, with a thickness of 50mm, was laid on a teak backing 120mm thick. The second armoured deck had the same thickness as the first. The iron conning tower, 250mm thick, was secured to a 400mm teak backing; the roof was 30mm. A steam capstan, rated at 12,500 tonnes, was located beneath the armoured deck.

The solitary superstructure block was only 2.50m wide, and supported a platform of moderate dimensions with a navigation bridge constructed of wood at its forward end. There was a single funnel immediately abaft the bridge, and a pole mast and a compass platform. Boats were stowed athwartships on crutches fore and aft of the mast. The steel mast was 21.39m high; at its base was a derrick to handle the boats atop the superstructure.

The ship's unusual silhouette led her to be nicknamed the 'flat iron' (*le fer à repasser*). Total cost of the ship was 6.43 million francs.

Table 1: Dimensions & Displacement

Length, oa:	73.60m
Length of keel:	70.80m
Beam at waterline:	17.60m
Depth of keel:	5.10m
Draught at deep load:	
forward	5.33m
aft	5.42m
Freeboard aft:	0.70m
Displacement:	
normal	4531 tonnes
deep load	4908 tonnes

Table 2: Equipment & Complement

Searchlight projectors:	three Mangin 40cm
Boat outfit:	8.5-metre cutter
	7.65-metre White cutter
	7-metre cutter
	8.5-metre whaler
	7-metre whaler
	5-metre dinghy
	3.5-metre punt
Anchors:	two 3700kg stocked anchors
	one 1270kg stocked anchor
	one 1140kg stocked anchor
Complement:	7 officers, 165 men

Armament

The two 274.4mm Modèle 1875 guns were housed in an enclosed turret which could train through 330 degrees. They were 4.5m above the waterline.

The turret armour comprised 300mm of iron secured to a teak backing 550mm thick. The roof was 50mm on 400mm of teak. The weight of the turret protection was just over 300 tonnes.

The fire of the main guns was controlled from a small cabin on the roof of the turret, which could be fully trained in 55 seconds. The guns could be elevated to +8°, corresponding with a range of 4700 metres; when fully depressed to -3° they could hit a target 50 metres away. The guns were elevated hydraulically using an articulated arm whose axis of oscillation was beneath the turret

Table 3: Armament

Calibre	No. of Guns	Rounds
274.4mm	2	160
47mm QF	4	2000
37mm QF	4	3000

commander's platform. The firing cycle was one round every three minutes when the gun was returned to the loading position, or one round every 1min. 33secs. with the ready-use ammunition. The weight of the 274.4mm projectile was 216kg.

The armament of the *Tempête* was completed by four 47mm and four 37mm QF guns. The 47mm projectile weighed 2.46kg, that of the 37mm 0.75kg.

Orders were transmitted from the conning tower to the guns via voice-pipes. A defect which was discovered only after completion was that the gun embrasures were so small that the gun commander, who was positioned behind the breech, was unable to see the target prior to opening fire.

Tempête was not fitted with torpedo tubes, but as a command ship for 'mobile defence' (*défense mobile*) forces she could embark and stow 18 torpedoes. The total weight of ammunition was 41.46 tonnes.

Machinery

There was a single machinery room. Four boilers built by

The coast defence battleship *Tempête* on the slipway at Brest (SHM)

Tempête: Firing Arcs

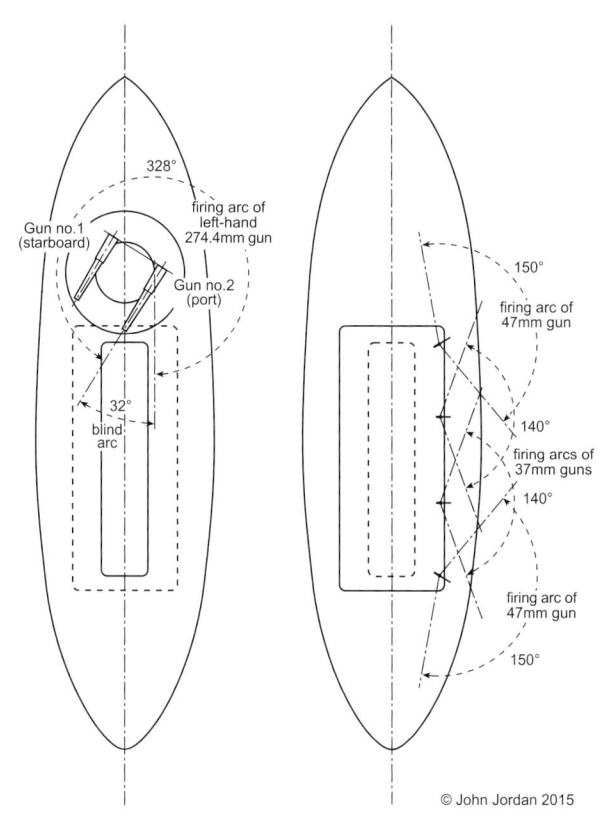

Table 4: Machinery

Boilers:	four Indret 4-bodied with two grates
Engines:	one Wolf horizontal 6-cylinder
Horsepower:	2164ihp
Coal:	168 tonnes
Endurance:	527nm at 10 knots; 1103nm at 5 knots
Trial speed (max.):	11.68 knots
Propellers:	one bronze 4-bladed 4.61m diameter
Rudder:	one 12.76m^2

A C Loire at Nantes provided steam for a six-cylinder horizontal reciprocating engine powering a single shaft. The transmission of orders from the bridge to the engine room was via voice pipes.

At cruise speed fuel consumption was 1800kg of coal per hour. In addition to the 168 tonnes of coal, 2.1 tonnes of wood was embarked to fuel the galleys.

Electricity was supplied by a dynamo rated at 100A/75V, a dynamo rated at 45A/66V, and a Meritons machine for signalling rated at 15A/30V.

At the maximum speed of 10 knots, with full rudder, the turning circle was only 150m with 2° of heel. Unusually, the servo-motor for the rudder was located forward of the boiler room at the upper level. Any leakage of steam in the boiler room rendered this compartment inaccessible. And water which flooded the forward compartments in rough weather could be only be evacuated via the heads!

An Uneventful Career

The launch ceremony took place on 18 August 1876 and was a success. During fitting out there were a number of modifications. The breakwater was demolished in 1878 and replaced by a more substantial one. The longitudinal deck which connected it to the forecastle was reconstructed. The heads for the crew, which were previously on the outer sides of the hull, were relocated directly abaft the breakwater. The forward part of the super-

Tempête fitting out at Brest; inboard of her is the central battery ship *Redoutable*, built by Lorient Naval Dockyard and completed in December 1878. (SHM)

The sister-ship of *Tempête*, the coast defence battleship *Vengeur*. (DR)

structure was suppressed to permit a complete rotation of the turret. The powder magazines underwent successive modifications, and a biscuit store was transformed into a magazine.

The machinery was in place from 12 July 1877 and fitting out completed on 9 January 1879. *Tempête* was manned for trials on 29 November 1879, and on 1st March 1880 sortied for her turning trials. The crew were at their posts, but the boilers were operated by specialists from *Constructions navales* (CN). For some unknown reason one of the lower taps in a tube which controlled the water level was closed, and the stokers failed to notice this. Three rows of tubes in the starboard after boiler were exposed and one of them split. Steam filled the corresponding grate, causing a huge flashback. All the stokers were safely evacuated, although three men suffered burns. At the same moment the rudder servo-motor failed and the ship no longer responded to the helm. The order was given to stop engines. Engineer Grolous rushed to check on the damage to the servo-motor but as he was entering the access trunk to the boilers he was pushed back by the engine room personnel fleeing the flames. Grolous decided to stem the steam pressure using a device which redirected the steam directly into the condenser. It was less than 15 minutes before the *Tempête* resumed her

Tempête during her prolonged period of trials and final fitting-out. (DR)

THE 'FLAT-IRON': THE COAST DEFENCE BATTLESHIP *TEMPÊTE*

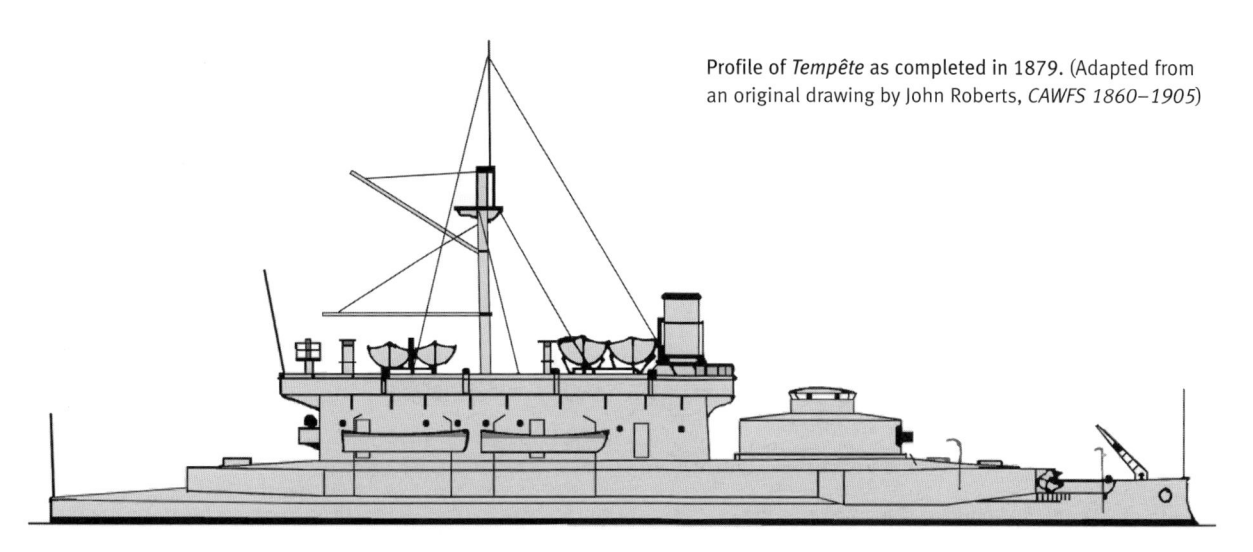

Profile of *Tempête* as completed in 1879. (Adapted from an original drawing by John Roberts, *CAWFS 1860–1905*)

Table 5: Commanding Officers

LV Corrard:	3 March 1887
LV Gourjon du Lac:	19 June 1888
CF Lecocq:	1st September 1889
CF Destephen:	1st September 1891
CF Motet:	18 June 1893
CF Thierry:	20 September 1893
CF Allys:	20 September 1895
CF Dufaure de Lajarte:	1st March 1897
CF Voiellaud:	3 March 1898
CV Merleaux Ponty:	5 March 1899
CF Morier:	7 January 1901
CF Aubert:	9 November 1902
CF Rey:	19 February 1904

This image highlights the unusual configuration of *Tempête*. (Bougault photo)

course at low speed; she subsequently entered the dockyard for repairs.

Sea trials were completed 26 March 1880. In 1898, the sea-keeping qualities of the ship were described as follows:

> The ship is exceptionally manoeuvrable. She is slow to work up to speed but even slower to lose momentum. When the engines are stopped steering is virtually impossible, making mooring difficult.

In 1902 a further report stated:

> The sea-keeping qualities of the ship are much better than might be expected considering the configuration of the

Tempête in the lake of Bizerte. (DR)

hull and her silhouette. Initial stability is exceptionally high. She rolls very little and recovers well. During a rough-weather transit to Algiers in May 1902 she behaved well with a head sea. Manoeuvrability is exceptional as long as the propeller is turning

Tempête was commissioned (*armement définitif*) on 4 July 1883 and the ship was assigned to the Trials Division; the latter was dissolved ten days later.

A cartoon of *Tempête* published in the magazine *Armée et Marine*; the bow view of the ship is decorated by two flat irons (*fers à repasser*), from which her nickname was derived.

On 10 July 1883, *Tempête* sailed for the Iroise Sea off Brest for tests to determine whether the blast from the main guns would damage the wooden service boats. Four 274mm rounds were fired and the boats were undamaged; however, the superstructure suffered blast damage. It was decided that major strengthening would be necessary before further trials with the guns trained aft were undertaken. In the same year, while exiting the Brest inlet in rough seas, the ship took on so much water that there were concerns for her safety. The incident was blamed on a broken turnbuckle and insufficient height of the coamings of the lower deck. The Maritime Prefect decided that in future *Tempête* would sail only in fine weather!

In July 1886 *Tempête* was docked, and advantage was taken to replace some of the deck plates, which had rusted badly. On 4 August 1888, while manoeuvring in the outer harbour at Brest, the ship grounded and had to await the next tide to be refloated. A docking revealed no damage. The next few years were occupied with training sorties, punctuated by periods in reserve for maintenance or modifications.

On 20 July 1892, during night manœuvres in the anchorage at Brest, torpedo-boat 76 was rammed by *Tempête* and sank.[1] On the 23rd, her 24-hour trials were pronounced a complete success, but it is difficult not to see this as an indication of how long it took to bring the ship to complete readiness.

First Category Reserve status was followed by 2nd, then 3rd Category status, a period marked by constant repairs and modifications: boiler and electrical problems, the replacement of the order transmitters, etc. During the first half of the ship's career, the engineers of the

THE 'FLAT-IRON': THE COAST DEFENCE BATTLESHIP *TEMPÊTE*

Constructors Corps would be permanently at her bedside. *Tempête* would deploy for the first time fourteen years after her entry into service!

In 1896 the four boilers were disembarked and replaced by new models brought to Brest from Cherbourg aboard the transport *Isère*. *Tempête* would again be operational from 15 June 1897. Soon she would be appointed flagship of the Naval Division of Tunisia, and she left Brest for Bizerte on 6 July, arriving on 25 July. As the key unit of the local 'mobile defence' forces, she was expected to make at least one sortie per month. During an uneventful career at Bizerte, her schedule comprised a variety of activities in the ports of North Africa. On 28 April 1898, she towed torpedo-boat *122* to Algiers, where she was docked on 9 May. The dockyard at Algiers was not particularly well-equipped to handle a vessel of this type. Echoes of this can be found in the local press of the period:

> Special keel blocks had to be prepared for the big graving dock in order to receive the *Tempête* because of her size and configuration; her beam is such that the ship could only just be accommodated in the dock, which is only 13.83m wide at its base. A bed of timber beams, placed longitudinally and perfectly levelled had to be constructed over a length of 25 metres to support the hull in the area of the armoured turret. *Tempête* is only the second battleship ever to be docked at Algiers, the only previous ship being the *Duguesclin*.

A second docking at Algiers took place the following year; *Tempête* then returned to Toulon on 22 May 1900

Part of the crew returning to the 'Flat-Iron'. (Armée et Marine)

to undergo repairs and maintenance in a major dockyard. On 29 May she entered Castigneau no.1 dock. On 21 June water was admitted into the dock to a height of 2.5m to enable the hull to be moved 50cm aft. This was to allow the keel to be painted in the areas previously obscured by the supporting timbers. At that moment a large number of the timbers gave way and the hull was damaged by the stone steps at the base and sides of the dock, which were crushed by its weight. An ingress of

Tempête, the keystone of the local 'mobile defence' forces, moored at Bizerte. She served there from 1897 until 1906, when she was recalled to France for deactivation. (DR)

167

Tempête docked at Algiers in 1898. (DR)

water through the damaged plating was partially stemmed with Makaroff mats, and the tank vessel *Aqueduc*, aided by the auxiliary machinery of the *Tempête*, pumped all the water from the hull. An enquiry set up on 22 June found that the insufficient width of the dock for a vessel the size of the *Tempête* had necessitated raising the timbers supporting the sides of the ship by 70cm. The timbers had shifted when the ship was

Viewed from this angle the ship still appears threatening, but by this stage in her career *Tempête* would have fared badly in a battle on the high seas against any modern warship. (Marius Bar)

Tempête fires a farewell salute in honour of Rear-Admiral Merleaux-Ponty, C-in-C of the Naval Division of Tunisia, following his death on 29 August 1902. *(Armée et Marine)*

Tempête at Toulon in 1900. (Marius Bar)

A rare view of *Tempête* showing the unusual configuration of the superstructure block, which was fined in order to maximise the arcs for the main guns on after bearings. (DR)

Tempête dressed overall. (Marius Bar)

refloated. As a result of the incident *Tempête* was transferred to the Missiessy docks with a slight list to port. Following a month of repairs, during which the plates on both sides of the hull were replaced, the ship again left Toulon for Bizerte on 8 August. During the repairs, 47mm guns from the battleship *Colbert* had been installed atop the superstructures.

On 4 May 1901, during a passage from Philippeville to Algiers, the counterweight wheel of the steering servo-motor shattered. This was subsequently repaired without a problem.

The next four years were uneventful, the customary gunnery exercises being interspersed with maintenance periods. The 29 August 1902 saw the death of Rear-Admiral Merleaux-Ponty, C-in-C of the Naval Division of Tunisia; his flag was brought aboard the *Tempête*, which would host the commemorative services held by the officers and the troops of Bizerte.

During this period the limited usefulness of the *Tempête* became more and more readily apparent; she was slow and her guns were outdated. Her age meant that in the event of mobilisation she would have served as a floating fortress moored behind the harbour walls at Bizerte. On 30 December 1905 a dispatch from the Navy Ministry ordered the *Tempête* to return to Toulon to pay off. She duly departed on 18 January 1906, arriving at Toulon four days later. She would be replaced in the Naval Division of Tunisia by the armoured gunboat *Phlégéton*.

Tempête hoisted her colours for the last time on 15 February. On 15 March a cost estimate was drawn up for maintaining the ship in a seaworthy condition with a view to her employment as a target ship.

On 26 April 1907, Navy Minister Gaston Thomson sent a telegram to the C-in-C of the Mediterranean Squadron instructing that *Tempête* should be stricken from the active list and her machinery condemned. The ship was to be employed as a target for practice shoots.

Target Ship

Tempête was the target of the Mediterranean Squadron throughout the year 1908, sustaining considerable damage in the process. On 31 August the Maritime Prefect for the region, Rear-Admiral Forestier, asked the Naval General Staff whether sufficient repairs should be carried out so that the ship could be used for further gunnery exercises, or whether she should be sold for scrap. The C-in-C of the Squadron considered that the ship could still be employed as a target, but that her superstructures had been so badly damaged that they would have to be removed and replaced by the wooden panels generally used on floating targets. In the event, *Tempête* would not give her executioners time to carry out the recommended amputation.

The battleship *Gaulois*, which administered the *coup de grâce* to the *Tempête* on 18 March 1909. (DR)

The following year she was anchored in Alicastre Bay at Porquerolles, where she endured the fire of the battleships *Justice*, *Liberté* and *Saint-Louis* from 17 March 1909. A 305mm shell from the *Justice* severed pipework in the double hull next to the starboard engine room. The resulting ingress of water was stemmed sufficiently to keep the former coast defence battleship afloat provided she was pumped out every twelve hours. On the 18th it was the turn of the *Gaulois* to batter the wreck to a point where she had to be towed to Brégançon in an effort to

Tempête displays the extensive damage sustained from shellfire during gunnery exercises by the Mediterranean Squadron. (DR)

keep her afloat. Following rudimentary repairs, *Tempête* returned to Alicastre at 2300 the following day listing heavily to port.

As a safety measure the cruiser *Galilée* despatched a boarding party to check the level of flooding. It quickly became apparent that the ship was doomed, and at 1000 the following telegram was sent to the admiral commanding the Squadron and to the Maritime Prefect:

> *Tempête* has been struck by several shells and can no longer serve as a target. She is currently moored at Porquerolles, but is taking on water. She could be towed to Toulon but only with a flat sea.

An auxiliary vessel, *Goudoux*, remained alongside the sinking ship while the *Galilée* returned to Toulon. Orders were given to keep the *Tempête* illuminated at night, to pump out regularly the double bottom, and if she looked like sinking to tow her into the shallows, evacuating all the materiel located in the turret. Unfortunately, the weather deteriorated to such an extent that the *Goudoux* could no longer maintain her station and took shelter behind Maubousquet Head. The inevitable occurred on Saturday 20 March when the semaphore station at Porquerolles signalled at 0750: '*Tempête* appears to be capsizing at her moorings. Very strong east wind, rain and fog'. At 1115 the Naval Staff received the following telegram: '*Tempête* sunk at her moorings as result of bad weather and serious shell damage'.

The harbour barge *Polyphème* arrived at the location of *Tempête* the same day and reported that the forward part of the keel was still above water over a length of some 40 metres, At 0700 the following day, when this same vessel made her way to Hyères to secure the moorings of the targets for the fleet, her crew observed that *Tempête* had completely disappeared. A dinghy was lowered and proceeded to the spot where she had sunk, took a one-metre sounding on the bow and moored two cork buoys. The following Monday, bubbles of air were still escaping from the hull, and it was judged inadvisable to investigate using divers.

On 23 March 1909 the commission investigating the wreck reported:

> The divers have conducted a complete survey of the wreck. The ship sank stern first, and the stern is buried in the sea bed up to the propeller. The mud is up to a height of 1.5-2m around this part of the ship, and the currents and tides mean that the after part of the ship will soon be covered. The forward part of the ship is not so deeply buried; the hull in this area is partially filled with air which is unable to escape and remains buoyant, so that the bow is some 80cm above the bottom. The keel is intact and relatively undamaged. The vertical part of the superstructure has collapsed under the weight of the hull, which has capsized to port. The depth of water over the stern is 13 metres and over the bow 2.10m metres. A red and black tun-buoy has been secured to the rudder with a chain: the entire body of the wreck is therefore to the east of the buoy; the light

The upper deck of *Tempête*, showing the effects of a multitude of shell hits; the photo was taken shortly before her loss.

> green of the underside of the hull is clearly visible from the surface. The wreck is a danger to shipping because the bow is only 2.10m beneath the surface, but is not on the route taken by steamers, and sailing ships, even when tacking, are unlikely to come that close to Cape Mèdes. However, it is recommended that a warning be posted for shipping.

Although the wreck had now been rendered relatively safe, it would still be necessary to clear the sea bed off Porquerolles of this large chunk of scrap metal, which hindered navigation. The same commission proposed the following solutions:

> Using sand pumps or jets to free up the entrance to one of the panels on the armoured deck – or possibly one of the shell passages on the deck, which would be more straightforward – then to insert a pump using compressed air and expel the water from the hull. The integrity of the underside of the hull should ensure the success of this operation. The *Tempête* should then be able to float and be towed, although righting the ship will almost certainly be impractical.

However, this operation could not be undertaken immediately as the Port Management Agency had no means of raising the hull.

In 1910 a certain M. Giardino of Marseille offered to break up the *Tempête* where she lay within three years for the sum of 30,000 francs. The offer was not taken up. On 4 February 1910 a first attempt to dispose of the hull by the Defence Estates Agency was made at Toulon, but the bids were unsuccessful.

The remains of the *Tempête* were abandoned for more than three years before a directive dated 4 November 1912 stipulated that there should be no sale to a private company. The Defence Estates Agency then handed over the wreck on 21 December of the same year, and the

Tempête was sold to M. Nicolini[2] of Toulon for the sum of 8755 francs. Demolition work began straight away but was interrupted by the outbreak of war. It resumed at a reduced pace on 18 September 1915. A substantial part of the hull and superstructures was removed, and the breakers duly left the site in the mid-1920s.

From April 1959, the Serra Company agreed to recover the last plates of armour and the 274mm guns; the wreck was then finally abandoned.

Diving on the Wreck

Even when ships are broken up on the spot, dives on the battleships *Iéna* and the *Tirpitz* have demonstrated that with a large ship there will always be some items remaining on the sea bed; it was therefore decided that our diving team would look for the remains of the *Tempête*.

In fact, our interest in this ship was only heightened by the discovery that the wreck could not be found, despite knowing the approximate location. During 2008 four of our expeditions failed to discover the remains of the ship. The sonar equipment remained silent, and dives revealed only an empty sea bed. The winter of 2009 was unusually harsh and we had to suspend operations for several weeks before leaving again for Alicastre Bay. Finally, on 31 July, we were able to resume our search using an underwater vehicle towed by a powerful boat. After several fruitless passages of the sea bed, we came across a large quantity of debris scattered over the sandy bottom. The site of the wreck was extensive, and was scarred by huge craters which almost certainly mark out the original location of the hull, now completely removed.

As with the battleship *Iéna*, which sank a few hundred metres away, little was left of the *Tempête*. The bow and the stern had disappeared; only the superstructures embedded in the sea bed remained. A second wreck, that of a sailing boat ten metres long, is in the centre of the site. The position of the *Tempête* according to WGS-84: 43°01'056N, 06°13'480W.

The Serra Company salvages one of the 274mm guns in 1959. (DR)

Acknowledgement:
This article was translated from the original French by John Jordan. The author also wishes to thank Rainer Schittenhelm, who took him in his boat to survey the wreck of *Tempête*.

Footnotes:
[1] The wreck of torpedo-boat 76 was found, stricken and sold for breaking up in 1894.
[2] M Lazare Nicolini was President of the Chamber of Commerce of the Var region; on the same day that he purchased *Tempête* he bought the wreck of the battleship *Iéna*.

WARSHIP NOTES

This section comprises a number of short articles and notes, generally highlighting little known aspects of warship history.

DERFFLINGER: AN INVERTED LIFE

Aidan Dodson looks at the twilight career of the German large cruiser (battlecruiser) *Derfflinger* during the Second World War, during which she spent longer afloat upside-down than she had done the right way up, despite a number of attempts at arranging for her scrapping.

Along with the other battlecruisers of the German High Seas Fleet, SMS *Derfflinger* (launched at Hamburg on 17 July 1913) was scuttled at Scapa Flow on 21 June 1919, capsizing and sinking at 1445 hours. During the 1920s and 1930s, most of the fleet was refloated, the capital ships (and some destroyers) being towed to Rosyth Dockyard, where they were demolished.[1] All but one of the big ships had been raised upside down and had thus to be scrapped in dry dock, of which Rosyth had three. *Derfflinger* was the last ship to be raised, rising to the surface in late August 1939,[2] having had all openings sealed and been pumped full of compressed air (see drawing). Under normal circumstances, she would then have been prepared for the tow to Rosyth and for her docking there.[3]

However, before she could leave Scapa, the Second World War had broken out, meaning that the Rosyth dry docks were all now earmarked to support the Royal Navy. Thus, they were no longer available to Metal Industries, since 1933 responsible for salvage operations at Scapa, once they had completed cutting up the wreck of the battleship *Grosser Kurfürst* (work commenced on 24 August 1938) in No.2 Dock. *Derfflinger* was therefore towed behind the island of Rysa Little, opposite Hoy, within Scapa Flow, and moored there with ten 7.5-ton anchors, awaiting events.[4]

The war effort's demand for scrap metal meant that the question of her future was reopened a few months later, in February 1940. It was argued that the 25,000 tons of scrap built into the hulk, which included armour plate and non-ferrous metals, justified the risk of tying up a dock for the four months that was cited as necessary for scrapping the battlecruiser.[5] The Director of Dockyards pointed out, however, that the Rosyth docks were the only docks on the East coast capable of taking capital ships, that they had been almost continuously occupied since the previous October, and that blocking one for even four months was therefore too much of a risk, particularly in light of the expected German offensive in the Spring. In addition, Commander-in-Chief Rosyth also noted the potential value of *Derfflinger* as a blockship for sealing the secondary entrances of Scapa Flow, although she proved to have too great a draught to be considered.

On the other hand, it soon became clear that two other big ships which might have yielded significant quantities of scrap, the ex-battleships *Iron Duke* (gunnery training ship) and *Centurion* (radio-controlled target), had ongoing war roles.[6] In addition, since the ability to salvage the hull of the sunken training ship (ex-56,551grt liner) *Caledonia* (ex-*Majestic*, ex-*Bismarck*), was unproven – although 7,000 tons was soon removed from her upperworks, the hull was not finally raised until July 1943 – the issue of *Derfflinger* was revisited in March. A tentative proposal was then made that she might be docked after the cruiser *Belfast* was removed from No.1 dock in June, following emergency repairs resulting from her mining on 21 November 1939, prior to proceeding to Devonport for reconstruction (she left dock on the 27th). Metal Industries stated that it was possible to break up the ship in 16–20 working weeks, and that it might be done in such a way that the hulk could be removed from the dock during the first 6–8 weeks, should an extreme emergency arise.

However, nothing was done to act on this, presumably due to the fall-out from the German attack on the Low Countries and France in May 1940. The idea remained alive, however, until the Admiral Superintendent at

Derfflinger at sea, following her 1917 refit, during which she received a tripod foremast. (Author's collection)

Derfflinger slips under the waters of Scapa Flow at 1445 on 21 June 1919. (C W Burrows, *Scapa and a Camera* [London, 1921], p.34)

175

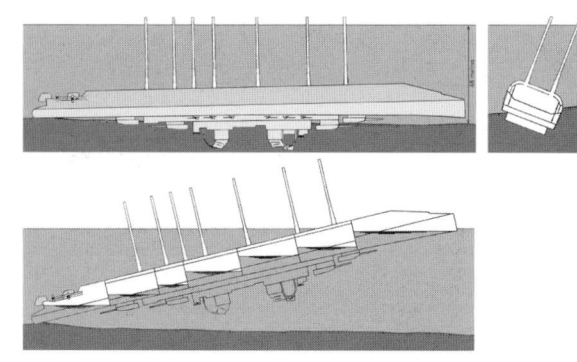

Sketches showing key stages in the raising of *Derfflinger* during 1938/39. The first stage involved the fitting of airlocks of between 28 and 40 metres in height on the bottom of the capsized hulk, allowing water to be displaced by compressed air and workmen to enter the ship to gradually seal openings and restore its watertight integrity. This was done in such a way that the hull was divided into seven airtight transverse compartments, while the longitudinal torpedo-bulkheads were also sealed to create similar longitudinal compartments along the flanks. These allowed the actual raising to be fine-tuned, in particular in removing the ship's 20.5º list to starboard – a technique that had been honed by more than a decade's experience of raising such inverted wrecks. The bow was raised first, to allow divers to inspect the underside (ie superstructure!) of the ship and check for any issues prior to bringing the stern of the ship to the surface. (Adapted from McKenzie, *Trans. Inst. Eng. & Sbdrs. Scot.* 93, figs. 2 & 3)

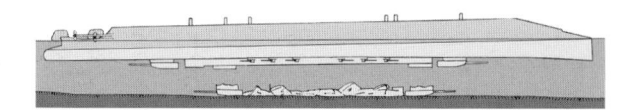

When raised, *Derfflinger* still retained all four turrets and the remains of her funnels and superstructure. To allow her to be docked for scrapping, her draught needed to be significantly reduced; thus 'B' and 'C' turrets were removed at her wartime berth behind the island of Rysa Little, together with all remaining top-hamper. (Adapted from McKenzie, *Trans. Inst. Eng. & Sbdrs. Scot.* 93, fig. 4)

Rosyth produced a minute in August that undermined the case. He pointed out that the break-up period cited was shorter than the time previously taken to scrap such vessels, and that cranage and storage space previously available would not now be. In addition, the time to remove and reinstate the support blocks in the dock would add six or seven weeks to the timeline. He thus considered that if *Derfflinger* were to be broken up at Rosyth, the dock would probably be unavailable for naval purposes for 6–7 months at the very least. He also doubted that the hulk would indeed remain floatable during the first weeks of scrapping. There were also manpower issues and questions as to where the hulk could be moored prior to docking. This killed the Rosyth option, while a proposal to use a dry dock at Southampton was also dropped for operational reasons – although the idea would be revived later.

Nevertheless, the scrap shortage continued, and in March 1942 the British Iron and Steel Corporation (Salvage), which had now requisitioned the hulk from Metal Industries, proposed that a new effort be made to break-up *Derfflinger*, and that Peterhead harbour, half way between Scapa and Rosyth, could be used. Approval was given on 5 April, the original conception being that she would be moored inside the breakwater and scrapped there.

In the subsequent discussions, Metal Industries pointed out that the non-use of a dry dock would present challenges, as the moment that any cut was made into the hull, compressed air would be released and the hulk would sink: it would not be possible to conduct the usual afloat scrapping process, whereby the removal of material gradually lightened the hull, allowing it to float ever higher in the water, until it could be beached for final dissolution. Accordingly, the concept was changed to beaching the hulk in the shallows on the opposite side of harbour, where the highest section would be broken up, the remainder of the hulk then having appropriate decks and bulkheads resealed to allow refloating and re-beaching.

The move of the hulk from Scapa to Peterhead was planned for the end of May, which then slipped to 7 June. Four days prior to departure, however, Metal Industries reported that having gone into the matter further, they had concluded that the break-up plan was not after all practicable. The tow was thus cancelled.

Derfflinger's steel continued to exert a draw (as did that of *Caledonia*, significant effort being expended on the difficult work of her salvage). Accordingly, the idea of breaking up the ship at Southampton was revived in January 1943, and approved in February. But in March the need for a strong escort between the Orkneys and the south coast, the lack of depth in the Southampton approach channel, and the availability of labour were raised as concerns by C-in-C Portsmouth. There were also worries as regards tying up the only two dry docks big enough to take her at Southampton, the King George V and Trafalgar, which were particularly used for merchant ships and armed merchant cruisers. Crucially, Chief Salvage Officer, Scapa, was of the view that, given experience with the much shorter tows of other High Seas Fleet inverted hulks to Rosyth, *Derfflinger* was insufficiently seaworthy to survive a tow all the way to Southampton. This and other objections led the project to being cancelled on 22 March 1943.

In June 1943, the Director of Sea Transport suggested that the hulk could be moved to the Firth of Forth against the possibility of dock availability, or that it might be possible to break her up on a beach or in a tidal basin in the Bristol Channel. In favour of the latter location was that the exceptionally large rise and fall of tide there would make the process easier than at Peterhead; it was also closer to the South Wales steel works. The Bristol Channel option was considered further, but dropped in August, although some ideas put forward for camouflaging the hulk as a group of barges were kept in case any move were to take place in the future.

WARSHIP NOTES

The hulk of *Derfflinger* in the process of being docked in *AFD.4* in November 1946. (CPL)

AFD.4, with the hulk of *Derfflinger*, arrives at Faslane in tow from the mouth of the Clyde, where the docking process had been carried out. The structures on the underside of the battlecruiser housed the passage crew and also the compressors necessary for periodically topping-up the air inside the hull that kept it afloat. The chimney-like pipes seen protruding at a slight angle from the hull are the airlocks giving access to the interior of the hull, with the extensions used during the raising removed. (Author's collection)

AFD.4 and *Derfflinger* coming alongside at Faslane, where the battlecruiser would be finally broken up. Behind them is *Iron Duke*, her after part already partially dismantled (the mainmast is in the process of being removed), following her own arrival from Scapa on 19 August 1946. (Author's collection)

This was seemingly the last wartime attempt to scrap *Derfflinger*, which continued to lie at her berth in Scapa Flow. With the end of the war, her final disposal could go ahead, and two turrets and the aft conning tower were removed (the turrets being released from inside the ship) to reduce the minimum draught of the hulk from 12.3 metres to the maximum allowable figure of 12 metres in the Rosyth dry docks. However, in March 1946, the Admiralty informed Metal Industries that Rosyth docks remained unavailable, meaning that an alternative dock had to be found, the wartime studies of the Peterhead option having made clear that a dry berth was needed to properly dismantle such an upturned vessel.

A solution was found when Metal Industries obtained in July 1946 the lease of the former Military Port No.1 at Faslane, on the Gare Loch in the Clyde estuary, as its new principal shipbreaking yard, replacing Rosyth. Nearby, at Rosneath at the mouth of the Gare Loch, was the 32,000-ton-capacity Admiralty Floating Dock (AFD) 4, which had been built by Swan Hunter in 1912, and served in the Medway until 1915. It had subsequently been operated in the Tyne, at Portland and at Devonport before being moved to Scotland in September 1941. Now surplus to requirements, a careful study indicated that it could accommodate *Derfflinger*; it was therefore acquired at the same time as the Faslane lease.

Derfflinger was then towed to the Clyde, where she was docked with some difficulty, the dock having to be sunk 2.5 metres below its normal maximum depth to accommodate its unusual contents, leaving it with a freeboard of only 2 metres. The dock and battlecruiser were then towed to Faslane, arriving towards the end of the year, where they were berthed against the wharf. They lay just aft of *Iron Duke*, which had also spent the war at Scapa, grounded after bomb damage, and had been sold to Metal Industries in March 1946, refloated in April and arrived at Faslane in August. It took some 15 months for *Derfflinger* to be entirely dismantled. It was briefly considered whether *AFD.4* might be used to break up the remains of the liner *Berengaria* (ex-*Imperator*), which had been partly dismantled at Jarrow between 1938 and 1940 and had since languished in the Tyne in two pieces, but instead the dock was sold in June 1948 for further service abroad: it still survives in use at Bergen-Laksevåg, after over a century of service.[7]

Footnotes:

[1] The key sources for the salvage and scrapping of the High Seas Fleet are S C George, *Jutland to Junkyard* (Cambridge: Patrick Stevens Ltd., 1973; reprinted Edinburgh: Birlinn, 1999) and I Buxton, *Metal Industries: Shipbreaking at Rosyth and Charlestown* (Kendal: World Ship Society, 1992). See also G Bowman, *The Man Who Bought a Navy: The Story of the World's Greatest Salvage Achievement at Scapa Flow* (London: Harrap, 1964; republished by Peter Rowlands & Stephen Birchall, 1998) and T Booth, *Cox's Navy: salvaging the German High Seas Fleet at Scapa Flow, 1924–1931* (Barnsley: Pen & Sword, 2005) for the first period of work.

[2] Curiously, many standard reference works state in error that she was raised in 1934 and broken up at Rosyth during 1935–36.

[3] The salvage and scrapping of *Derfflinger* are described in detail in T McKenzie, 'Marine Salvage in Peace and War', *Transactions of the Institution of Engineers and Shipbuilders of Scotland* 93 (1949–50), pp.124–39, with a summary in the works cited in Note 1.

[4] See also *Warship 2008*, p.149.

[5] For this and other aspects of the history of the hulk between 1939 and 1943, documentation is contained in National Archives file ADM1/13330.

[6] As base/accommodation/harbour defence and base ship/training vessels, respectively.

[7] For a history of Admiralty floating docks, see Ian Buxton, *Warship 2010*, pp.27–42.

WARSHIP NOTES

THE SINKING OF HM SUBMARINE *K13*

Ian Johnston provides an eye-witness account of the accidental sinking of the submarine *K13* during her trials on the Gareloch in 1917.

On 29 January 1917 the 'K'-class submarine *K13*, built by the Fairfield Shipbuilding & Engineering Co. Ltd. of Govan, Glasgow, sank while running trials on the Gareloch with the loss of 32 lives. The sinking set in motion frantic rescue attempts assisted by Fairfield personnel who were rushed to the scene. This first-hand account written by Thomas Dey, assistant foreman shipwright at the Fairfield yard, was written some years after the event, and describes the tragedy as he remembers it.

'HM Submarine *K13*' by Thomas Dey

It was on 29 January 1917 that the above vessel left Fairfield Basin, Govan, to go to the Gareloch for diving trials. On the way down she had two minor mishaps. First, when leaving the builder's basin a mooring wire got foul of her propellers and she was delayed for half an hour till the yard diver cleared it. Secondly, when about 300 yards below Whiteinch Ferry, she grounded for a few moments, but as she was going downriver on a rising tide she soon freed herself. Proceeding to the Gareloch she completed all her diving tests very satisfactorily. The Admiralty and Fairfield officials were highly delighted with the manner in which she had behaved. They all praised the grand new machine. Included in the Fairfield's official party were Sir Alexander Gracie, the Chairman of the firm, Mr Hugh M McMillan, Managing Director, and Mr Percy Hillhouse the famous naval architect.

The submarine was under the command of Commander Herbert, a very efficient naval officer who I believe escaped from a sunken submarine off the coast of Harwich. On her trials on the Firth [of Clyde] and Gareloch she was under the charge of an old experienced Clyde pilot, Captain Duncan. All her trials and tests were completed shortly after 2pm and the signal was passed around: 'All diving trials finished for the day'. Several of the Fairfield officials left the vessel about this time and Captain Goodhart, the commanding officer of another submarine, came aboard. Just before 3pm she submerged in deep water above Shandon and failed to come to the surface.[1] Through some mistake, which is unknown to the writer, she had dived with two ventilators in the after part of the vessel wide open. The engine room and after part of the vessel immediately filled with water. Luckily for the men in the forward portion the Engine Room watertight door was closed and it confined the water to the after compartments. I believe Mr Wm Struthers, who was assistant manager, took the work of going along the passage to see that the door was properly bolted, a very

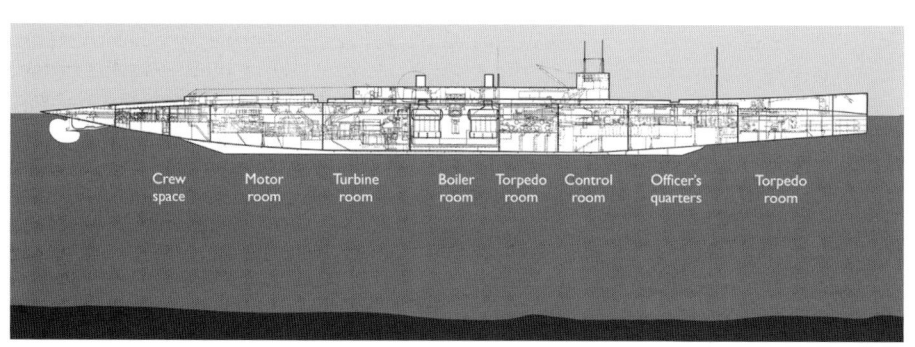

The 'K' class submarine had a length overall of 339ft (103m). The Gareloch off Shandon where *K13* made her final dive that Monday has a depth of about 60ft (18m). (All drawings by the author)

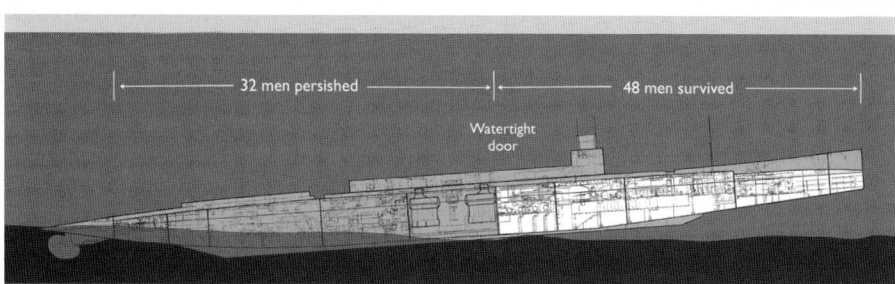

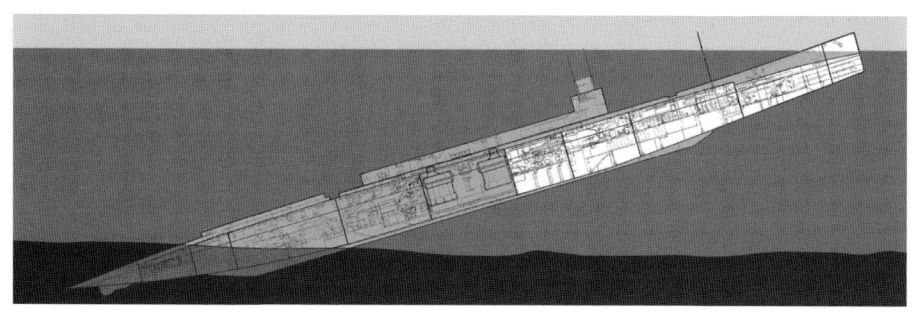

K13 came to rest on the bottom of the Gareloch at a four-degree angle with the after part of the vessel completely flooded, drowning 32 men. To prevent the entire vessel from flooding, the watertight door in the bulkhead between the boiler room and the torpedo room was closed.

As air filled the tanks, *K13*'s bow began to rise and was immediately supported by wire hawsers passed under the submarine by salvage vessels. A hole was cut in the bow and by 10pm on Wednesday the men had been released, 55 hours after *K13* had dived.

179

gallant action under critical conditions. Unfortunately 39 [32] men lost their lives: they included Fairfield foremen and engineers and naval personnel, amongst them a very clever and highly promising Engineer Manager Mr Steel. Two engineers had a very lucky escape: Donald Hood of Fairfield's staff, and an engineer from Messrs Chadburn the telegraph people whose name in the meantime I have forgotten. They had gone forward to grease and oil the forward torpedo shutter. It had been a little bit stiff on trials and they went to adjust it. It was certainly a lucky bit of trouble for them, for they were the only members of the Engine Room to come out alive.

When she made her fatal dive an officer on another ship had watched and he thought all was not right in the way she had gone under, so with great thought and wonderful intuition he lashed a weight on to a ship's grating and put it above the spot where he saw her submerge. It was a good thing he did so for it saved valuable time in looking for the stricken ship.

I had gone to bed early telling my wife I wanted a good sleep, for I was working excessive overtime and was feeling pretty tired. At that time I had been for a few years Assistant Foreman Shipwright on warship construction. Shortly after midnight I was roused by a messenger at the door telling me to report to the yard immediately. I had no idea what was wrong or why I was wanted and the man at the door could not tell me. Hurrying to the yard I was given the fatal news but I was warned to keep it secret.[2]

A rescue party was collected comprising of two shipwrights, two riggers and three labourers; they were the pick of the best men in the shipyard. We collected tools, wire hawsers, lumber, etc. and put them aboard an armed tug boat that had been ordered to call at Fairfield for them. We set off for the Gareloch shortly after 6am with this exhortation from Mr Allan Aitken, Head Foreman Shipwright: 'Do your best to save these fellows' lives and I know you won't let them down.' On the way down the River we called in at Clydebank and collected a heavy log of timber and took it in tow. When we arrived at the spot in the Loch some distance above Shandon pier we found two salvage vessels moving into position. They were the *Ranger* and her sister ship under the command of Lieut.-Commander McKay. Commodore Young was the Naval Salvage expert and chief of staff, and what a man he was. Nothing was too trying or difficult for him and I never saw a man with such boundless energy; he never seemed to tire. In no time he had his naval divers over the side and on the top of the sunken submarine. We could follow every movement of them by the bubbles on the water created by escaping air in the head gear.

I have often wondered if the survivors of *K13* realised how much they owed to these brave, resourceful men. I often look back and think on these gallant fellows. They hardly took time off for meals. They had a job to do and they were determined to do it. One of them took his heavy hand hammer down and tapped on the hull to let the imprisoned men know (if any were alive) that help was on the way. They then proceeded to make the only outboard connection on the hull that was made, namely the Divers Connection. This is a hole about 6in diameter with a flange connection inside the hull of the submarine. It allows small bottles of stimulants, chocolates, food, etc. to be passed down to the imprisoned men and this was eventually done in this case.

Sometime around 11am I saw a great commotion in the water just like a pot boiling. I knew it was air escaping that was causing it and it was in such volume that I thought they had at last managed to release the Drop Keel Blocks on her Keel and that the ship was freeing herself, but my conclusions were wrong. It was no such luck, for in the midst of the commotion a man's head and shoulders suddenly appeared; it was Commander Herbert. His first words to us before we pulled him into a boat were 'Have you seen Commander Goodhart?' and of course we had to say no. He replied 'Well he should have been here, we both left the submarine together'.

Captain Herbert was clad only in a shirt and a pair of socks, a very scanty dress for the very cold water in the Gareloch in the month of January. Poor Goodhart failed to come to the surface, for in making the bold attempt to escape, his head had come in contact with a beam and he perished. The escape was made through the conning tower hatch. Herbert got clear but Goodhart had missed the opening. In less than a quarter of an hour Commander Herbert was out working like a Trojan in a suit of borrowed clothes. His help and advice was invaluable. He gave us an indication of how many men were alive, their actual position and how things were in general in the interior of the vessel. It was [had been] decided after a consultation inside the submarine with the officers and the leading ratings that the two commanders would try to come to the surface to give the rescue party the data they so badly required.

By this time the Clyde Trust Hoppers Nos. 7 and 10 had arrived on the scene to lend a helping hand. Wires were passed from ship to ship under the bow of the *K13* and when all were in position and made fast they were hauled taut, [and] they began to keep a very heavy hauling strain on them. This was kept up the remainder of the day on the Tuesday and all night, and it was heartbreaking; no sign of any movement. We were beginning to think it was a hopeless task, every man jack doing his utmost and all of us were praying that some kind of miracle would happen and set the stricken ship free. But miracles just don't happen that way and our quarry seemed to be sucked down too deep in the mud plus the weight of water inside her. None of our men left their posts, so keen were they to see the men saved. We kept up our heaving strain on the wires all day on the Wednesday; in fact so great was the strain that some of the wires snapped.

On the Wednesday evening, as dusk was coming down, I thought I could detect a little cross swell on the water but it just may have been imagination only. I have a pretty keen eye where movement of ships is taking place. I was taking depth sounding with my sounding rod when

Tom Dey, third from right, sitting with other shipwrights. In the hierarchical structure of the Clyde shipyards, bowler hats were worn by foremen and managers, thus distinguishing them from the flat caps worn by workmen.

I realised there was a slight movement taking place. I did not say anything till I was absolute certain and most of the naval people were away to mess rooms for a little food. When I realised that *K13* was gradually coming to the surface it was the only time in my busy life in shipbuilding that I got really excited. I got so excited and also intoxicated with sheer joy that I nearly fell overboard into the Loch. I shouted at the pitch of my voice: 'Boys, she's coming up'. I think they thought the strain had taken my senses but no, there she was, 6in of her nose showing, now it was feet and finally we got her to an angle of I would roughly say 10 degrees. She was showing about 16 or 18 feet of her hull along the water line. Now was the time when great care had to be taken in steadying the supporting ships because a slight movement would have meant wires slipping and the vessel plunging to her doom, but Commander McKay knew his job thoroughly and he soon had her just the way he wanted her. There was no time to lose and [it was] a blessing there was no wind nor swell on the water, a very unusual thing on the Gareloch in the winter months. A hole was immediately cut in the top of the hull and thro' the bulkhead and shortly after 11pm the 46 survivors escaped after being entombed for 57 hours. There was no sign of panic or jostling to get out. It was a great pleasure to see all our friends, Mr Percy Hillhouse, Naval architect, Mr Wm McLean, the greatest shipwright the Clyde has ever produced, Mr Ted Skinner, Electrical Manager, Mr Bullen, Mr Struthers, Asst. Managers.

It was amazing to see the cool, calm heroism shown by these survivors. Captain Duncan the 64-year-old Clyde Pilot's first words to me when he got out were 'Hello, what part of the Loch are we actually in?' Mr Edward Skinner, Fairfield's Electrical Manager, said to me when he came out: 'Hello Tam, you will be thinking it was you chaps outboard that saved us but it was my wee machines.' I replied, 'That's alright, Mr Skinner, but I wonder what all those heavy wires are on the fore end for?'

It was amazing to see a pair of boots coming up through the hole and then a naval rating would appear, out would come an alarm clock and then another naval rating, the Admiralty overseer in white overalls with his torch and notebook in his breast pocket and his chamois gloves in his hand just as you see them coming out of a tank in the shipyard after an inspection.

During the time we were so busy in the Gareloch, the Fairfield yard in Govan was also busy constructing and building an escape shaft or tube. A squad of men under the leadership of the late Tom Paul, Iron manager, worked night and day without rest to complete it and send it down to us but it never got past Helensburgh. When I was told by Mr McMillan, Managing Director, who was on the spot all the time, that such an escape tube was coming down I just could not see how it could be fitted. The vessel was in 98 feet of murky water. Time and tide was our greatest factor in the men's salvation. Steadying such a tube, keeping it in position, drilling holes to fix and keeping it watertight, and then to cut a hole to let the men have access to it, and if water got into the Battery Tank it meant that gassing would take place and it would gather a volume of water in the erection that we could not get rid off. It was utterly impossible to attempt to do anything with it. In theory on the drawing office bench it was alright, but in the murky swelling waters of the Gareloch it was a different matter. I think the idea was Sir Alexander Grace's and some naval experts. On the way up to Helensburgh on a puffer I explained and pointed out all the difficulties to Mr McMillan and I was delighted to hear him say: 'You are quite right, Dey, it can't be done' and he ordered it to be returned to the yard without even taking the fastenings off it.

It was a good job that we had a man with the experience and the courage to condemn what his boss and so-called experts had created. Recently I read in Glasgow papers of a diver who claimed to have fitted a shaft on *K13* and thereby saved 40 lives. It was an untruth. I have no recollection of him being nearer the scene than at least 20 miles away. Quite a number of the men who survived this unhappy ordeal have passed away, but some are now enjoying a well-earned retirement.

Although it's nearly 31 years since it happened I never forget the scenes in the Gareloch on those three eventful days: the wonderful work of the salvage people, especially Commander McKay and the divers, the grand help by the Clyde Trust Hoppers nos. 7 and 10; they also had their tug 'Clyde' and their diver Mr McKenzie in attendance. Mr Gush [?] the Greenock diver also played a grand part. We of the Fairfield Rescue Party were not called on to play a big part, but what they did do they did it well. They say the onlooker sees the most of the game well; I was the onlooker.

One peculiar incident happened when we went to the scene of the disaster in the Gareloch: a servant girl in one of the big houses reported that she had seen two men coming out of the submarine as she sank. She said that they both swam some distance before disappearing from view. No one paid any heed to her thinking it was only her imagination running away with her. But when *K13* was finally brought to the surface nearly seven weeks later it was discovered that Mr Steel, Manager of the Engine Dept and the Chief Naval Engineer of *K13* were both amissing [*sic*]. One of their bodies was recovered and identified several months later.

Although it was supposed to be kept secret, when I came into Craigendoran Station late on Thursday afternoon on my way home, I met an old lady from the extreme south of England, wanting to know if her only son was saved. I told her the truth when I said I did not know, as we were not acquainted with the names of the naval ratings, but I learned later that he had been drowned. But how did she get to know of the accident? *K13* was raised and brought back to the yard; she was thoroughly overhauled and rechristened *K22*, and I believe she done [*sic*] grand service before finally being broken up.[3]

As long as life lasts I will remember those brave cool courageous men coming through that small hole. And as long as we have men like Willie McLean, Mr Bullen, a great Englishman, Wm Struthers and Ted Skinner, Britain will be able to keep her old flag flying.[4]

Acknowledgements:
This hand-written account was given to Ian Johnston by the Reverend James Dey (retd.), Thomas Dey's son. The drawings are based on those in Percy Hillhouse's paper 'The Accident to the Submarine K13' published in *The Channel*, November 1919.

Editor's note: As far as possible Dey's original wording has been retained, but some punctuation has been updated for ease of reading. Corrections and amendments to Dey's (handwritten) account by Ian Johnston are in square brackets.

Footnotes:
1. On 29 January 1917, *K13* made her way down river from Fairfield's to the Gareloch to conduct diving trials. After successfully completing a two-hour dive, it was decided to make a short dive of 15 minutes duration off Shandon. The dive took place at 3.00pm with 80 persons on board including 14 Fairfield men.
2. *K13* was detected early on Tuesday morning. Various rescue options were discussed in the knowledge that air in the submarine was running low. By 4am on Wednesday, a compressed air line was connected providing air not only for the men but for the air bottles used to blow the tanks.
3. It was not until March 1917 that *K13* was finally raised and taken back to Fairfield where she was refitted and put into service and renamed *K22*.
4. A memorial to *K13* and the six Fairfield men who were drowned when she sank was erected in Elder Park adjacent to the shipyard. The men together with the 26 others who lost their lives are buried in Faslane Cemetery.

THE NETHERLANDS TYPE 47A DESTROYER

Henk Visser writes about the first destroyers built for the Royal Netherlands Navy during the post-war era.

The Type 47A destroyers of the *Holland* class were designed by the Chief Naval Constructor to the Netherlands Navy, K. de Munter. Displacement was 2765 tons full load and dimensions 111.3m x 11.3m x 3.8m. Their geared turbines produced 45,000shp for a maximum speed of 32 knots, and they had a complement of 247 officers and men.

Design
In broad outline the design brief stipulated three basic requirements:

– to provide defence for other surface forces and convoys of merchant vessels against submarine attack and air attack;
– to track down and destroy enemy submarines;
– to combat light enemy surface forces.

The first meeting of the project group charged with the task of fulfilling the Naval Staff requirements took place in January 1947. Among the topics discussed was the design of the ship's weather deck in the event of an atomic bomb attack. It was decided that the hull would not be penetrated by scuttles; instead the group recommended the adoption of an entirely smooth all-welded hull constructed of high-tensile Type A52 steel. At this early stage it was envisaged that the Type 47A would carry two twin 4.5in guns of British design, together with five 40mm AA guns. In deciding the overall dimensions of the proposed ships former British designs, and particularly the 'N' class, were considered as well as earlier Dutch destroyer classes. The construction of the class was approved by the government on 27 December 1948.

It soon became apparent that increases in top-weight were causing problems and would begin to affect stability, maximum speed and radius of action. Concerns were raised about these matters in August 1949 in a report to the Flag Officer Shipbuilding and Construction. The report stated that, due to added weight, at deep load displacement the metacentric height (GM) was down from 0.83 to 0.73 metres. Correspondence of the time reveals that a principal concern was the weight of the twin 12cm (4.7-in) gun mounting which, it was asserted, should not be allowed to exceed 55 tons as 'the tonnage of the ship does not permit much more'.

In the event the total weight of the mounting (plus crew and ammunition) was 67 tons, a heavy burden for the Type 47A with its lightweight hull. When the ships were in commission and in a lively sea, it was possible to detect working of the hull amidships by up to an inch. However, this was not considered to be structurally dangerous and similar movement was also perceptible on board the larger ships of the following Type 47B. Aluminium was used for parts of the superstructure, such as the after deckhouse and the funnel cowls, as a further weight-saving measure.

Machinery

At the time of the German invasion of the Netherlands in May 1940, the Parsons turbines for the new destroyers of the *Gerard Callenburgh* class, which had been laid down in 1938-39, were in various stages of manufacture.[1] Subsequently the Kriegsmarine ordered four turbine sets to be constructed for their projected *Flottentorpedobooten*. Despite the fact that work on the turbines was deliberately delayed during the years of occupation, these turbines were practically complete when the Low Countries were liberated. They provided first-rate propulsion power for the *Holland* class. There were two boiler rooms, located fore and aft, each housing one Admiralty 3-drum type boiler, together with fore and after engine rooms in a 'unit' arrangement.

Gunnery

The main armament consisted of two twin 12cm Mk 10 dual-purpose guns. These weapons were fully automatic and radar controlled. Rate of fire ranged from a minimum of twelve rounds per minute to a maximum of 42 rounds per minute, with a maximum theoretical range of 21,500 metres and a ceiling of 12,500 metres; in practice, of course, the effective range was considerably less. Ammunition stowage was 720 rounds per magazine. The projectile weighed 23kg; the weight of the fixed round (including cartridge) was 42 kg. All the crews liked these fine, modern-looking guns and praised their handling, reliability and accuracy. Credit needs to be given to those who made the decision to adopt the Bofors design in the face of an opposing group who advocated the adoption of the battle-proven British 4.5in weapon. The British design suffered from being 'over-engineered', requiring a very large 19-strong gun crew to operate it while the mounting machinery was dependent on hydraulics which were prone to malfunction caused by leakages.

HrMs *Zeeland*. This is an early view of D809, as indicated by the old-model Carley Floats below the signal deck. The letter and figures used for the pennant number are different from the standard design introduced in the 1960s.

The dual-purpose guns were supported in the anti-aircraft defence role by a solitary 40mm AA gun which was sited above the galley between the funnels. This excellent 70-calibre Bofors Mk 6 mount had a firing cycle of 240 rpm, a maximum range of 10,500 metres and an effective range of 4,500 metres. The reduction in the number of 40mm guns from five (as originally envisaged) to one was to save topweight, and contradicted wartime practice when destroyers often bristled with AA weaponry. It was a risky strategy as, despite the fact that the Cold War enemy lacked carrier air power, there was a real threat from land-based bombers which could reach warships' operational areas in time of war: the Southern Norwegian Sea, the North Sea and, most importantly, the Greenland/Iceland/UK (GIUK) gap.

ASW Weapons

Two Bofors four-tube 375mm Mk 1 antisubmarine rocket launchers were positioned on 'F' Deck [RN: 02 deck] just forward of the bridge. The weapons were built under licence by Wilton-Fijenoord. The projectiles had a range of between 520 and 820 metres and were propelled by two solid-fuel cordite motors. The range could be adjusted by the employment of either one or two of the motors and by varying the elevation of the tubes up to a maximum of 60 degrees. The maximum velocity using both motors was 130m/sec and the final sinking speed roughly 11m/sec. When practice rounds were fired only one motor was ignited, and the body of the rocket would float to the surface afterwards for subsequent recovery. The launch crew disliked live firings because the mass of flames and cordite powder smoke which erupted from the launcher required the entire installation to be washed, brushed, scoured and repainted!

HrMs *Holland*. The red-white-red signal flag which is hoisted indicates 'I am restricted in manoeuvring'. The coloured shapes of the flag show that the photograph was taken before 1976 as from that year the shapes would have been black. A replenishment at sea (RAS[F]) is underway; pumping Dieso F76 has commenced, as indicated by Flag Bravo close up at the starboard inner yardarm.

Initially, two DC racks were fitted at the stern, each with a capacity for six British-type depth charges. Later, one was removed to make space for a noise-maker anti-torpedo device and a BT/XBT set used for providing bathythermographic information.

Radar and Sonar

The radars were of indigenous design, being manufactured by Hollandse Signaal Apparaten (HSA). The ZW-01 high-definition surface and low-flying air warning and navigation set had a range of 100km, while the DA-01 medium range surveillance and target acquisition radar had a range of 200km. The third set, the LW-02 long range surveillance set had a range of 160km. There were two fire control sets, the GA-03 and the KA-01, which had ranges of 35km and 24km respectively.

The sonars were also developed by Netherlands companies. The Van der Heem PAE-1N was a 17–35kHz 'searchlight'-type attack sonar set with an output of 250W. The 'warning' sonar, developed by the same company, was the CWE-10, a 10kW set with a range of 10,000 metres. These sonars were subsequently acquired by the German, Swedish and other navies.

Brief Service Histories

Holland

Laid down 21 April 1950; launched 11 April 1953; commissioned 30 December 1954.
Builder: Rotterdamsche Droogdok Maatschappij, Rotterdam; Yard No.266.
Visual call sign: D808 (pennant number); international call sign: PAOP (signal letters).
1955: after a maiden voyage around the African continent (via the Suez Canal and Capetown) the ship was laid up in reserve until re-commissioned in 1962.
1965: In collision with the Danish MV *Mayumba* (Dansk-French Line 1955, 3899grt) 3nm ENE of Texel Light Vessel.
1968: *Holland* joined the first STANAVFORLANT squadron which commissioned at Portland on 13 January under the command of Captain G C Mitchell RN. Ships of the squadron included HMS *Brighton* (F106), HNoMS *Narvik* (F304) and USS *Holder* (DD-819).
1969: the ship, with two sisters *Zeeland* (D809) and *Noord-Brabant* (D810), assembled at Spithead for the NATO Naval Review. Other Netherlands Navy warships present included *De Ruyter* (C801), *Van Nes* (F805) and *Evertsen* (F815).
Holland also took part in several NATO exercises including 'Razor Sharp' (1969), 'Northern Merger' and 'Safe Pass' (1974).
The ship was decommissioned for the last time on 9 November 1977. She was transferred to Peru and commissioned as BAP (Buque Armada Peruana) *García y García* (75). She was stricken in 1986.

Zeeland

Laid down 12 January 1951; launched 27 June 1953; commissioned 1 March 1955.
Builder: Koninklijke Maatschappij De Schelde, Vlissingen; Yard No.269.
Visual call sign: D809 (pennant number); international call sign: PAAU (signal letters).
1955: after trials and work up and a maiden voyage in the North Sea and English Channel, the ship visited Portland before further fitting-out at Vlissingen.
1956: in July *Zeeland* made a ground-breaking visit to Leningrad in the company of *De Zeven Provincien* (C802) and *Friesland* (D812). She then decommissioned into the reserve until November 1965.
1966: joined the Netherlands Task Group (NlTG).
1968: succeeded *Holland* in STANAVFORLANT, joining HMS *Argonaut* (F56), FGS *Bayern* (D183) and USS *Glennon* (DD-840) in August.
1969: took part in the NATO Naval Review at Spithead.
Zeeland took part in NATO exercises 'Grande Chance', 'Cut Loose' (1956), 'Silent Rain' (1966), 'Perfect Play' (1967), 'Silver Tower' (1968), 'Razor Sharp' and 'Peace Keeper' (1969).
1976: after two years in reserve *Zeeland*, in company with *Tromp* (F801) sailed for New York in order to

participate in the Bicentennial Naval Review held in the Hudson River.

1978: decommissioned for the last time on 29 September, stricken, sold for scrap and towed to breakers in Bilbao, Spain in December 1979.

Noord-Brabant

Laid down 1 March 1951; launched 28 November 1953; commissioned 1 June 1955.

Builder: Koninklijke Maatschappij De Schelde, Vlissingen; Yard No. 270.

Visual call sign: D810 (pennant number); international call sign: PAIP (signal letters).

1955: after visiting Amsterdam to participate in the *Vlootweek* (Navy Days) the ship undertook further trials before decommissioning into the reserve.

1957: was temporarily re-activated to attend the 350th anniversary of the birthday of Admiral De Ruyter at Flushing in company with *De Zeven Provincien* (C802) and HMS *Russell* (F97), amongst others.

1965: recommissioned, working up with the NlTG. In 1969 at Spithead.

Noord-Brabant took part in NATO exercises 'Sailors Pride' & 'Silent Rain' (1966), 'Wicked Lady' & 'Perfect Play' (1967), 'Silver Tower' (1968), 'Razor Sharp' (1969), 'Strong Express' (1972), 'Sunny Seas', 'Quick Save' & 'Swift Move' (1973).

1974: on 9 January *Noord-Brabant* collided with the British-flagged MV *Tacoma City* (Reardon-Smith Line, 1972, 16,700grt) in a position off Flushing (buoy W10 – Wielingen). *Noord-Brabant* suffered heavy damage on the port side close to the after boiler room, which rapidly filled with sea-water, and also the after engine room. One rating was killed and despite a brave rescue attempt by two POs another rating died later. Damaged beyond repair, the destroyer was decommissioned in March of that year. She was towed to the breakers in Ghent, Belgium, the following year.

HrMs *Noord-Brabant*. The photograph shows the destroyer sailing close by the esplanade at Vlissingen (Flushing) where she was built; Vlissingen was also the birthplace of Admiral De Ruyter. She is still on trials and her LW-02 Air Surveillance radar has yet to be fitted. She lacks the standard design of pennant number, which tells us that the photograph was taken in the 1950s.

HrMs *Gelderland*. The second funnel on the ship is covered, indicating that the after boiler room is not in operation. In the first years she carried, besides a motor boat, a whaler on the upper deck which was later removed. The extent of the visible boot-topping indicates that *Gelderland* requires re-fuelling. Flag Romeo is flying at the yardarm; it indicates 'commencing approach' with sometimes 3–5, even 7 knots extra called for on the run-in. Replenishment speed was often 12 knots.
(All photographs from the author's collection)

Gelderland

Laid down 10 March 1951; launched 19 September 1953; commissioned 17 August 1955.

Builder: Wilton-Fijenoord, Schiedam; Yard No. 725.

Visual call sign: D811 (pennant number); international call sign: PARY (signal letters).

1955: maiden voyage to Lisbon; decommissioned on return and thence to Wilton-Fijenoord to rectify defects during the 'guarantee period'.

1956: participated in various exercises.

1958: attended Amsterdam *Vlootweek*.

1959: visited New York to commemorate the 350th anniversary of the voyage of Henry Hudson.

1960: assisted in disaster relief with other units of the NlTG following an earthquake in Agadir, Morocco.

1964: placed in reserve until 1969 with weapon systems and machinery in a state of preservation.

1973: decommissioned 29 March to be based in Amsterdam as a static training ship and berthed in front of the TOKM (Technical School, Royal Netherlands Navy). *Gelderland*'s two twin Bofors mountings were refurbished and transferred to the new guided missile frigates *Tromp* (F801) and *De Ruyter* (F806), which were nicknamed 'Kojaks' because of the distinctive dome housing the 3-D (three dimensional) radar.

Stricken and sold in 1993, and broken up in 1994 at a former shipyard in Amsterdam.

Concluding Remarks

The bridges of the Type 47A Class were more spacious than their larger 47B (*Friesland*) class half-sisters. Purely by coincidence, none of the ships was deployed to the Netherlands New Guinea or acted as West Indies guard ships in the Caribbean. The *Holland* class spent approximately 30% of their service lives laid up in reserve.

Footnote:

[1.] The nameship of the *Gerard Callenburgh* class and her sister *Tjerk Hiddes* were built at Rotterdamsche Droogdok Maatschappij shipyard. The *Phillips van Almonde* and *Isaac Sweers* were built at Koninklijke Maatschappij De Schelde. The *Gerard Callenburgh* commissioned into the Kriegsmarine in October 1942 as *ZH1* and the incomplete *Isaac Sweers* was towed to the UK and later commissioned in May 1941 with the pennant number G83.

Bibliography:

Nooteboom, S.G, *DeugdelijkeSchepen* (Zaltbommel 2001).
Quispel, H.V, *The Job and the Tools* ('s-Gravenhage 1960).
Veenstra, A J C, *Onderzeebootjagers van de Holland-en-Friesland-klasse* (Meppel 1987).
Jaarboek voor de Koninklijke Marine (various editions).

A's & A's

Night Action, Malaya 1942 (*Warship* 2015, pp.62-80)

Pages 73–74 and 80 of Peter Cannon's article outlined the 1942 gunnery and torpedo control equipment of the 'V' class destroyer HMAS *Vampire*. Subsequent research by the author has unearthed a 26 June 1941 Admiralty Fleet Order authorising sound-powered telephone communications to supplement the original voice pipe and mechanical transmitter/receiver arrangements in ships of the class remaining as fleet destroyers and officially designated as 'Vees' from 1940.

Gun control: Two Mark X head-set type phones on the upper bridge, port and starboard, for the Gun Control Officer to the TS. Another Mark X phone in the TS communicated with identical phones at the director and each gun along with a Mark XI handset on the bridge.

Torpedo Control: Port and starboard upper bridge to the torpedo tube mounting, all Mark X.

Having hitherto relied exclusively upon early powered four-wire (Graham's 'Navyphone') exchanges superseded between the wars by parallel two-wire direct and exchange phone systems, the RN introduced sound-powered telephones for action and armament control purposes during the Second World War, many years after the US Navy. Although supporting records appear not to have survived, it is quite possible that *Vampire* received these modifications during her final June-November 1941 refit in Singapore and fought her action off Malaya utilising telephone communications with older systems employed as alternative backups.

Sources:

National Archives of Australia (NAA), Item BOX 5, Admiralty Fleet Orders – 1941: 1 to 5696, AFO 2734. – Telephone Communications – Fire and Torpedo Control 'Vees' and Leaders (D. 0959/41. – 26.6.1941).
B.R. 157/1933, *Naval Electrical Pocket Book*, London, Admiralty, 1933, pp.251–70.
B.R. 157/1953, *Naval Electrical Pocket Book*, London, Admiralty, 1957, pp.224–38.

Warship Prefixes

As hoped (and indeed invited), a number of people contacted *Warship* with corrections and additions to the Warship Note in 2015 on the subject of national prefixes. The following is a digest of the more important or interesting of these:

Brunei: KDB – *Kapal Di-raja Brunei*.
Germany: In a fine example of the sort of prefixes that can sometimes appear even in official documents, after the famous battle between HMAS *Sydney* and SMS *Emden* in 1914, Captain Glossop of the *Sydney* refers in his report to 'HIGMS *Emden*'.
Indonesia: KRI – *Kapal di Republik Indonesia*.
Malaysia: KD – *Kapal Diraja* (Royal Ship).
Netherlands: Queen Wilhelmina actually reigned from 1890 to 1948 – during her minority (1890–98), her mother, Dowager Queen Emma, was regent. Since 2013 (as well as pre-1890) Zr Ms (*Zijner Majesteits*) should be substituted for Hr Ms (*Harer Majesteits*) since the Netherlands has been ruled by a king. During the years of exile (1940–45) HNMS was used. In addition, up until about 1955, it was sometimes the custom to insert the type between the prefix and name, for example 'Hr Ms *kruiser Java*'.
Philippines: BRP – *Barko Republika Pilipinas*.
Turkey: TCG – *Türkiye Cumhuriyeti Gemisi* (Ship of the Turkish Republic). This first came into use in 1946 for the destroyers TCG *Gayret* and TCG *Muavenet* and the submarine TCG *Burakreis*, bought from the UK. It may have been a result of a desire amongst senior Turkish naval officers to clearly distinguish their ships from those of their new allies, Britain and America.
United Kingdom: The comment in column 3 on p.179 is not strictly accurate in that Great Britain became the official title of the union of England and Scotland in 1707, though not including Ireland. This was succeeded by the United Kingdom of Great Britain and Ireland from 1801 to 1922, and then the United Kingdom of Great Britain and Northern Ireland from 1922 to date. Britain is an unofficial name, usually denoting the UK.
WWII Allied navies in exile: Of interest are the prefixes used in the official Admiralty 'Lists of stations and movements of Allied and Royal Naval Ships' (otherwise known as the 'Pink Lists'): Netherlands – HNMS; Norway – NS; Poland – PS; Yugoslavia – YS; Greece – GS; Belgium – BS.

Thanks to Ian Sturton, Henk Visser, Conrad Waters, Devrim Yaylali and Kenneth Fraser for information used in the above.

NAVAL BOOKS OF THE YEAR

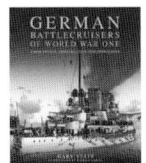

Gary Staff, with illustrations by Marsden Samuel
German Battlecruisers of World War One: their design, construction and operations

Seaforth Publishing, Barnsley 2014; hardback, 336 pages, approximately 600 photographs, drawings and maps in B&W and colour; price £45.00.
ISBN 978-1-84832-213-4

In contrast with the ships of the much smaller *Kriegsmarine* of the Second World War, the ships of the *Kaiserliche Marine* (Imperial Navy) are poorly served in English-language works. Accordingly, this large-scale treatment of the German equivalents of the British battlecruisers – never so classified in the German navy, where they were *Große Kreuzer* ('large cruisers') – is to be warmly welcomed.

The structure of the book is simple, with a chapter dedicated to each of the seven completed ships and a further two covering the two uncompleted classes, bookended by introductory matter and a set of appendices, a glossary, a bibliography and an index. The Introduction has an account of the development of earlier German 'large cruisers' (an official designation enshrined in the 1898 Fleet Law) but, given that the later ships were very much their lineal successors, one would have liked to see much more detail given here – perhaps even some photographs! As it is, we have an account of the development of the last of the 'pre-dreadnought' large cruisers, *Blücher*, but nothing on her career, and only two line drawings to give an idea of a ship that bridged the gap between what in British terminology would be termed 'armoured cruisers' and 'battlecruisers'.

In contrast, the seven completed ships are treated in exhaustive detail. There is an account of their genesis and design (derived from German archival sources, though it is not clear whether this is the result of new research or based on the work of Axel Grießmer), with sketches of options considered, followed by sections dealing with the ship's technical characteristics: headings include: 'hull', 'armour', armament', 'ship's boats', 'anchors', 'electrical plant', 'searchlights', 'machinery', '"leak" pumps' (an odd terminology: cf. below), 'manoeuvrability and seakeeping', 'general characteristics and changes', 'namesake' (all large cruisers were named after military heroes), 'trials' and, finally, 'operational history' (supported by detailed maps). Large sections of the 'operational histories' are clearly taken from the ships' war diaries (cited in the Bibliography), but nowhere is it indicated what is verbatim quotation, what is author's *précis* and what is author's comment. The 'ship-by-ship' approach also means that coverage of the overall war at sea is fragmented, leading to multiple views of Jutland, all from slightly different perspectives, with significant repetition and no coherent picture of the whole action. Nor is there any kind of overview of German capital ship design of the period: rather more on the battleships which were constructed in parallel with each generation of battlecruiser would have been useful.

Copies of official 'as fitted' drawings are also included, plus others showing internal arrangements, weaponry and other details. Drawings showing armour arrangements are also provided, but most of these are reproduced so small that the armour thickness figures are illegible, even with a powerful magnifying glass. On the other hand, there are superb computer-generated full-colour images by Marsden Samuel, providing a range of perspectives on the ships.

There are also many large-format photographs of the ships throughout (almost) their entire careers, many of which are new to the reviewer and make the book worth having almost for the photographs alone. Curiously, while there are a number of images of the ships on the stocks, there are none of them during the final stages of their careers – in almost all cases scuttled at Scapa Flow, and then raised for scrap in the 1920s and 1930s (of which many images exist). The text also skims over these episodes extremely briefly, in contrast with the detailed accounts of the ships' earlier histories. In particular, the failure to follow the last half-century of the career of *Goeben* (as the Turkish *Yavuz*) in any substantive way seems most odd.

Equally strange is much of the idiom used in the text. As noted above, considerable use is made of German archival sources, but rather than translating into idiomatic English the author often opts for as a more 'direct' translation. Thus, when talking about magazine capacity, he talks about 'shots per gun' (rounds per gun); we also find the great sea battle of 31 May/1 June 1916 referred to as the 'Skagerrak Battle' (directly translated from 'Skagerrakschlacht'), rather than the normal English 'Battle of Jutland' (or even the idiomatically correct 'Battle of the Skagerrak'). These idiosycracies extend to the author's use of German words where there are perfectly acceptable English equivalents (eg *offizier* for 'officer').

There are other stylistic issues that detract from the overall experience of reading Mr Staff's book, some of which derive from the 'one ship per chapter' approach, which leads to paragraphs being reproduced almost unchanged in chapters dedicated to sister-ships. One wonders whether an overarching introduction dealing with issues of commonality in German battlecruiser design might have been preferable. However, these, and other points noted above, do not detract from the immense amount of effort that has gone into producing this book, which is a mine of information and will remain the fundamental source for these ships for the foreseeable future.

Aidan Dodson

John D Grainger
The British Navy in the Baltic
Boydell & Brewer, 2014; hardback, 294 pages, 3 maps; price £65.00.
ISBN 978 1 84383 947 7

Published with the support of the Swedish Society for Maritime History and forming a part of the Forum Navale series, this keenly anticipated book is essentially an overview of the Royal Navy's involvement in the Baltic Sea from Anglo-Saxon times until the 20th century. It is based mainly on Navy Records Society volumes and other secondary sources, and throughout aims to highlight naval actions in the context of wider diplomacy and geography.

Grainger makes the point that from 1650 to 1801, the year Nelson bombarded Copenhagen, no Royal Navy warship was engaged in aggressive action in the Baltic, despite the importance of Britain's Baltic trade. Naval presence, and with it the implied threat of naval combat, sufficed to achieve British aims, and it was only between 1801 and 1920 that any European war in which Britain was involved brought the Royal Navy into action in the Baltic.

It is the book's final three chapters which may be of most interest to readers of *Warship*. These chapters cover the war against Russia of 1854–55 and the bombardments of Bomarsund and Sveaborg, and the attacks of the Coastal Motor Boat Force on Kronstadt in 1919, which Grainger describes as a minor action with major results. As Grainger points out, the 1854 war was one of the few occasions when British and Russian policies actually led to armed conflict. Although Russia possessed the largest Baltic fleet, it was confined to its harbours, while the bombardment of Sveaborg was a significant demonstration to the Russians that both their fleet at Kronstadt and the capital, St Petersburg, were vulnerable to allied sea power. It was this knowledge and not the fall of Sevastopol that brought Russia to the peace table.

Before the outbreak of war in 1914, British ships had frequently visited the Baltic but the implicit threat of British naval action, including Admiral Fisher's plans for landings in the Baltic, were only a tentative part of British war planning, and by the time war actually occurred Germany's strategic position effectively precluded any British naval incursions by the surface fleet into the Baltic, although a small force of submarines operated there until 1917.

British ships did enter the Baltic after the war, the subject for Chapter 13. Admiral Sinclair and his successor, Admiral Cowan, were sent to the region with vague instructions to show the flag and take appropriate action in what was a confused and complex political situation. Both Sinclair and Cowan, with light naval forces, had to use their initiative in deciding on the best course of action. Grainger points out that the Royal Navy, despite its actions against Russian naval forces, was only one of many factors to be considered when considering the freeing of the Baltic states.

From 1921 onwards, the Baltic became less of a key concern for the Royal Navy. The growth of air power made big ships extremely vulnerable, and again, Churchill's aborted ideas for 'Operation Catherine' notwithstanding, Germany's strategic position made any thoughts of operations in the Baltic unrealistic. Post 1945, apart from the occasional courtesy visit, the Soviet Navy dominated the Baltic Sea. British strategic priorities now lay elsewhere.

Grainger has successfully set out to illustrate that naval power does not always have to depend on battle to be effective, and that the implicit threat of battle can often be sufficient to secure broader policy aims. The author has provided an enjoyable, readable, and often entertaining overview of Britain's involvement in the Baltic Sea. Production values are high although, as in most books by this academic publisher, the only illustration comprises three general maps of the Baltic region.

Andy Field

Nick Childs
Britain's Future Navy (Revised Edition)
Pen & Sword Maritime, Barnsley 2014; softback, 172 pages, 16 colour illustrations; price £14.99.
ISBN 978-1-47382-324-2

A revised edition of a book first published in 2012, *Britain's Future Navy* looks at potential directions for the Royal Navy's structure and equipment in the first half of the 21st Century. Given its origins in the immediate aftermath of the controversial 2010 Strategic Defence and Security Review (SDSR), its analysis is inevitably heavily influenced by key decisions taken at that time. However, the book's overall scope is much broader than this, encompassing themes such as Britain's own view of its place in the world and the growing influence of the emergent Asian naval powers on the global maritime balance.

After a foreword from Admiral Sir Jock Slater, First Sea Lord at the time of the acclaimed 1998 Strategic Defence Review, Nick Childs commences with an introduction summarising some of the key questions the Royal Navy currently faces. The opening chapter outlines the Navy's development from the end of the Cold War to the SDSR, encompassing key events such as the 1966 cancellation of the CVA-01 fleet carrier and the 'Nott navy cuts' that immediately preceded the 1982 Falklands conflict. Subsequent chapters cover broader issues, including the role of navies and the impact of developing technologies. The most significant chapters, however, focus on the Royal Navy's key equipment. These include separate discussions on aircraft carriers, nuclear submarines, the Type 45 destroyer programme, the proposed global combat ship, and amphibious and minor warfighting vessels. The concluding chapters look at the case for maintaining British maritime influence and set out a number of possible force structures under different scenarios.

The author's credentials as a well-established and authoritative journalist are clearly evidenced by his insightful and widely-sourced analysis. He succinctly summarises a broad range of often complex considerations whilst demonstrating in-depth knowledge of the underlying subjects. He has the ability to describe, for example, the key design features of the Type 45 destroyer and the inter-service infighting that impacted SDSR 2010 with equal clarity and credibility.

There are some minor criticisms. The small number of photographs and lack of other supporting illustrations and diagrams mean the unbroken text can be a 'heavy' read. Perhaps more significantly, there is relatively little about current manning and support factors that can be just as important as equipment to fleet effectiveness. The revolution in unmanned technologies is accorded just three short pages. It would also have been helpful to see more analysis of maritime capabilities provided by other elements of the armed forces, particularly with respect to maritime patrol aircraft.

All in all, however, *Britain's Future Navy* is a thoroughly researched and well-thought-out review which should be required reading for anybody interested in the modern Royal Navy. In due course it would be good to see a further edition, this time updated for the conclusions of the 2015 SDSR.

Conrad Waters

Adam Thompson.
Kustenflieger: The Operational History of the German Coastal Air Service 1935–1944
Published by Fonthill Media, Oxford, 2014; soft back, 192 pages, 118 B&W illustrations; £16.99.
ISBN 978 1 78155 283 4

Adam Thompson's book is a history of the German Coastal Air Service up to its eventual absorption into the Luftwaffe. It comprises fifteen chapters, together with acknowledgements, notes, bibliography and index.

The first two chapters offer a chronological history of the service from 1918 to 1944, followed by a chapter on colour schemes and markings. The following nine chapters are individual *Gruppe* histories, each taken from the establishment of the *Gruppe* to its final dissolution; this is a problem in that each chapter tends to duplicate material from earlier chapters. There is then a chapter covering German involvement in the Spanish Civil War, which effectively disrupts the chronological sequence. The penultimate chapter covers the various special squadrons (*Sonderstaffeln*), and the final one the Stavanger *Gruppe*.

The *Gruppen* chapters are well illustrated and give an insight into the trials and tribulations of the service and the crews, trapped between short-sighted admirals and a grasping Goering. The early training was clearly well thought through, but the lack of modern aircraft was always going to hamper even the limited objectives of Admiral Raeder. When more modern aircraft finally became available the Kustenflieger was always going to be accorded a lower priority than the Luftwaffe. Of particular interest in the author's account is the small number of mines and torpedoes available for use, the former often badly delivered, the latter rarely working, initially at least. (The reviewer's father-in-law spent his war rendering safe mines dropped in some strange places, Lyndhurst in the New Forest being one.)

The author has chosen some interesting illustrations for his text, many of which are new, but the book would have benefited from better editing: 'intact' is spelt as two words, the white summer crown of an officer's cap is referred to as its 'peak', and there is repeated use of 'mute' in place of 'moot'. However, the book has covered a gap in the history of the Second World War, and is clearly the result of significant research.

W B Davies

Anthony J Cumming
The Battle for Britain: Interservice Rivalry between the Royal Air Force and the Royal Navy, 1909–40
US Naval Institute Press, 2015; hardback, 306 pages, a few B&W images, maps; price $39.95.
ISBN 978-1-61251-834-3

The publisher's summary describes this book as 'a provocative reinterpretation of both British air and naval power from 1909 to 1940'. Provocative it certainly is: *The Battle for Britain* is essentially a 'hatchet job' on the RAF, and indeed on the entire concept of 'independent' air power. According to the author, the so-called 'Battle of Britain' was an inflated piece of propaganda dreamed up by Churchill to impress the American public, and the strategic bombing concept which dominated RAF thinking during the interwar period was simply the result of scare-mongering by politicians who were fed lies and exaggerations by the aviation lobby. The RAF failed to provide suitable aircraft in sufficient numbers for the Royal Navy and for Coastal Command, or suitable support aircraft for land operations. And so it goes on...

At least Cumming doesn't make any claims to objectivity. That much is clear from the outset: the foreword is written by an Admiral of the Fleet, and the author's first sentence includes the words 'a shining new branch of the British armed forces' (in reference to formation of the RAF of 1918). Evidence for the author's thesis is selective. Thus the 'Battles' [sic] of Mers el-Kebir and Taranto, both of which involved surprise attacks by RN forces on stationary ships in harbour, get extensive coverage, while the crippling of *Illustrious* (at sea) by land-based Stuka dive-bombers and the sinking of *Prince of Wales* and *Repulse* (also at sea) by Japanese land-based torpedo bombers merit only a cursory mention. The author is of the opinion that the RN's wariness of operations close to the Norwegian coast, in the Channel and in the central Mediterranean, was largely unjustified, as the aircraft which posed the threat had unsuitable bombs and gunsights, and their pilots were insufficiently trained for attacks on mobile naval targets.

The main problem with this book, however, leaving aside the author's insecure understanding of naval hardware and tactics – of which there are too many examples to mention in a brief review – is that it does not do what it says on the tin. We get potted histories of the First World War and the first years of the Second which are by no means confined to an analysis of the impact of aircraft, plus an overview of interwar thinking about air power, but very little about the institutional conflict between the Royal Navy and the Royal Air Force between the wars and its consequences for the Navy in terms of 'air mentality', personnel and hardware. There are many missed opportunities: the author mentions the Navy's 'failure to deploy aerial reconnaissance' at Jutland, but makes no attempt to explain why this occurred; in a similar vein, he criticises (briefly) the Admiralty's predilection for 'multi-role aircraft' during the 1930s, but fails to explain or to analyse these decisions, which were often taken against Air Ministry advice.

Cumming is so intent on demolishing the case for independent air power that he fails to provide satisfactory answers to a number of key questions. If the RAF had not become an independent service, would homeland air defence for the British Isles have been provided by the Navy (which was primarily interested in aircraft for reconnaissance, spotting and strike at sea) or the Army (which was primarily interested in air superiority over the battlefield and close support)? And if the Admiralty had been allocated a proportion of the defence funding redirected to the RAF during the interwar period, would it have spent it on developing semi-independent carrier aviation (as was the case with the IJN and the US Navy), or would it have blown the money on cruisers? As the air lobby pointed out during the early 1920s, airfields in northern France were only minutes' flying time away from the cities and ports of Britain. Britain's geo-strategic position was therefore very different from that of the United States or Japan.

John Jordan

Henry 'Hank' Adlam
The Disastrous Fall and Triumphant Rise of the Fleet Air Arm from 1912 to 1945
Pen & Sword, Barnsley 2014; hardback, 216 pages, B&W photographs and maps; price £25.00.
ISBN 978-1-47382-113-2

Despite its broad title, much of this book is actually devoted to criticism of senior RN officers with no flying background who commanded carrier operations in the Second World War, a group the author refers to as 'Penguins' – creatures that have wings but cannot fly. The dust-jacket describes it as offering 'a degree of critical candour that sets it apart from other Fleet Air arm histories' and the author, who was a well-respected Fleet Air Arm fighter pilot, has to be admired for attempting this when he was over 90 years old and suffering from macular degeneration in both eyes. Sadly, 'Hank' Adlam died shortly after the book's publication.

Unfortunately the book contains so many errors of fact that it cannot be taken seriously as a work of history. On p.4 the author writes that '*Furious* served as a carrier until the second year of WWII when she was sunk'; she was actually paid off into reserve in 1944 and sold for scrap in 1948. The statement on p.53 that HMS *Thetis* was a light fleet carrier allocated to the British Pacific Fleet is incomprehensible; the ship in question was actually *Glory*, while *Thetis* was a submarine that sank off Anglesey in June 1939.

Errors of fact undermine every point Adlam tries to make, even though there is an element of truth in the criticisms he levels at senior naval officers, especially Admiral Vian, who was certainly not popular with many of the aircrew who served under him. However, some criticisms are unjustified. In his description of Operation 'Avalanche', the Allied landings at Salerno, the author states that 'our two Penguin Admirals [Willis and Vian] had not apparently given any thought to that vital need for preparation'. In fact Force V under Vian, comprising the maintenance carrier *Unicorn* and the escort carriers *Attacker*, *Battler*, *Hunter* and *Stalker* with a combined total of 11 Seafire squadrons embarked, had actually undergone more preparation than any previous RN carrier task force. The Seafire proved to have limitations, but it should have been explained that the carriers were only expected to provide air cover for the first 24 hours of the operation, during which Montecorvino airfield was to be captured for use by land-based fighters. The stiff resistance offered by the German Army was not anticipated by any Allied intelligence source, and there was no choice but to keep Force V in action despite the loss of 32 Seafires in deck landing accidents out of the 106 aircraft committed to the operation (both numbers far below those quoted in error by Adlam).

The later section on operations by the BPF against Japan again contains many errors which completely undermine confidence in the text. Many of these concern the positions held by various serving personnel. For some reason Adlam thinks that Admiral Fraser was in the Admiralty and that his second-in-command, Vice Admiral Rawlings, commanded the BPF. Moreover, the statement on p.203 that Fraser 'was rather boot-faced at this wholehearted submission of the BPF to the Americans' is a complete miss-representation of the man who, more than any other, had set out to form a close working relationship with Admiral Nimitz and who was determined to achieve the closest possible Anglo-American relationship using American signal procedures, codes and practices in order to learn and understand their strike warfare techniques.

I have no difficulty with books that are critical of accepted views of history, but to succeed they must be based on evidence that is both factually correct and supported by source notes and a bibliography. Neither of these are present in this book. I find it sad that the author was not given the degree of knowledgeable editorial or co-writing support that he obviously needed, and the result is that this book cannot be recommended.

David Hobbs

Arthur J Marder
From the Dardanelles to Oran: Studies of the Royal Navy in War and Peace 1915–40
Seaforth Publishing, Barnsley 2015; softback, 320 pages, B&W illustrations and maps; price £16.99.
ISBN 978-2-914017-86-2

This paperback edition of a Marder classic first published in 1974 is particularly welcome, as the book has been out of print for many years. *From the Dardanelles to Oran* is a collection of essays on the Royal Navy, some of which had previously been published as articles in academic journals, while others had their origins as lectures.

From a modern perspective it has to be said that some of the chapters have aged better than others. Chapter 2, a critique of RN strategic thinking between the wars, rightly highlights the Navy's obsession with re-fighting Jutland, its distaste for convoy and unjustified confidence in the capabilities of Asdic, and its institutional conservatism. However, Marder seriously underplays the extent to which, with Germany defeated and militarily 'neutered', the Royal Navy of the 1920s refocused its attention on the threat posed by Japan in the Far East. He also fails to give sufficient weight to the treaties and protocols (signed even by Hitler's Germany) which effectively prohibited unrestricted submarine warfare.

Chapter 3, a study of the Ethiopian crisis of 1935, makes some useful points. The Navy was less interested in Mussolini's aggressive colonial expansionism than in Italy's key strategic position astride its primary line of communication with the Far East, had no time for the League of Nations, and was seriously unprepared for conflict in terms of ammunition stocks and anti-aircraft weaponry. What the crisis demonstrated was that Britain could only realistically send a fleet to the Far East in the absence of a European threat; however, it did serve to kick off British naval rearmament, and the war planning against Italy developed in 1935 served as the basis for the Navy's strategy in the Mediterranean in 1940.

Chapter 4, which focuses on Winston Churchill's spell as First Lord of the Admiralty 1939–40, is arguably the weakest of the five chapters. At times it verges on the anecdotal, and at others it degenerates into an academic spat with Roskill concerning the degree of Churchill's interference in naval planning and operations. Roosevelt has the best line: 'Winston has a hundred ideas a day, of which at least four are good.' Operation 'Catherine' (up-armoured 'R' class battleships go on the rampage in the Baltic) was not one of them.

The two chapters which stand out through the clarity of their exposition and the cogency of their argument are Chapters 1 and 5. In Chapter 1 Marder reappraises the Dardanelles campaign in the light of the Mitchell Report of 1919. He puts a strong case for the original, exclusively naval strategy, and argues that with sufficient preparation, appropriate use of aerial spotting, and a force of properly-equipped minesweeping destroyers the Navy could have broken through the Narrows into the Sea of Marmara, resulting in the desired Turkish collapse.

Chapter 5, on the attempted neutralisation of the French Fleet at Oran and Mers el-Kebir on 3 July 1940, is an astonishing *tour de force*. Using official records on both sides of the Channel, together with the reminiscences of participants who were serving officers at the time, Marder pieces together an account of an event riddled with misunderstandings and miscalculations during a desperate period for France and an only slightly less desperate period for Britain, in which the genuine attempts by the naval officers on the ground to find a solution were over-ridden by Churchill's desire for a show of ruthlessness to impress American public opinion. Marder's account of the bitter aftermath of this unprovoked aggression, which resulted in the deaths of 1300 French sailors, is carefully nuanced, although his final endorsement of Churchill jars with much of his analysis.

John Jordan

Roger Parkinson
Dreadnought: The Ship that changed the World
I B Tauris, 2015; hardback, 306 pages, 26 B&W images, 55 diagrams, 1 cartoon; price £25.00.
ISBN 978-178076-826-7

The name 'Dreadnought' in the title catches the eye, and the author uses it to flag a reworking of naval history and technology for the 33 years from the Naval Defence Act of 1889 to the 1922 Washington Treaty; his alternative title was 'The Navalist Era in Defence'. The topic is familiar and well-worn, and the book has to squeeze into a niche between authoritative and well-known works, notably Marder (extended slightly by Roskill's *Naval Policy between the Wars*) for the naval side and Massie's majestic *Dreadnought: Britain, Germany and the Coming of the Great War* for the European scene. The ships themselves have been fully described in numerous publications.

In the 1890s the 'Pax Britannia' came to be challenged successively by France (1893, Siam), the USA (Venezuela, 1895–96), Germany (Kruger telegram, 1896), Russia (Port Arthur, 1898), France again (Fashoda, 1898) and Germany again (Boer War, 1899–1900). The strengthening of the Royal Navy by the 1889 Act and the Spencer Programme kept Britain ahead of any likely European grouping, and spurred a naval arms race. The Anglo-Japanese Alliance (1902) and ententes with France (1904) and Russia (1907) marked the end of 'Splendid Isolation', setting the stage for Anglo-German rivalry, the Dreadnought era, a new, greater arms race, the Great War and the Washington Treaty.

The book is fully referenced, with numerous primary archival files and a very long list of secondary sources; almost all references cite the latter. The text adopts the kitchen sink approach: everything possible is included, every situation is summarised, every box ticked. The result reads like expanded lecture notes, with anecdotes and snippets to keep the reader's attention. The photographs are familiar IWM stock, while the drawings, from

Brassey occasionally amplified from Parkes, are too small for clarity. There are no acknowledgements. The appendix of capital ships omits incomplete and projected German battlecruisers. There are irritating factual errors: Admiral Aube was Minister of Marine under Freycinet and Goblet, not Clemenceau; Fisher was never an earl; capital ships armed with 12in guns were dreadnoughts rather than super-dreadnoughts. Overall, the book might be considered a useful present for a favoured nephew or niece, but is not for the serious bookshelf.

Ian Sturton

Ian Johnston
A Shipyard at War: Unseen Photographs from John Brown's, Clydebank 1914–1918
Seaforth Publishing, Barnsley 2014; hardback, large format, 192 pages, many B&W photographs; price £30.00.
ISBN 978-1-84832-216-5

In many respects a sequel to the author's excellent *Clydebank Battlecruisers*, published by Seaforth in 2011 and reviewed in *Warship 2012*, this book focuses on the work of the John Brown Shipyard during the First World War. Like it's predecessor, it is much more than just a photo album, and uses the photographs, taken by in-house professional photographers using large-plate cameras, to tell a story.

The photographs are stunning, and have been beautifully reproduced in this large-format book. They cover the full range of ships built at the yard from just before the outbreak of war to just after its conclusion. Unlike certain other shipyards and naval dockyards which specialised in a limited range of ship types, John Brown's, which before the war had been Britain's premier builder of fast transatlantic liners, undertook the full range of naval construction, from battleships and battlecruisers to destroyers, and supplemented these with orders for war-standard merchant ships. The photographs taken during the fitting out of the battleship *Barham*, the battlecruiser *Repulse* and the late-war destroyers of the 'S' class are particularly prominent, but there are also images of the liner *Aquitania*, the seaplane carrier *Pegasus*, light cruisers of the 'C' and 'E' types, and submarines of the 'E' and 'K' classes. The John Brown shipyard was located at a strategic point on the Clyde, and besides images of the ships built at the yard there others of ships passing down-river from other shipyards.

The images are placed in context by a concise and well-researched introduction on the shipyard itself and its experience of war. This is accompanied by the author's plan of the yard in 1918 and by clear graphics detailing the wartime programme and destroyer construction times. There is also a useful (and attractive) graphic on the endpapers which illustrates the ships themselves and their date of completion. Each of the photographs is accompanied by a caption which covers every aspect of what can be seen; the depth of field of these old plate negatives is extraordinary, so there is plenty of material worthy of comment.

Although the primary focus is on the technical side of warship construction, the social aspect is not neglected. The rigid hierarchical dress code which marked out the shipyard workers from the management and office staff is much in evidence, and there are a few images which depict the women who were specially recruited to the workforce as part of the wartime 'dilution' initiative. Because the photographs of the day had to be taken on long time exposures, these necessarily had to be 'posed'; anyone whose work involved movement ended up as a blurry figure in an otherwise sharp image. The inclusion of these images is important, however, as the author is keen that the photographs should constitute a social document of the period: the book is not just about the ships but about shipbuilding in Glasgow 1914–18. I cannot recommend it too highly.

John Jordan

Phil Carradice
The First World War at Sea in Photographs: 1915 and 1916
Amberley, 2014; paperback, each 144 pages, 240 illustrations incl. 16 pages of colour; price £14.95 each.
ISBNs 978 1 4456 2237 8 and 978 1 4456 2242 1.

These two compendia provide an illustrated chronological account of two years of the First World War at sea. Some of the images will be familiar to anyone with an interest in the subject; others are less commonplace. The colour sections are especially attractive. The one in the 1915 volume comprises a series of striking watercolours of naval aspects of the Gallipoli operation by the famous war artist Norman Wilkinson. However, not only is the artist not named, but the images are not captioned: subjects include the *River Clyde*, amphibious operations, shore bombardment, coaling ship, the ill-fated hospital ship HMS *Britannic*, HMS *Triumph*, seaplanes, a seaplane carrier, submarines and a host of large armoured warships.

The treatment of this campaign, the Battle of Jutland and other events is rather one-sided. The idea that early naval bombardments of the Dardanelles defences were 'hugely successful', and the great naval attack a 'success' would be contested both by Turkish commanders and the unfortunate troops landed to open the Straits after the naval attack had signally failed. Some of the captions are simply incorrect. The 1915 volume claims to show HMS *Formidable* on p.4, but the ship in view is one of the older French battleships!

The 1916 volume enters the contested world of gunnery fire control, using an image of *Warspite*, which performed superbly with the standard Dreyer system, claims the German had a computerised system, which they did not, and that the British suffered from not adopting Arthur Pollen's system, which is hotly disputed. The image of an 'American Dreadnought' on p.120 is in fact the old pre-Dreadnought USS *Kearsarge*, complete with her characteristic double turrets. On p.96 of the 1916 volume a caption dealing with the murder of

Captain Charles Fryatt conflates Bruges with Brest; the one below claims to show a U-Boat departing Dunkirk, but the submarine is French! The implication on p.113 that the Royal Navy refuelled at sea with coal during the war is inaccurate: the photograph shows an experimental effort with a Temperley Transporter at least a decade before 1914.

These handsome little volumes provide a useful introduction to the imagery of the conflict for non-experts, but those with a better-developed understanding of the subject may find their pleasure tempered with frustration.

Andrew Lambert

Edward Hampshire
From East of Suez to the Eastern Atlantic: British Naval Policy 1964-70
Ashgate, 2013; hardback, 266 pages, 7 tables; price £85.00.
ISBN 9780754669722

In 1953, the year of the Coronation, the Royal Navy could claim (if the reserve fleet were included) to be the world's second navy. Three years later came the Suez operation. Defence cuts (the Sandys 'Axe') followed, but Britain as a global power continued to police the Indian Ocean. In the early 1960s, Kuwait was protected from Iraqi threats, legitimate regimes were restored or sustained in Tanzania and Kenya, and Indonesia was confronted in the defence of Malaysia. Yet in 1982 it was only with the utmost effort that the Navy could mount an amphibious operation to recover the Falkland Islands from a minor power.

The reasons for this gradual naval decline are familiar: exhaustion after a ruinous world war and loss of empire was characterised by recurrent financial and economic crises. An understanding of the Labour years from 1964 to 1970, which saw the most drastic and fundamental cuts, is vital to detailed study of the decline. The Wilson administration cancelled the proposed TSR-2 strike/reconnaissance aircraft (April 1965) and the strike carrier CVA-01 (February 1966) and, following the sterling devaluation of November 1967, decided to withdraw from east of Suez (July 1968). An order for the American F-111, the designated TSR-2 and CVA-01 replacement, was also cancelled.

The three core questions debated by the author are: why and how was CVA-01 cancelled, how did the naval leadership recover, and how did the navy change in the years after cancellation? Was this trauma really the end of the world role for the Royal Navy? The author follows in the wake of Eric Grove's *Vanguard to Trident*, but covers the ground in more detail, unhindered by the Thirty Years' Rule. A deep and extensive examination of official papers gives a fascinating insight into the strategic, political, economic and financial issues, and of the leading protagonists. By 1964 Mountbatten (a peer since 1946, so no longer 'Lord Louis Mountbatten', as the author styles him) was a spent force, supplanted by the new Defence Secretary Denis Healey. Personalities aside, new ships had to be designed and ordered to offset the lost carrier. As the major role east of Suez was relinquished, the Mediterranean emerged unexpectedly (and briefly) as an important theatre for UK naval deployments. The Suez Canal was closed between 1967 and 1975, affecting logistics and strategy.

This is a serious academic study, unrelieved by maps or photos; it is to be hoped that the publishers will bring out a paperback edition with a less exorbitant price tag.

Ian Sturton

Malcolm Wright
British and Commonwealth Warship Camouflage of WWII
Seaforth Publishing, Barnsley 2014; hardback, 160 pages, over 700 colour illustrations; price £30.00.
ISBN 978-1-84832-205-9

The undertaking of any accurate description of warship camouflage schemes during the Second World War is not for the faint-hearted. This is particularly the case with respect to the British Royal Navy and the fleets of its Commonwealth allies. A prevalence of unofficial schemes (at least in the early years of the war), the post-war destruction of records, lack of reliable photographic material, and conflicting memories amongst those who served all make any analysis fraught with difficulties. An additional complication is the sheer volume of different camouflage patterns that were used.

This last factor is clearly evidenced in Malcolm Wright's book, which focuses exclusively on the schemes used by destroyers, escort vessels, minesweepers, coastal craft, submarines and auxiliaries. The author commences with a short introduction outlining the basis of his approach and a frank description of some of the challenges facing his work. Lists of references, abbreviations and the main colours used follow this. The bulk of the work – some 145 pages in 14 separate chapters – is given over to around 740 colour drawings of more than 600 individual ships at various dates in their careers. The illustrations generally comprise a single starboard profile of a specific ship at an approximate date. However, a few ships are accorded more than one drawing to represent their appearance at different stages in their careers. There are also a small number of plan views. Chapters are allocated to ship types rather than to the different types of camouflage pattern used. Each commences with a full-page sized painting by the author, who is a well-known maritime artist. The book concludes with a three-page index of the various ships depicted.

The book has both strengths and weaknesses. To his credit, Wright is very clear about the factors limiting his work. He makes some good points about the difficulties of determining accurate colours and shades. This includes the impact of the 'that looks about right' (TLAR) approach that must have inevitably been adopted by crews when painting ships under the stress of wartime conditions. The extent of his knowledge is evident in his coverage of a very wide range of minor warships, many of which have not been depicted in other published works on camouflage. The layout of the book is good and the drawings themselves – whilst relatively simple – depict the patterns with clarity.

Unfortunately, this reviewer also holds some significant reservations about the value of the book as a reference. These largely relate to the accuracy of some of the drawings presented. Little is said as to the sources used to produce each specific profile; it is therefore difficult to know how firm the evidence is to justify the various patterns depicted. An associated problem relates to lack of precision as to exactly which period each drawing represents, given the tendency of camouflage patterns to be frequently repainted.

Concerns grow with a cursory comparison of a few of the book's drawings against readily available photographic evidence. To cite an early example of a questionable conclusion, the book portrays the destroyer HMS *Tenedos* wearing a light and dark grey camouflage pattern in 1942. However, a number of photographs exist which demonstrate that she wore a scheme of light upperworks and dark hull from at least the start of February 1942 to her loss at Colombo in April of that year. Similarly, the preserved destroyer HMS *Cavalier* is depicted in what is described as an 'Admiralty Home Fleet pattern' (otherwise known as a 'Special Emergency Fleet Scheme') of green and blue on white in 1944 despite photographs showing her in one of the new standard schemes around that time. Whilst these may well be isolated examples, they suggest it may be unwise to rely too heavily on the book without augmenting it with further research.

The author asks his readership to take his work on trust, stating that, '…his deductions are probably as good as any.' However, given the underlying complexity of the subject, this reviewer would like to have seen more evidence to support Mr Wright's conclusions.

Conrad Waters

Matthew Seligmann, Frank Nägler and Michael Epkenhans, eds.
The Naval Route to the Abyss: The Anglo-German Naval Race 1895–1914
Navy Records Society, Aldershot 2015; hardback, 510 pages; price £90.00.
ISBN 9781472440938

'The Naval Route to the Abyss' follows the standard Navy Records Society format, combining extensive publication of original documents with concise, effective scholarly introductions, both to the subject as a whole, and to specific sections within it. While most volumes have been restricted to material from a single nation this one, edited by two distinguished German historians, Frank Nägler and Michael Epkenhans, and Matthew Seligmann, a leading British expert on the Royal Navy in the Fisher era, looks at the subject from both sides. There are 153 documents from both British and German sources, the latter translated into English.

As befits a scholarly work in a contested area the editors set their volume in the context of older interpretations and current debates. The subject has attracted considerable attention in both countries, and sparked some serious arguments about the meaning of the past. In Germany inter-war scholarship stressed the domestic agendas served by the new Imperial Fleet of Wilhelm II and Admiral Tirpitz, while Anglophone scholars focused on the link between arms races and the outbreak of war in 1914, creating a false link between the very real naval arms race and the events of July and August 1914, by which time the 'race' was long over and Anglo-German relations were improving. In the past three decades the British side of the subject, best known through Arthur Marder's five-volume work of the 1960s, *From the Dreadnought to Scapa Flow*, has been contested by self-styled 'revisionist' historians who have put forward an alternative assessment of the thinking of Admiral Lord Fisher, in which Germany was relatively unimportant, making the arms race more of a domestic political tool for enhanced naval budgets than a serious competition, with battleships less significant than battlecruisers, submarines and destroyers. These theses have been valuable in developing scholarship but have not stood up to sustained archival research. On too many occasions the lack of supporting evidence in the archives has been explained by conspiracy theories, while those who contested the new version were ignored.

This volume demonstrates beyond doubt that the Anglo-German naval arms race was a genuine contest. The documents clearly show that British and German naval authorities viewed each other with suspicion, concerned by the threat that the other navy posed to their own aims and policy. By 1901 at the latest, Germany was a major factor in British planning, while the Royal Navy had always dominated the thinking of Alfred Tirpitz as he struggled to build a fleet capable of sustaining the world power ambitions of Imperial Germany. The latest German scholarship has done much to reconnect the High Sea Fleet with national ambition. The fact that the Royal Navy recognised this threat at an early stage, and responded effectively in strategic, technical and policy terms, demonstrates that the Edwardian Admiralty was a highly professional body staffed by intelligent, sophisticated officers, among whom Jacky Fisher was the leading light but was by no means alone.

The arms race was conducted in annual programmes: those of Germany were fixed by Tirpitz's Navy Laws, the British retained the freedom to set fresh targets annually. The unit of measurement was the Dreadnought capital ship, and the British had a massive advantage over their rivals, with far greater construction capacity and a profound commitment to sustain naval mastery which was shared by a very large proportion of the population. Ultimately the race was settled by financial pressure. Germany simply could not afford to build a fleet to face Britain and an army ready to fight France and Russia, and Britain would not surrender her sea power. The British offered opportunities to compromise down to 1912, but Tirpitz managed to block any meaningful agreement. In 1914 the consequences of the political stand-off created by the arms race and the construction of the High Sea Fleet became clear. While naval armaments did not cause the First World War, they ensured that Britain would be on the side of France and Russia.

Pre-war thinking about economic warfare in both countries demonstrated considerable insight. The Germans sought to use the growing framework of international law created at The Hague and London Conferences to exploit British vulnerability. Tirpitz saw cruiser warfare as a useful weapon, even if he was determined to complete his battle fleet project. The British also examined the role of economic warfare in a war with Germany, but they did not advance a radical doomsday scenario of the type posited in Nicholas Lambert's recent book *Armageddon*, in which he claimed the Admiralty was planning to collapse the global economy in order to destroy Germany's economic base in a matter of weeks. British assessments were more sober, recognising that, as history indicated, economic warfare would take a long time to work and that Britain would also be affected by heavy maritime losses.

This volume, expertly introduced and strikingly well-balanced, and with its arguments backed by an exceptionally rich and varied collection of material, will be essential reading for students of the arms race and the ships that were built to conduct it. The range of archives consulted offers a useful primer for anyone essaying a project in this period. The Navy Records Society has been publishing high-quality collections of naval documents since 1893, and this volume may be the best yet. Many serious students of naval history will already be members of the Society, which has done more to sustain the subject through the decades that any other organisation. Those who are not will find this volume the ideal occasion to take up the opportunity.

Andrew Lambert

James Goldrick
Before Jutland: The Naval War in Northern European Waters, August 1914 – February 1915

Naval Institute Press, 2015; paperback, 382 pages, 32 black & white photographs and 11 maps; price £36.50/$44.95.
ISBN 978-1-59114-349-9

In 1984, Lieutenant James Goldrick's *The King's Ships Were at Sea* was published, and would soon be recognised as the definitive account of the early months of the First World War in the North Sea. During his distinguished career, Goldrick has gone on to serve with the Royal Navy twice in these waters and also in the Baltic; his experiences there and as a senior commander have led him to look with different eyes at the events he described, and to realise that we no longer properly understand how ships were worked and fought in the Great War. Now retired, Rear-Admiral Goldrick draws on these perceptions for a new and expanded edition in which he analyses the challenges facing navies in 1914, some of which he barely recognised in 1984 but which now seem fundamental. He has also included the Baltic theatre, not least because understanding it is necessary for a fair assessment of the activities in the North Sea.

Before Jutland, which is 15% longer than its predecessor, has a similar overall plan but three new chapters on the Russian Navy and the war in the Baltic. There is also a new chapter entitled 'War Plans' covering the dispositions taken up by the three navies as they put these plans into effect. For this reviewer, the outstanding addition is the chapter on 'Operational Challenges'. Among many topics, it emphasises the problems facing navigators due to fog and mist, rain and snow, strong North Sea currents, the unreliability of the recently-introduced gyro-compass, and the often grim conditions on open bridges, especially in destroyers. Positional errors of five miles were expected after eight hours steaming while, if a destroyer had been in action, her reckoning might be out by 25 miles. Positions were still checked by the traditional methods of depth sounding and sampling the seabed. Forces that remained out of sight of one another had no shared frame of reference, which may explain why no navy had yet attempted to keep anything resembling the later tactical plot. Poor visibility also interfered with the accurate enemy identification so necessary for tactical decision-making; examples are given of both British and German ships mistaking light cruisers for armoured cruisers.

Goldrick casts new light on the difficulties of operating coal-fired ships. Apart from the bane of regular coaling, high-speed steaming could not be maintained once it became necessary to draw coal from bunkers more remote from the boilers. The Germans were especially handicapped by the poor quality of the available coal and it is possible that this restricted their battlecruisers to a speed of no more than 24 knots. A further difficulty was that their heavy ships could not traverse the Kiel Canal without discharging coal; thus it took four to five days to complete a move between the Baltic and the Bight.

August 1914 did indeed see the beginning of modern naval warfare but many of the new technologies, as well as being untried in war, were still being assimilated. Both the British and the German fleets had evolved into 'grand fleets of battle', comprising capital ships, cruisers and destroyers, and effective methods of control were hotly debated. Mines had shown their potential during the Russo-Japanese War and they would be used effectively by the Russians in the Baltic, both defensively and offensively. At the start: 'Few submarine commanders could successfully attack a surface ship'. However, the overseas-type submarines on both sides soon proved capable of operating near enemy coast and bases and of hitting with their torpedoes. Aircraft were at an even earlier stage of development, but Goldrick notably describes no fewer than three pioneering seaplane raids by the Royal Navy on the Zeppelin base at Cuxhaven. When scouting, the Zeppelins themselves were frequently frustrated by poor visibility, while they proved too unmanoeuvrable – and lacked sufficiently heavy bombs – to be effective against warships.

Goldrick recognises the tension between the promise and the actual achievements of the new technologies. He accepts that British specialist officers were generally at ease with technology but, in stating that their education 'tended to produce mathematicians and mechanics' and that only mechanical antisubmarine devices were being developed, he overlooks both the electrical expertise of the Torpedo Branch and fails to acknowledge that acoustic submarine detection could make little progress until electronic amplification using valves had been fully developed. In discussing ships' machinery, he shows that problems were by no means confined to British ships; German engines were frequently under repair, there were many cases of 'condenseritis' and the small-tube boilers were unreliable.

The chapters of *Before Jutland* that describe operations and engagements remain as lucid and perceptive as before and incorporate important new material. A case in point is that Goldrick reveals the part played by Beatty's flag-commander, Reginald Plunkett, in the confused conclusion to the Battle of the Dogger Bank. By his own account, Plunkett recommended the signal ('Engage the rear of the enemy') that seemed to confirm that the other battlecruisers should attack the already-crippled *Blücher*; on this occasion Seymour, the flag-lieutenant, seems to have been largely blameless.

This brief review has only been able to mention some of this work's many qualities. For those who already possess a copy of its predecessor, *Before Jutland* offers many new insights derived from its author's own experience of command at sea, while new readers can be confident that the new edition will again become the standard reference for the early months of the Great War at sea in northern waters.

John Brooks

Malcolm Smith
Voices in Flight: The Royal Naval Air Service During the Great War
Pen & Sword, Barnsley, 2014; hardback, 210 pages,
27 B&W images; price £25.00.
ISBN 978 1 78346 383 1

This book was produced in association with the Society of Friends of the Fleet Air Arm Museum (now part of the National Museum of the Royal Navy). The author, a former air engineering officer in the Royal Navy, has drawn on the Museum's archive to produce a wide range of material concerning the early years of British naval aviation.

There are sections headed 'Early Days', 'Home Waters', 'Secret History', 'Western Front', 'Mediterranean and Middle East', 'My Time as a Prisoner of War' and 'Armoured Cars in Gallipoli and Russia', containing material drawn from diaries, transcripts of letters, log books and other documents. However, the source documents are not identified, while some sections have been written by modern writers, with the result that the origin of what is printed here is not always obvious. Ideally, each source document should have been clearly identified so that it could be used by the reader as the basis for further study. The lack of a bibliography further weakens the work's value as a reference source and this is a pity, as it represents a lost opportunity on the part of the NMRN to demonstrate the size and importance of its archive. There are no maps either, but there is a small glossary of abbreviations and naval terms and a six-page index.

Malcolm Smith is referred to on the dust jacket as the author but, apart from a two-page epilogue, his role could better be described as that of editor, selecting the archive material to be reproduced and the documents to be re-written. He has certainly drawn together material which gives a clear insight into the wide-ranging activities and people of the RNAS, but there is no explanation as to why men in the RNAS found themselves in armoured car units fighting in Russia, or indeed why their story has been included in a book entitled 'Voices in Flight'. While any serious historian will know why they were there, for the wider readership an explanation would have been helpful. The book is well presented in hard-back but there are some surprising errors which should not have escaped the proof-reading process. Typical of these is the mis-captioned photograph of a Sopwith One-and-a-Half Strutter taking off over the 12-inch guns of a battlecruiser's wing turret. It is a well-known image but the ship in question is not *New Zealand* as captioned; it actually shows Flight Lieutenant Donald RNAS taking off from 'Q' turret of HMAS *Australia* on 7 March 1918 – the ship in the background is *New Zealand*.

Malcolm Smith has produced a workmanlike book which generally succeeds in its aim of describing a variety of RNAS activities, but which would have benefited from some thoughtful editing. Although not aimed at the serious historian, the book will hopefully still have the effect of stimulating interest in the RNAS, for which it is a good starting point.

David Hobbs

Anthony Clayton
Three Republics One Navy: A Naval History of France 1870–1999
Helion & Company Limited; hardback, 215 pages,
small profiles; price £29.95.
ISBN 978-1-90998-229-4

This slim volume attempts to compress 130 years of naval history into 196 pages. The scope is ambitious, the author well-read with a good understanding of French language and culture. The organisation of the book is logical, and the text is illustrated by small profiles by Peter French of key French warships of the period covered; these are more decorative than anything but do provide a flavour of French warship architecture and paint schemes.

Unfortunately, the author's account of the naval politics of the period is too superficial to be useful, and the degree of compression results in a lack of clarity and 'direction' in the narrative. Clayton's understanding of

France's position during the naval arms limitation process of the interwar period is seriously flawed, and his account of French warship development contains a multitude of errors. The *Courbet* class were not the 'first ships to be oil-fuelled' (they had coal-fired boilers); the torpedo attack on *Jean Bart* in December 1914 by an Austrian submarine was successful (the author claims it missed – in fact the ship spent nearly four months under repair at Malta); and there were twelve (not eight) ships of the *Bourrasque* (not *Simoun*) class. This is just a sample; there are so many factual errors here that they tend to undermine the reader's confidence in the political *aperçus* with which the narrative is littered.

The book is also a difficult read because of the idiosyncratic punctuation (of which the title is a prime example): commas are used in place of full stops or colons, and some sentences self-propagate in a stream of subordinate clauses without a pause for breath. There were numerous occasions when this reviewer needed to read a sentence two or three times in order to grasp its meaning. Mr Clayton is an author is search of an editor.

John Jordan

Martin Stephen
Scapegoat: The Death of *Prince of Wales* and *Repulse*

Pen & Sword, Barnsley 2014; 194 pages; 26 B&W photographs, map, endnotes, select bibliography, index; price £19.99.
ISBN 1783821784

Martin Stephen lists no fewer than 18 books in his bibliography which are wholly concerned with the events leading to the sinking of HMS *Prince of Wales* and *Repulse* or to the fall of Singapore in December 1941. A new perspective is not unwelcome, and in this new work the author seeks to counter the received wisdom that Admiral Sir Tom Phillips, the commander of Force Z, was largely responsible for the loss of the two capital ships, claiming instead that he should be wholly exonerated by history. In Stephen's eyes, Phillips has been made the scapegoat for arguably the most serious naval blunder of the Second World War involving the Royal Navy.

Conjecture and debate over who was ultimately responsible for this disaster have exercised the minds of historians throughout the post-war era. Wide-ranging historical factors have been considered, including the problems of Britain defending an over-stretched worldwide empire with increasingly limited and ageing naval resources, the 'peace dividend' sought by many in the interwar years, Britain's strict adherence to successive naval treaties, and the myth of the impregnability of the Singapore island 'fortress'. More specifically, the wisdom of the decision to send the force to defend Britain's Far East territories without shipborne air cover has been widely criticised. Inevitably, the decision of Admiral Phillips to venture out in search of the enemy and his handling of events after leaving Singapore have been minutely scrutinised and almost universally condemned.

Stephen mounts his defence by contradicting each and every accusation made against Admiral Phillips and placing the blame instead on the political elite, other senior naval officers, an incompetent naval intelligence organisation and even the official naval historian. With one small exception, he can find no fault in Admiral Phillips' tactical handling of Force Z after the squadron left Singapore.

Winston Churchill probably receives the greatest share of the author's opprobrium, which centres on the Prime Minister's perceived 'meddling' in naval matters and his failure to intervene once the decision had been taken by Phillips to leave Singapore without air cover. Moreover, Admirals Cunningham, Layton, Palliser and Somerville are variously accused of a range of misdemeanours which contributed either directly or indirectly to the disaster itself or to the subsequent besmirching of Phillips' reputation. Naval intelligence is blamed for failing properly to brief the Admiral about the capabilities of the Japanese aerial threat. Finally, Stephen Roskill, who wrote the official history, is deemed to have allowed personal enmity to cloud his judgement with regard to Phillips.

Is Martin Stephen presenting us with a serious piece of revisionist history or is he merely 'flying a kite'? Much of the evidence he presents in support of his theories is prefaced with 'ifs', 'buts' and 'maybes'. He was allowed access to the personal papers belonging to the Phillips' family, which must count as a significant piece of new evidence. However, it appears that apart from contradicting the accepted norm that Tom Phillips was a difficult man to work with or to like, the papers do not seem to offer anything substantial in defence of his decision-making prior to the loss of his ships.

Stephen condemns both the 'armchair historian' and those who succumb to putting 'a spin' on events in order to prove their point. Unfortunately, he is guilty himself on both these counts. One would expect the author to be able to support his opinions with objective evidence from primary sources, yet the endnotes reveal only a scattering of this kind of documentation; the vast majority of his references are derived from other books on the subject.

While Pen & Sword publish a wide range of books on naval history for which there is an enthusiastic audience, their standard of editing remains a concern. It does not exactly fill one with confidence to read here that Force Z searched 'the length of what was then Malaya and is now Sri Lanka'. Sadly, this is not an isolated error.

Jon Wise

John Jordan and Jean Moulin
French Destroyers: *Torpilleurs d'Escadre* & *Contre-Torpilleurs* 1922–1956

Seaforth Publishing, 2015; hardback, 296 pages, illustrated with numerous B&W photos and line drawings, and colour paintings; price £40.00.
ISBN 978-1-84832-198-4

The third in a series of books co-authored by John Jordan on the *Marine Nationale*'s major surface combatants over the 1922–1956 timescale, *French Destroyers* examines the *torpilleurs d'escadre* ('fleet torpedo boats') and *contre-torpilleurs* ('torpedo boat destroyers') of this period. Although the two types were designed for different roles – the former were primarily torpedo attack craft whilst the latter had a broader series of functions but with an emphasis on fleet scouting – they shared many developmental and technological characteristics. Hence, the authors considered it logical to consider both types of flotilla craft together.

The book is sixty-four pages longer than its predecessors, properly reflecting the much greater extent of the subject matter covered. It is of credit to publisher Seaforth that this expansion has been achieved with no increase in the cover price. Otherwise, *French Destroyers* has a very similar feel to previous books in the series, benefitting from large, clear photographs and numerous, detailed line drawings.

Structurally, the book adopts broadly the same approach taken in *French Cruisers 1922–56*, being split into two principal parts. The first, predominantly technical portion is largely the work of John Jordan. It commences with a brief introduction outlining the development of French destroyers prior to 1922. The nine following chapters each focus on the development and design characteristics of a specific class, commencing with the *Jaguar* class of the 1922 programme and concluding with the *Le Hardi* type *torpilleurs d'escadre* of 1932–38. A short colour section between chapters 6 and 7 provides scope to display a small selection of watercolours from the extensive work of Surgeon General Jean Bladé, whose paintings also appeared in *French Cruisers*. The 'technical' part concludes with a chapter, new to this book, describing paint schemes and identification markings.

Part 2 is written by Jean Moulin and comprises the book's historical section. It encompasses four main chapters set out in chronological order, commencing with the pre-war period. The war years are split between two chapters focused respectively on 1939–1943 and 1943–1945, the latter corresponding with the return of France to the Allied side and the modernisation of many of its remaining warships. The concluding, post-war chapter includes brief summaries on the former German and Italian warships acquired by the *Marine Nationale* and operated in destroyer-type roles. A number of additional appendices, largely contained in the second part, provide more detail on specific topics, such as the careers of *Léopard* and *Le Triomphant* with the Free French Naval Forces and details of the modernisation work carried on the four surviving members of the *Le Fantasque* class in the United States between 1943 and 1945.

French Destroyers is distinguished by the quantity and quality of information provided in the chapters describing specific destroyer classes. These benefit from following a largely common arrangement. An introduction to each class's construction programme is followed by an in-depth technical description ordered into standardised sub-headings. Information on subsequent modifications complete each description. Supporting drawings include exterior profiles and plans, general arrangement and sectional views. These are supplemented by detailed drawings covering such areas as key weapons systems, propulsion plants, bridge layouts and accommodation arrangements. Tables encompass building dates, general characteristics, power trials and more. It is hard to envisage what additional information a reader might want about these ships.

The historical chapters are set out in a somewhat more economical fashion. They provide a clear and accurate overview of the diverse range of operations the destroyers participated in throughout their careers. It is notable that much of their active service was undertaken in particularly adverse circumstances, with the result that they were rarely able to demonstrate their true potential. Nevertheless, it would be have been interesting to see more analysis on what these actions revealed about their relative strengths and weaknesses in operational conditions. A tabular summary of the ships' ultimate fates would also have been welcome. However, these are very minor criticisms.

In conclusion, *French Destroyers* more than matches the exemplary standards of technical detail and presentational excellence achieved by the preceding volumes in a series that is inevitably now considered the standard reference work on a fascinating subject.

Conrad Waters

Andrew Faltum
The Supercarriers: The *Forrestal* and *Kitty Hawk* Classes
US Naval Institute Press, Annapolis MD, 2014; hardback, 288 pages, 81 photos, 6 maps; price $42.95/£29.95.
ISBN 978-1-59114-180-8

This is the third book in a series by Andrew Faltum, who formerly served as an air intelligence officer on board USS *Midway* (CV-41); the earlier books had as their subjects the light carriers of the *Independence* class and the fleet carriers of the *Essex* class. The subject of the present book is the series of conventionally-powered 'supercarriers' built post-war for the US Navy beginning with the *Forrestal* (CVA-59, laid down July 1952) and ending with the *John F Kennedy* (CVA-67, commissioned 1968).

The first half of the book, comprising an Introduction and 14 chapters, is an account of the design and construction of the ships, of the aircraft which operated from them and the associated weaponry, and of their operational service lives. It is a competent, if slightly bland account which, as far as the historical record is concerned, concerns itself with reporting the facts rather than attempting any deep technical or political analysis. Faltum twice makes the point – once in the Introduction and then again in the Epitaph – about the omnipresence of these ships in every politico-military crisis of the period, and twice quotes the often-reported presidential question 'Where are the carriers?', but does not get drawn into the more pertinent question as to whether

American power projection resolved any of these conflicts except in the narrowest and most temporary military sense. It is sobering to read through the accounts of these past conflicts and to realise how many of the areas of intervention – Southeast Asia, East Africa, the Middle East, Libya – are still in turmoil, and arguably even less stable than before. However, this is probably the subject of another, very different book.

There are six theatre maps and many black & white photographs, but these are grouped into two 'plate' sections, which means that they are not used to best advantage to illustrate the text. It is unclear why USNIP opted for this format, as the entire book (including the photo sections) is printed on plain paper so there is no compelling reason why the photos need to be grouped together. There are port and starboard profiles, together with a plan view, of *Saratoga* on the front endpaper, and of *Constellation* on the rear endpaper; these are reasonably well executed, but suffer from being reproduced across the gutter.

Arguably the most valuable part of the book is the Appendices section, which occupies the second half of the book. There are technical data tables for the carriers themselves and for their aircraft and electronic systems, together with tables listing the composition of the air wings which operated from each of the ships at every stage of their respective careers. This is an admirable piece of work, and Faltum has provided clear and useful explanations for the designations of US naval aircraft and the air wings to which they were assigned. The book is probably worth purchasing for these alone.

John Jordan

Brian Lavery
Nelson's *Victory*: 250 years of War and Peace
Seaforth Publishing, 2015; hardback, 208 pages,
illustrated throughout in full colour; price £30.00.
ISBN 978 184832 232 5

Published to accompany a special exhibition at Chatham Historic Dockyard marking the 250th anniversary of *Victory*'s launch in May 1765, this handsome volume tells the full story of Britain's most famous preserved ship, from her inception right up to the present day, and also charts the parallel career of Britain's best-known sailor. These parallels are often quite remarkable: for example a young Nelson learned much of his early navigation and seamanship skills in small boats in the tidal reaches of the river Medway, full of various warships lying 'in ordinary', of which one was *Victory*.

The early part of the book looks at the various issues resulting from placing wooden warships in this state, for it posed some serious challenges to their structural longevity. *Victory* spent the first dozen years of her life thus, and it was only when she was finally commissioned in 1778 that various problems came to light. However, once these had been resolved and she was in service, *Victory*, as well as being big and heavily armed, turned out to be notably fast for a first rate, and this meant she soon became popular with the more aggressively-minded admirals such as Hood, Jervis and, in due course, Nelson.

Since it was not apparent during the early careers of ship and man that either was destined for future fame and glory there is something of a dearth of records, but Lavery turns this to advantage, taking the opportunity to quote from a wide range of sources and thus placing both into their proper historical context. For example, in the chapter on life on board during blockade of Toulon he delves into the backgrounds of some of those present; the picture he paints is not one of a particularly happy ship, though no doubt the long, exhausting and apparently fruitless mission had some influence on this.

The middle part of the book is, perhaps inevitably, a straightforward account of the Trafalgar campaign, but it is the final section which may well be of most interest to *Warship* readers, since it covers *Victory*'s long, varied and not entirely incident-free post-Napoleonic career, first in harbour service and then as a museum ship. Quite apart from the problems of her increasing age, there were various calls for her to be scrapped; she was rammed by the battleship *Neptune* in 1903, and bombed during the Second World War. Among the illustrations are a number of WL Wyllie's fine paintings, including an evocative one of Portsmouth Harbour reminiscent of Turner. In contrast, earlier in the book there is a watercolour sketch by Britain's most celebrated painter showing *Victory*'s quarterdeck after Trafalgar, stripped practically bare and without the mass of cabins, fittings, and decoration with which we are familiar today.

Written in Brian Lavery's customary highly readable style, the book is clearly intended as much for the informed mainstream reader as for the true specialist. Numerous books, both technical and historical, have been written about *Victory*, but for an overall picture of the ship and her remarkable story this volume is hard to beat.

Stephen Dent

Gérard Garier
Les avisos dragueurs de 630 TW du type Elan: Vol.1
Lela Presse, Le Vigen 2015; hardback, large format, 336 pages,
heavily illustrated; price €49.00.
ISBN 978-2-914017-86-2

The 'minesweeping sloops' (*avisos dragueurs*) of the *Elan* class were ordered as replacements for the series of sloops built during the last two years of the Great War. Powered by diesels, robust, and with good sea-keeping qualities, they proved well-suited to convoy escort duties and were rarely used in their designed minesweeping role. Four had been completed by the end of 1939 and a further eight were rushed into service before the Armistice of June 1940. Four were seized by the Royal Navy, which was quick to appreciate their potential. One would be manned by the RN and three were handed over to the FNFL. From 1943 onwards the weapons and sensors of these units would be extensively upgraded: they were fitted with the distinctive 'lantern' for the Type 271 radar, British-model depth charge rails and DCTs, and given a close-range AA armament of 40mm Bofors and 20mm Oerlikon guns. Many surviving ships of the class continued in service well into the post-war era, with further upgrades to their equipment.

This new French-language study by Gérard Garier looks at the origins of the design, the technical characteristics (including the plethora of modifications), and the service history of the first six ships of the class. A second volume, due to be published in 2016, will cover the seven later ships ('B' type).

The strength of the book lies in its attention to detail. This is no broad-brush account, but an attempt to assemble all that is known about these ships under one cover. The technical section is illustrated by the customary plans of ships and weapons held by the CAA. These are well-reproduced, but at a scale which inevitably means that labels are not always legible. The layouts are extraordinarily 'busy', and the cursive typeface used for some sections of the introduction is not always easy to read, particularly when printed on a coloured background. The author's decision to publish excerpts from official specifications and commanding officers' reports for different ships also means that there is a degree of duplication in the descriptions of equipment and evaluation of the ships' qualities. The author has clearly spent many years collecting photographs of these ships, and all of them are published here, good, bad or indifferent, interspersed with beautiful watercolours by the distinguished naval artist Jean Bladé and photographs of other ships relevant to the technical and historical accounts. There are some excellent close-up views taken on board, and the comprehensive nature of the coverage means that there are photographs which illustrate all six ships throughout their service lives, accompanied by detailed captions which highlight the modifications shown.

The final section of the book is a real *tour de force*. Using his extensive photograph collection the author has produced artwork colour profiles (generally seven/eight for each ship) showing each of the six ships at key points in their career. These are clear, beautifully crafted and reproduced at a good size which enables the reader to see every detail noted in the captions.

Despite the uneven production qualities this is a stunning book, and one which sets a new standard for monographs of this type.

John Jordan

Brian Izzard
Yangtze Showdown: China and the Ordeal of HMS *Amethyst*
Seaforth Publishing, Barnsley 2015; 276 pages;
35 B&W photographs, map; price £25.00.
ISBN 978-1-84832-224-0

It could be argued that the *Amethyst* incident, like the defence of Rorke's Drift during the Zulu wars, was blown out of all proportion by the government of the day for political gain. *Amethyst*'s escape from the Yangtze in 1949 provided a tonic for a public in the grip of post-war austerity and was a welcome re-affirmation of the status of the Royal Navy. The over-riding question which remains, however, is what this lone British frigate was doing on a foreign river amidst the climax of a long civil war.

Brian Izzard's book provides a well-written, very readable account of this incident. His research has been thorough and includes both first-hand accounts by survivors and reference to primary source material not expected to be released before 2030. The ordeal for those members of the crew who remained on board lasted from 19 April to 31 July 1949 and is recounted in detail. Izzard also describes the failed interventions by the cruiser *London*, the destroyer *Consort* and the frigate *Black Swan*, each of which cost dear in terms of casualties.

Wisely, quite a lot of space is given over to the broader aspects of this unusual engagement: to the political ramifications, to the historical reasons for the international presence on the Yangtze River, and to the parts played by the RN senior command most closely involved, Admirals Fraser, Brind and Madden. Much of this background material has been examined before in detail, as the incident offers a prime example of the problems faced by Britain as it struggled to come to terms with its place in the world in the aftermath of war, the inexorable rise of the two superpowers USA and the USSR, and the beginning of the end of colonial domination.

Anyone with knowledge of the *Amethyst* incident will understand the central part played by the stand-in commanding officer, Lt. Commander John Kerans. Brian Izzard describes him as being 'the wrong man in the right place' when he was chosen to take command in place of *Amethyst*'s CO Lieutenant Weston, who had been badly wounded and had to leave the ship (he would later die of his wounds). A heavy drinker and a womaniser, Kerans had made a mess of his naval career thus far and was at the time filling the role of assistant naval attaché in nearby Nanking. Izzard describes Kerans' tireless efforts throughout the months of fruitless negotiations with the Communists to ensure the safe passage of the ship, but also his bad behaviour on the long journey home, freed at last from the stresses of his experience on the Yangtze. Despite his failings, Kerans did bear the brunt of the negotiations, and it was his decision alone to make the bid for freedom which he managed without further loss of life or damage to HMS *Amethyst*.

While there is little new evidence presented in this book, the reviewer was unaware that it was the Australian frigate *Shoalhaven*, not *Amethyst*, which was originally scheduled to sail to Nanking to assume guardship duties,. The Australian Government refused to allow *Shoalhaven* to sail because of the internal situation pertaining in China. The unwillingness of the Labor Prime Minister at the time, Ben Chifley, to kow-tow automatically to the wishes of the mother country was as significant as the Americans' own decision not to send their warships up-river.

In pre-war Shanghai the most central berth on the Woosung River was traditionally taken by a 10,000-ton British cruiser. In the few short years after the Second World War, before the Chinese Communist government drove the Western nations away, it was American warships which occupied the same buoys. The old order was changing, as *Yangtze Showdown* demonstrates so well.

Jon Wise

Richard Beale
One Man's War: an actor's life at sea, 1940–1945
Conway, 2015; hardback, 264 pages, 15 B&W photographs, 2 maps; price £16.99.
ISBN 978-1-84486-333-4

One of the first titles to appear from Conway under the new owners, *One Man's War* is authored by Richard Beale, who has been a successful television, film and stage actor for more than fifty years. Back in the spring of 1940, not long out of school, Beale volunteered for service in the Royal Navy, and the book is his account of what ensued over the next five years.

Beale had a narrow escape not long after going to sea. While at Scapa Flow he was temporarily transferred from his first ship, the light cruiser *Arethusa*, to HMS *Hood* to replace a sick rating, during which time he was surprised to witness the ship's officers letting off steam in a thoroughly juvenile large-scale prank. A short while later he was back on board *Arethusa*; *Hood*, and her officers, meanwhile, were at the bottom of the Denmark Strait…

Commissioned shortly afterwards under the CW scheme, Beale was very much a product of his class and era, and his confident (some might say headstrong) nature sometimes brought him into conflict with members of the 'regular' Navy. The deliberate ramming and sinking of an abandoned drifter which happened to be occupying a mooring he wished to use off Greenock resulted in a court martial, which he describes as a 'pantomime'. He survived this with a reprimand but was then packed off to the Mediterranean, possibly with a view to placing him where his approach was more likely to harm the enemy than his own side. Here, in command of first *HDML 1070* and then *ML 135*, he took part in the campaigns to liberate Italy, Greece and the Balkans. Direct encounters with the enemy were few and far between; when *ML 135* did have to open fire with her main armament it was on Greek bandits (who promptly fled) rather than on the armed forces of the Third Reich. This is very much a tale of ordinary life in the small craft of the RN: patrol, convoy escort, minesweeping, ferrying troops or refugees, and so on. When Beale does get injured (and nearly loses his sight) it isn't in action, but as a result of the explosion of a faulty smoke generator.

Throughout the narrative he comes across as straightforward, sensible and practical, keen to get on with whatever job is to hand, with no great love of the alcohol which seemed to keep many of his fellow officers going.

He is not a great one for contemplation or reflection; however, encountering the body of a young German, floating in the Adriatic, clearly had a great impact on him. He conveys well the sense of anti-climax that the end of hostilities brought about: suddenly he, his crew, his ship, and their mission, were all no longer wanted, and were just shunted back into 'ordinary' life as quickly as possible. No wonder so many found the transition difficult to cope with.

Beale is happy to concede that occasionally his recollection of events seventy years ago isn't what it might be, and there is the odd minor factual error (*Arethusa* wasn't a sister of *Calypso*, and had six 6in guns, not four). He is also honest in his admission that he behaved pretty disgracefully while on leave in Italy (though he met a GI who behaved even worse), while if there is one constant in the tale it is the concern with food: what provisions were available, and what the cook would have done with them. It is an engaging, honest, readable book, played straight – neither for horror nor for laughs – with the result that those moments of genuine fear or humour stand out all the more.

Stephen Dent

Rear Admiral Dave Oliver, USN (Rtd.)
Against the Tide: Rickover's Leadership Principles and the Rise of the Nuclear Navy
Naval Institute Press, 2014; 178 pages; price $27.95/£21.95.
ISBN 9781612517971

Admiral Hyman G Rickover, dubbed the 'father of the nuclear navy', master-minded the development of naval nuclear propulsion and spent more than three decades in control of US naval nuclear reactors.

This book is not a biography. The clue is in the title: it sets out to explore the principles which underpinned Rickover's unique brand of leadership, which many deemed to be abrasive, intrusive and alienating. The author, who served under Rickover in nuclear submarines, clearly thinks otherwise and believes that aspiring managers from across the business spectrum might learn important lessons from this unique character.

Rickover had a curiously inauspicious career prior to his involvement with the naval nuclear programme. Although he volunteered for submarine service in 1930, he subsequently failed to be selected for command. Indeed his only experience in that respect was captaining a minesweeper off China. He spent most of the Second World War as Head of the Electrical Section at the Bureau of Ships. However when the wat ended, and with the prospect of forced retirement beckoning, Rickover was one of a handful of naval officers and civilians despatched to Oak Ridge, Tennessee, to familiarise themselves with the possible naval applications of the findings of the Manhattan Project. The US Navy's original intention was to fit nuclear reactors to surface ships. However, Rickover realised the potential of the nuclear submarine and, with the aid of a small team, overcame obstacles, gained the support of Presidents Truman and

Eisenhower, and was the driving force behind the commissioning of USS *Nautilus* in January 1954.

Oliver uses successive chapters of this book to describe different aspects of Rickover's leadership style. To this end, his own experiences in senior business management after leaving the Navy lend credence to the topics he selects to discuss. At the end of each section he asks the reader various questions which might prompt them to apply Rickover's theories to their own workplace. Thus, Oliver shows how Rickover was quite ruthless with potential COs: if there were doubts about their abilities to shoulder the burden of responsibility they would be sacked. His nuclear captains had to demonstrate that they were aware of the need for continual improvement. He actively encouraged innovation and often preferred eager and thrusting young minds rather than experienced veterans. From a broader perspective, Rickover wanted the men who commanded these nuclear boats not only to accept the great responsibilities they held but also to be aware of the higher purposes they served.

Safety was of paramount importance. One of the defining moments of Rickover's career was the loss of USS *Thresher* off Cape Cod in 1963 when a faulty weld allowed an overwhelming ingress of seawater. The entire crew was lost. If it had been the boat's maiden voyage, Rickover himself would have also perished. The admiral was always on board for the inaugural voyage of vessels of his nuclear fleet; he insisted on supervising the programmes personally and on testing the submarine and its crew to the limits.

Against the Tide is not exactly a panegyric, but nor could it be described as an objective study of the methods Rickover used to ensure that the nuclear submarine fleet, both ballistic and attack, fulfilled its burdensome duties during the long years of the Cold War. Oliver clearly had enormous regard for his superior officer and shares his values. He regards the men, more specifically the officer corps who commanded the boats and supervised the precious reactors, as men apart – alpha males, both highly intelligent and highly motivated. One might criticise this self-acclaimed 'elite' service for producing narrow-minded individuals, but the drive and dedication shown by Admiral Rickover was undeniably a vital factor in the successful conclusion of the Cold War.

Jon Wise

Roger Branfill-Cook
Torpedo: The Complete History of the World's Most Revolutionary Naval Weapon
Seaforth Publishing, 2014; hardback, 256 pages, 461 illustrations; price £35.00.
ISBN 978-1-84832-215-8

There have been many books written on the history and development of submarines, but only a handful dedicated entirely to the subject of the weapons which made them one of the most significant naval developments of the last 200 years.

In this large format hardback, the author covers the history of the torpedo from when the idea of underwater explosive devices was first conceived through to the sophisticated weapons of the modern day. It was not until the *Nautilus* of around 1800 that the search for a suitable new underwater weapon really began, and only in 1866 that Robert Whitehead created the first self-propelled torpedo. The book is divided into four main sections (24 chapters in total): 'Inventors and their Torpedoes', 'Torpedo Launchers and Delivery Systems', 'Anti-torpedo Defence' and 'Notable Torpedo Actions and Incidents'. Each section is laid out in a sensible chronological order and abundantly illustrated with photographs, drawings and diagrams which ably support the text.

The section covering the evolution of the torpedo from the 'infernal machines' of the American Civil War to the technologically advanced weapons of today is detailed and engaging, as is the study of some of the notable torpedo actions. While some of these incidents have been covered in greater detail elsewhere, many are less well known and worth discovering. The section about the various torpedo launchers and delivery systems that have been used by different navies in the attempt to get torpedoes to their intended targets, from early rail systems to underwater or deck-mounted torpedo tubes, covers in some detail a subject that is often overlooked.

The book ends with a useful appendix listing details of the various torpedoes that have been designed and built over the last 150 years. Arranged by country, it includes data such as speed, size, weight, year of first production, method of propulsion and type of warhead carried, and will be valuable for future reference. There is a comprehensive bibliography and a detailed index.

In his introduction Branfill-Cook acknowledges the size of the subject and states that 'to tell the whole story in all its minor details would take a multi-volume work running to several thousand pages.' Thankfully the author avoids this by maintaining a clear, readable style; he knows when to go into detail and when to keep to the main facts and key points. This results in an interesting, informative and enjoyable book which will appeal to the general reader. It may feel oversimplified in places to the specialist looking for a more academic approach to the subject, but in general the author maintains a good balance between the technical side and 'readability'. With a cover price of £35.00 *Torpedo* is on the expensive side, but a great deal of work has gone into this book and those with an interest in the subject will not be disappointed.

John Peterson

Benjamin Armstrong
21st Century Sims
Naval Institute Press, 2015; paperback, 176 pages; price £14.49.
ISBN 978-1-61251-810-7

Throughout his naval career Admiral William Sims, commander of American naval forces in Europe during the First World War, battled against reactionary forces within the US Navy. A noted thinker and innovator, Sims first came to prominence when serving as a lieutenant with the Asiatic Fleet in 1901 where, following the improvements achieved in Britain by Percy Scott, he endeavoured to introduce continuous aim into the Navy, thereby revolutionising naval gunnery. This was no easy task, and it took a direct approach to President Roosevelt to overcome the institutional torpor of the Bureau of Ordnance. Sims repaid the faith subsequently placed in him by increasing the overall efficiency of US gunnery by some 500%, and earned himself the title of the 'Gun Doctor', the man who taught the Navy how to shoot.

Sims continued to push the boundaries throughout his career. He was an aggressive and vocal proponent of the all-big-gun battleship and developed effective methods of fire control. As commander of the Atlantic Fleet Destroyers he encouraged and motivated his commanding officers (who included the future Admiral Halsey) to develop and refine fleet tactics. Later, as president of the Naval War College, he established strategic wargaming, which in turn facilitated the development of naval aviation and ASW.

Sims realised early on the importance of communicating his ideas, and became a prolific writer in support of his controversial opinions both within the US Navy and outside it via the Proceedings of the Naval Institute. In this book Commander Ben Armstrong USN, employing a similar format to that of his previous work *21st Century Mahan*, has assembled six of Sims' essays which, although written 100 years ago, have a particular relevance to contemporary military challenges.

The essays cover: Professionalism and Military Innovation, the Military Mindset, Preparing for Command and Preparing for War, The Forces of the Status Quo, The Peace Dividend and the Promotion System (which before 1900 was solely based on seniority). Each essay is as important for the structure of the argument as it is for its content. In the first, Sims defends 'the big gun single calibre battleship' as the future of naval warfare. Still a junior officer, he was writing in direct opposition to an essay by Admiral Alfred T Mahan, generally regarded as the greatest naval thinker of the day. Through precise detail and painstaking research Sims skilfully deconstructs Mahan's analysis, disproving his assertion that large numbers of small, low-calibre warships were preferable to a smaller number of heavier, higher-calibre battleships.

Each of the essays is preceded by a short, well-written introduction by the author which highlights and reinforces the message of the lasting value of Sims' contributions. As Armstrong points out, the problems of today are not unique, and the lessons of history remain pertinent for a modern military in an age of budget cuts and conflicting priorities.

The essays are followed by an epilogue in which Armstrong provides an insight into what Sims' contemporaries thought of him. It reinforces the view of Sims as a man who was idolised by junior officers as the man who 'attacked cherished beliefs as shams'. *21st Century Sims* is essential reading for the student of naval strategy;

it is also highly recommended for those interested in innovation and civil-military relations.

Philip Russell

James Bender
Dutch Warships in the Age of Sail 1600–1714: Design, Construction, Careers and Fates

Seaforth Publishing, Barnsley 2014; 328 pages, fully illustrated with maps, paintings and plans; price £50.00.
ISBN 978-184832-157-1

The Dutch Navy has a long and proud history; even today its warships remain technologically of the highest quality and are crewed by a strong professional cadre. This large-format book, well illustrated with paintings and sketches from the period, has as its subject what might be described as the 'golden age' of the Dutch Navy. Dutch warships of the 17th century proved equal in battle to those of the major European naval powers and the crews, mostly drawn from the ranks of experienced merchant seamen, were led by some of the greatest commanders of the day, including Admirals Michiel de Ruyter and Maarten Tromp.

The book is the result of practically a quarter century of research by James Bender. It was the author's intention, amongst other things, to address the errors and omissions in two studies which hitherto had been considered seminal works on the subject. Essentially, it is a reference book. There is a long and excellent introduction by J D Davies which describes how the disparate collection of Dutch provinces, each with its own admiralty, managed to unite and, for a short while, become a formidable naval power. The bulk of the work brings together what is known about the individual warships employed, together with details of the composition of successive fleet lists. Bender should be congratulated for his organisational skills in arranging a mass of data in a way that can be readily understood.

The standard naval reference book layout features a fixed fleet list and warships with neatly defined 'careers'. However, in this instance there was not an established fleet which could be called 'The Dutch Navy'; instead, ships were drawn together from the different admiralties for a short while for specific purposes. Perhaps a more cohesive understanding of the evolution of the navy over time can be gained from the 'Ships Listed by Period and Admiralty' section, which is divided into roughly 20-25 year periods. Known details about the ships employed by each of the admiralties within a particular time-slot are listed. Typically, the information includes dimensions, ordnance carried, the commanding officer's name and the crew size, but this can vary according to what has survived in the records.

Any specialist study which relies heavily on meticulous research based on primary sources in a language other than English is to be applauded. *Dutch Warships in the Age of Sail* is certainly that. It will be welcomed by those with an interest in the various campaigns and battles fought during a complex period when allies quickly became enemies and vice versa, while those who simply gaze at the wonderful examples of marine art in museums depicting the warships and engagements of the time will find the factual information contained here very useful in understanding this important era of naval history.

Jon Wise

Robin Brodhurst
Churchill's Anchor: The Biography of Admiral of the Fleet Sir Dudley Pound OM, GCB, GCVO
Pen & Sword, Barnsley 2015; paperback, 320 pages, 31 B&W photographs, maps; price £14.99.
ISBN 9781473841833

This biography of Admiral of the Fleet Sir Dudley Pound, first published by Leo Cooper in 2000, has been reprinted by Pen and Sword and is to be welcomed as it sheds light on that somewhat shadowy figure in naval uniform who is often seen in photographs trailing in Winston Churchill's wake. It also provides a fresh angle on some of the hugely controversial decisions taken at the highest level during the Second World War.

However, that is not all. As the title of the book suggests, the book delves into the relationship between the First Lord of the Admiralty, later Prime Minister and Defence Minister, and the First Sea Lord. In this respect, one feels for Dudley Pound. He inherited the legacy of the many ill-conceived policy decisions of the 1920s and 1930s which left the Royal Navy with inadequate naval aviation, air defence and antisubmarine warfare capabilities during the first years of the war. He also had to cope with an immensely demanding leader whose bizarre working hours were both physically and mentally exhausting for all who worked closely with him. In one important respect, though, he learned to manage the great man's many hare-brained schemes. Pound's tactic was to refrain from saying 'no', but instead to show an initial interest, to offer to investigate the matter further, and later to present Churchill with a series of reasoned arguments as to why the idea was impracticable (if not downright reckless).

Naturally this biography covers not only the 1939–43 period and Pound's untimely death, but his entire naval career. His first command, the battleship HMS *Colossus*, took part in the Battle of Jutland. The account of his service during the interwar years is an exemplar for a naval officer destined for the highest rank: Chief of Staff to C-in-C Mediterranean, Commander Battle Cruiser Squadron, Assistant Chief of Naval Staff, Second Sea Lord, C-in-C Mediterranean Fleet, First Sea Lord.

Dudley Pound proved to be a highly successful sea-going officer, but his greatest talents lay in his organisational skills and his single-minded, unsparing work ethic. Although he loved his 'hunting, shooting and fishing', his socialising and especially his dancing, it is readily apparent that Pound's 'down-time' was neatly rationed, compartmentalised and totally subservient to his service to the Navy.

He is often criticised for his role in some of the controversial decisions and disasters which afflicted the Royal Navy during the war: the Norwegian campaign, the destruction of the French Fleet at Mers-el-Kebir, the loss of HMS *Prince of Wales* and *Repulse*, and Convoy PQ17. As the most senior officer, of course, he had to accept ultimate responsibility. Robin Brodhurst does not attempt to exonerate Pound at every turn but instead presents a broad and fair picture of the pressure he was under (not least from Churchill), a pressure which was ultimately to kill him. Through all this, and despite the author's best efforts, which are supported by meticulous research, Pound the man remains stubbornly elusive, but that may well have been part of his character.

Jon Wise

Bernard Edwards
Dönitz and the Wolf Packs
Pen and Sword Maritime, 2014; hardback, 240 pages, 12 photographs, 11 maps; price £19.99.
ISBN 978-1-47382-293-1

Dönitz and the Wolf Packs was originally published by Arms and Armour Press in 1996. It has now been republished by Pen & Sword with a striking new cover, and is one of a number of informative and readable books on the war at sea 1939–45 by Bernard Edwards.

Following a brief introduction, seventeen chapters tell the story of the Battle of the Atlantic focusing on the wolf packs, the flexible U-boat hunting groups devised by Admiral Dönitz to attack Allied convoys. It was a tactic which had proven itself during the First World War when Dönitz himself was a submariner, and as Commander of the U-boats in 1939 he reintroduced it with devastating effect.

Edwards brings together the details of several significant convoy battles to help tell the story, while the inclusion of several useful maps of the North Atlantic illustrate the progress of both the convoys and the wolf packs and detail the unfolding action when the two sides clashed. There is also a small selection of mostly standard photographs from the Imperial War Museum collection, a useful bibliography and a comprehensive index.

If there is a criticism to be made it is that the cut-off point of the story comes shortly after May 1943, the date widely acknowledged as the turning point in the Battle of the Atlantic. Edwards' primary focus is on the early years of the war up to the notorious battle for convoy ONS-5, when the U-boats suffered unprecedented losses, compelling Dönitz to temporarily withdraw them from the Atlantic. The remaining two years are therefore only summarised in the last chapter, which makes it feel as though a second volume is needed to explore how the wolf packs which caused such devastation in the early years of the war were then devastated themselves as the conflict drew to a close. That said, *Dönitz and the Wolf Packs* is a worthwhile addition to the existing canon of works about the Battle of the Atlantic.

John Peterson

WARSHIP GALLERY

HMS *Colossus*, 1911–1923

A D Baker III, former editor of *Combat Fleets* and illustrator for many of the classic books by Norman Friedman, has kindly agreed to provide a series of detailed drawings for *Warship* based on the official plans, together with a brief technical description of the featured ship. This year's contribution features the British dreadnought HMS *Colossus*.

The first two capital ships of the Royal Navy's 1909 Programme, *Colossus* and *Hercules*, were ordered as a result of the Navy League's 'We want eight and we won't wait' campaign for increased naval construction to counter the perceived expansion of Germany's fleet. The disposition of their main battery of five twin 12-inch 50-cal. Mk XI guns, the last of that calibre installed on a British battleship, followed the same general layout as on their single-unit predecessor, HMS *Neptune*, with the wing turrets mounted en echelon in order to provide a full 10-gun salvo to either beam, albeit with a very small 60-degree arc of fire athwartships for the mount on the disengaged side. The turrets, however, were farther apart laterally than on *Neptune*, requiring a different distribution for the sixteen 4-inch 50-cal. single-mounted secondary battery guns, with ten within or exposed atop the forward superstructure, and the remaining six atop or within the after structure. *Colossus*' 4-inch guns were not provided with any protection, wheareas on her sister *Hercules* they were shielded within casemates. One hundred projectiles per 12-inch gun and 150 rounds per 4-inch gun were carried. For whatever reason, the hull of *Colossus* was 1ft broader in beam than that of *Hercules* (see characteristics table on p.207).

The major visual distinction between the *Neptune* and *Colossus* designs was the omission of the after tripod mast, whose control top in *Neptune* was generally useless because of funnel smoke. Thus, in order to

Colossus departing Scotts yard at Greenock for trials on 11 March 1911, watched by some of the men that built her. (NRS, courtesy of Ian Johnston)

HMS COLOSSUS
1910

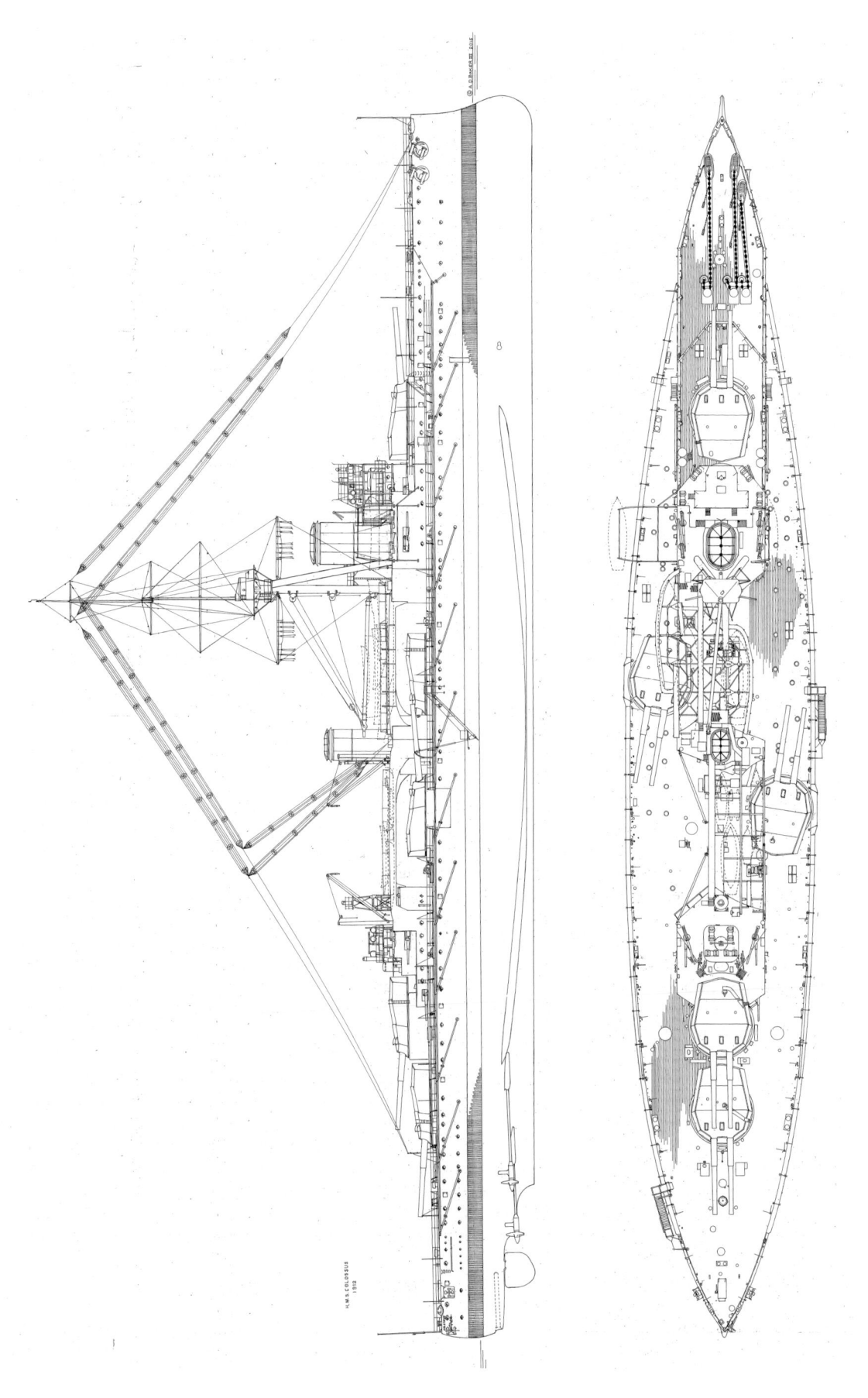

handle the ships' heavier boats, the main boat boom had to be stepped against the forward tripod's central leg, with the supporting legs splayed forward rather than aft. On both designs, all of the ships' boats except the two 32ft life cutters were stowed on skids running athwartships between the 'flying bridges' spanning the area between the three superstructure blocks. While on *Neptune* the heavier boats, which required a large boom for handling, were on the after boat structure, on *Colossus* and her sister they were forward. The flying bridges consisted of heavy, slightly arched longitudinal beams on their sides, connected by lighter transverse beams to support the boat skids; various narrow gangways ran between the boats and the superstructures at either end of the boat bridges to allow access to their crews. In the plan view of the accompanying drawing, which shows the boats as broken lines, it can be seen that the turrets and deck fittings on the upper deck could be seen through the open boat bridge structures. During the First World War, when boat complements were reduced in part to save weight, *Neptune* lost her forward stowage and the other two the after boat bridge (although, in 1919, Admiralty as-fitted plans showed *Colossus* to have been fitted post-war with portable supports for four boats added on the angled sides of the after two superstructure blocks).

There were other, internal differences as well. From *Dreadnought* to *Neptune* inclusive, side armour ran from bow to stern, although it was considerably thicker (8in) between the forward and after turrets, reducing to 2.5in at the extreme ends in *Neptune*. In the *Colossus* class, the sides at the ends were unarmoured, but protection over the midbody was strengthened by two inches and the rectangular patch of armour abreast the wing turrets was increased from 8in to 10in. In addition, the torpedo protection structure within their hulls was greatly improved, and deck thicknesses were increased to better protect against plunging fire. These and other changes increased full load displacement by about 690 tons over that of the *Neptune*.

Below the waterline lay yet another difference. While the torpedo tube layout of one firing aft at the extreme stern and one firing on each beam forward of 'A' turret was retained, the torpedoes were now of 21-inch diameter and had an effective range of 7000 yards; a total of six torpedoes per tube were carried, while six 14-inch torpedoes were on board for use in portable drop gear by the battleships' two 50ft steam pinnaces when they were employed on picket boat duties.

Both ships had numerous modifications prior to and during the First World War, primarily to their forward superstructures, which were greatly extended, and the eight 24-inch twin searchlight arrays were moved several times and ultimately replaced by single 36-inch projectors mounted atop control stations. *Colossus* acquired a 'manoeuvering compass platform' jutting 16ft forward from above the forward edge of the pilothouse and an enlarged charthouse added at the after end of the pilothouse roof. The original bridge deck was extended well

Characteristics:
Displacement: 20,030 tons normal; 23,266 tons deep
Dimensions: 545ft 9in overall (510ft pp) x 86ft 8in hull beam x 27ft draft (29ft 5in deep)
Armament (as completed): Five twin 12-inch/50-cal. (1,200 rounds); sixteen single 4-inch/50-cal. (2,400 rounds); one 12-pdr field gun; four single 3-pdr (saluting; removed during WWI); six Maxim machine guns; three 21-inch submerged torpedo tubes
Propulsion machinery: 18 Babcock & Wilcox watertube boilers (235–240psi); four sets Parsons steam turbines; four 3-bladed propellers; 25,000shp designed (29,000shp max. on trials)
Fuel: 900 tons coal normal (2900 tons max.) and 800 tons oil
Range: 4800nm at 10kts on coal; 6680nm at 10kts on coal and oil
Complement: 751 total as completed; 778 at Jutland
Boats as completed: two 50ft steam pinnaces; one 36ft sail pinnace; one 42ft steam launch; two 32ft life cutters (on davits abreast forefunnel); one 30ft gig; three 27ft whalers; one 16ft dinghy; and one 13ft 6in balsa raft

aft, and the original straight ladderways on either side replaced by spiral stairways.

Wartime changes included the removal of the torpedo net-supporting booms and their numerous brailing winches, while the net stowage shelves running down the sides of the upper deck were cut off. As a result of damage from the blast from cross-firing of the starboard wing turret during the Battle of Jutland, *Colossus* had several skylights, ventilators, and the forward deck winch removed to prevent future damage. All but 'A' turret had range-finding equipment added in hoods at their after ends, while even prior to the war a small cylindrical fire control station was added atop the gunnery control station atop the tripod mast, replacing the rangefinder originally fitted. Changes in the secondary armament came after Jutland as well, and by 1919 only ten of the original 4-inch guns remained, two aft and eight forward, while two anti-aircraft guns were added: a 4in mount at the extreme stern and a 12-pdr on a platform atop the after superstructure above the after compass platform.

The drawing shows *Colossus* in August 1912 after her forward funnel had been raised to reduce smoke disturbance to the activities in the control station atop the tripod. In the plan view, the wing turrets are shown trained out at their minimum 5-degree firing offset from the ship's centreline; in the elevation view, they are shown in stowed position fore and aft. The W/T antenna array's six-strand wire arrangement is shown in heavier line than it would have appeared; indeed, the arrays are almost invisible in photographs of the period. Note that the terminus for the array was in a pit, topped by a circular Faraday cage to starboard of the after funnel. The 4-inch secondary guns are shown in their stowed positions.

Operational history:

Colossus was laid down on 8 July 1909 at Scott's Shipbuilding & Engineering Co. Ltd., Greenock, about

A close-up of the midship section of HMS *Colossus* in 1912, with an excellent view of the distinctive 'flying bridges' for the ship's boats. The fore-funnel has already been raised. (Richard Perkins collection, © National Maritime Museum, Greenwich, London.)

25 miles west of Glasgow on the south bank of the River Clyde. Launched on 9 April of the following year, she ran sea trials on 28 February 1911 and was commissioned on 8 August 1911.

Having entered service with the 2nd Division, Home Fleet, she was assigned to the 2nd Battle Squadron in 1912, then the 1st Battle Squadron of the Home Fleet, renamed the Grand Fleet in August 1914. Flying the flag of Rear-Admiral Gaunt, *Colossus* led the 5th Division of battleships at Jutland and was 17th in the battle line. Opening fire at 1830 local time, her guns damaged the German light cruiser *Wiesbaden*; at 1912, her guns were directed at a range of about 9000 yards at the *Derfflinger*, which led the German battlecruiser line, causing the German ship severe damage. Four minutes later, *Colossus* became the only RN ship in the battle line to be hit by German fire; a 12-inch shell hit and exploded just abaft her forward funnel, starting fires on the port side amongst the exposed 4-inch guns and causing three injuries, two serious. A second shell was seen to hit forward but failed to detonate and skipped over the side, causing no damage, while the ship did experience splinter damage from some shells that missed entirely. *Colossus* also evaded three torpedoes fired during the battle.

The post-war career of the *Colossus* was more interesting than that of the other 12in-gun RN dreadnoughts. Although the ship was put on the disposal list on 30 June 1920, she was reprieved, rendered non-effective, and adapted as Boys' Training Ship at Portland, commissioning on 22 September 1921; in that guise she was repainted in Victorian livery, with black hull and white upperworks. With the drawdown of the British fleet after the war, and because of the reductions in fleet size mandated by the Washington naval disarmament treaty of 1922, the school was closed down during May of 1922. *Colossus* was then hulked at the HMS *Impregnable* establishment, commissioning on 31 January 1924, and was sold for scrap on 25 July 1928.

Sources:

Burt, R A, *British Battleships of World War One*, US Naval Institute Press (Annapolis 1986).

Friedman, Norman, *The British Battleship, 1906–1945*, Seaforth Publishing (London 2015).

Parkes, Oscar, *British Battleships, Warrior to Vanguard, 1860–1950*, Seeley Service & Co., Ltd. (London 1957).

Pears, CDR Randolph, RN (Ret'd.), *British Battleships 1892–1957, The Great Days of the Fleets*, Putnam (London 1957)

Winklareth, Robert I, *Naval Shipbuilders of the World, From the Age of Sail to the Present Day*; Chatham Publishing (London 2000).

Acknowledgments:

The author wishes to thank Dr Norman Friedman for providing a copy of his manuscript covering the origins of the *Colossus* and *Hercules*, and the invariably helpful staff of the National Maritime Museum's 'Brass Foundry' collection of warship plans for copies of the Admiralty 'as-fitted' plans set used to develop the drawing for this article. Thanks also to John Roberts for his help in clarifying details of the postwar career of *Colossus*.